ALLE · ZEIT · WACH · 1842

A. Bernd J. Bereiter-Hahn F. Hevert
H. Holzmann (Eds.)

Cell and Tissue Culture Models in Dermatological Research

Springer-Verlag
Berlin Heidelberg New York
London Paris Tokyo Hong Kong
Barcelona Budapest

Dr. August Bernd
Zentrum der Dermatologie und Venerologie
Abt. 1, Johann Wolfgang Goethe-Universität
Theodor-Stern-Kai 7, 6000 Frankfurt/Main 70, FRG

Professor Dr. Jürgen Bereiter-Hahn
AK Kinetische Zellforschung, Institut für Zoologie
Johann Wolfgang Goethe-Universität
Senckenberganlage 27, 6900 Frankfurt/Main 70, FRG

Priv.-Doz. Dr. Frank Hevert
Basotherm GmbH, Dermatologische Entwicklung
Eichendorffweg 5, 7950 Biberach a. d. Riss, FRG

Professor Dr. H. Holzmann
Zentrum der Dermatologie und Venerologie,
Abt. 1, Klinikum der Johann Wolfgang Goethe-Universität
Theodor-Stern-Kai 7, 6000 Frankfurt/Main 70, FRG

ISBN-13:978-3-642-77819-3 e-ISBN-13:978-3-642-77817-9
DOI: 10.1007/978-3-642-77817-9

Library of Congress Cataloging-in-Publication Data
Cell and tissue culture models in dermatological research / A. Bernd ... [et al.].
Includes bibliographical references and index.
ISBN-13:978-3-642-77819-3 : DM 168.00
1. Dermatology–Research–Methodology. 2. Cell culture.
3. Tissue culture. I. Bernd, A. (August)
RL79.C45 1993
616.5'0072–dc20

Softcover reprint of the hardcover 1st edition 1993

Typesetting: Cicero Lasersatz, Augsburg

27/3145-5 4 3 2 1 0 – Printed on acid-free paper

Preface

The *Frankfurter Gespräche* are a continuing series of scientific symposia dealing with interdisciplinary topics related in particular to dermatology. In the third of those, the impact of cell culture models on dermatological research was discussed by scientists from medicine, physiology, biology, pharmacology, and pharmacy.

The skin is the largest organ of the human body. Besides protection from external, mechanical, thermic, and chemical influences, skin has immunological and communicative functions. Accordingly, its cellular organization is complex. In skin, we find such different types of cells as keratinocytes, melanocytes, Langerhans cells, fibroblasts, Merkel's cells, nerve cells, and glandular cells. To understand the function and dysfunction of this complex structure, it is necessary to study partial aspects in models of manageable size and to reintegrate the results back into its context. Descartes formulated, 300 years ago, principles for the investigation of complex interactions. One of those is:

> *Diviser chacqune des difficultez que j'examinerois en autant de parcelles qui'il se pourroit, et qu'il seroit requis pour les mieux resoudre.* (If a problem is too difficult to be analyzed, then divide it into the smallest, possible pieces. That is the best way to resolve it).

In particular in the life sciences significant results have been achieved with this procedure. However, results obtained by reductive simplification become meaningful only when reintegrated in to an entity. This step may be more difficult than the investigation of the partial aspect itself. The poet, philosopher, and scientist Johann Wolfgang von Goethe, for whom our university has been named, stated this problem in the first part of his drama *Faust*:

> *Wer will was Lebendiges erkennen und beschreiben,*
> *Sucht erst den Geist herauszutreiben,*
> *Dann hat er die Teile in seiner Hand,*
> *Fehlt leider! nur das geistige Band.* (He who strives to comprehend and describe living things, should first try to extract the spirit, he then holds the pieces in his hands, unfortunately the spiritual link is missing).

With the contributions presented here, therefore, we have tried to elaborate the relation of tissue culture to the in vivo and to the clinical situation. We think this has been a successful endeavor.

The book consists of seven parts corresponding to the program of the meeting. In the first part, in vitro models suitable for investigations of the orthology and pathology of skin are presented. The second part deals with the differentiation of keratinocytes, and the third part is dedicated to the problems of cultivating melanocytes and to the biochemistry of pigment metabolism. In the fourth part, models for wound healing and tissue renewal are presented, while the fifth part treats fibroblast function and in particular the metabolism of collagen. The sixth part covers work on the problems of aging. Finally, in the seventh part, models for pharmacological and toxicological tests are presented.

Thus, a wide arch is spanned, from basic problems with cultivation and characterization of cell cultures to examples of application. The cooperation with the Basotherm company has proved to be especially favorable in this respect, since in combining the interests of both industry and university research application-oriented models have been considered.

We hope this book reflects the high scientific level of the meeting and elucidates the significance of cell and tissue cultures for dermatological research.

Frankfurt
July 1992

A. Bernd
J. Bereiter-Hahn
F. Hevert
H. Holzmann

Contents

4. Models in Wound Healing

5. Models in Fibroblast Function and Collagen Metabolism

6. Models in Skin Ageing

7. Models in Skin Pharmacology and Toxicology

Contributors

Bachmann, M., Priv.-Doz. Dr.
Institut für Physiologische Chemie, Abt. für Angewandte Molekularbiologie, Duesbergweg 6, W-6500 Mainz, FRG

Bayreuther, K., Prof. Dr.
Institut für Genetik, Universität Hohenheim, Emil-Wolff-Str. 14, W-7000 Stuttgart, FRG

Bereiter-Hahn, J., Prof. Dr.
Johann Wolfgang Goethe-Universität Frankfurt, Arbeitskreis Kinematische Zellforschung, Senckenberganlage 27, W-6000 Frankfurt/Main, FRG

Bernd, A., Dr.
Johann Wolfgang Goethe-Universität, Zentrum der Dermatologie und Venerologie, Abteilung für Dermatologie I, Theodor-Stern-Kai 7, W-6000 Frankfurt/Main 70, FRG

Blaheta, R.
Johann Wolfgang Goethe-Universität, Zentrum der Chirurgie, Transplantationsimmunologisches Labor, Theodor-Stern-Kai 7, W-6000 Frankfurt/Main 70, FRG

Brinckmann, J., Dr.
Institut für Molekularbiologie, Medizinische Universität Lübeck, Ratzeburger Allee 160, W-2400 Lübeck, FRG

Gast, W., Dr.
Universität Leipzig, Hautklinik, Liebigstr. 21, O-7010 Leipzig, FRG

Görmar, F. E.
Johann Wolfgang Goethe-Universität, Zentrum für Dermatologie und Venerologie, Abteilung für Dermatologie I, Theodor-Stern-Kai 7, W-6000 Frankfurt/Main 70

Hänel, H., Dr.
Hoechst AG, H811, Postfach 80 03 20, W-6230 Frankfurt/Main 80, FRG

Hegner, S.
Johann Wolfgang Goethe-Universität, Arbeitskreis Kinematische Zellforschung, Senckenberganlage 27, W-6000 Frankfurt/Main, FRG

Herrmann, K.
Universität Leipzig, Hautklinik, Liebigstr. 21, O-7010 Leipzig, FRG

Hoh, A.
Battelle Europe e.V., Am Römerhof 35, W-6000 Frankfurt/Main, FRG
Hoppe, W., Dr.
Universität Osnabrück, Medizinische Chemie (AGW), Albrechtstr. 28, W-4500 Osnabrück, FRG
Kaufmann, Dieter, Dr.
Universität Ulm, Humangenetik, Eberhardstr. 90, W-7900 Ulm, FRG
König, A., Dr.
Universitätsspital Zürich, Dermatologische Klinik, Gloriastr. 31, CH-8091 Zürich, Switzerland
Kramer, Michael D., Dr.
Institut für Immunologie und Serologie der Universität, Im Neuenheimer Feld 305, W-6900 Heidelberg, FRG
Lang, E.
Battelle Europe e.V., Abt. Molekulare und Zelluläre Biologie, Am Römerhof 35, W-6000 Frankfurt/Main 90, FRG
Maus, C.
Universität Leipzig, Hautklinik Liebigstr. 21,O-7010 Leipzig, FRG
Meyer, J. C., Dr.
Universitätsspital Zürich, Dermatologische Klinik, Gloriastr. 31, CH-8091 Zürich, Switzerland
Münnich, A.
Johann Wolfgang Goethe-Universität, Arbeitskreis Kinematische Zellforschung, Senckenberganlage 27, W-6000 Frankfurt/Main, FRG
Pfeifer, K., Dr.
Johannes Gutenberg-Universität, Institut für Physiologische Chemie, Abt. Angewandte Molekularbiologie, Duesbergweg 6, W-6500 Mainz, FRG
Ramirez-Bosca, A., Dr.
Johann Wolfgang Goethe-Universität, Zentrum der Dermatologie und Venerologie, Abteilung für Dermatologie I, Theodor-Stern-Kai 7, W-6000 Frankfurt/Main 70, FRG
Ramme, K.
Hautklinik des Klinikums Minden, Portastr. 7–9, W-4950 Minden, FRG
Reichrath, J.
Universitätshautklinik Poliklinik für Hautkrankheiten, W-6650 Homburg/Saar, FRG
Reinartz, J., Dr.
Institut für Immunologie und Serologie der Universität, Im Neuenheimer Feld 305, W-6900 Heidelberg, FRG
Rodemann, H. Peter, Priv.-Doz. Dr.
Sonderforschungsbereich 223, Entwicklungsbiologie, Hoppe-Seyler-Str.3, W-7400 Tübingen, FRG
Ruszczak, Zb., Dr.
Hautklinik des Klinikums Minden, Portastr. 7–9, W-4950 Minden/Westf., FRG

Schäfer, D., Dr.
Johann Wolfgang Goethe-Universität Zentrum für Frauenheilkunde und Geburtshilfe, Theodor-Stern-Kai 7, W-6000 Frankfurt/Main 70, FRG
Schachtschabel, D., Prof. Dr.
Institut für Physiologische Chemie, Philipps-Universität, Karl-von-Frisch-Straße, W-3550 Marburg, FRG
Scharffetter-Kochanek, Karin, Dr.
Universität Düsseldorf, Abt. Dermatologie, Moorenstr. 5, W-4000 Düsseldorf 1, FRG
Stadler, R., Prof. Dr.
Hautklinik des Klinikums Minden, Portastr. 7–9, W-4950 Minden, FRG
Strohmeier, R., Dr.
Johann Wolfgang Goethe-Universität, Zentrum für Frauenheilkunde, Abt. für Gynäkologische Onkologie, Theodor-Stern-Kai 7, W-6000 Frankfurt/Main 70, FRG
Stumpenhausen, G., Dr.
Universität München, Dermatologische Klinik, Frauenlobstr. 9–11, W-8000 München 71, FRG
Solano-Munoz, F., Prof. Dr.
Departamento de Bioquimica y Biologia Molecular, Facultad de Medicina, Universidad de Murcia, E-30100 Murcia, Espana
Theilig, C.
Johann Wolfgang Goethe-Universität, Zentrum der Dermatologie und Venerologie, Abteilung für Dermatologie 7, Theodor-Stern-Kai 7, W-6000 Frankfurt/Main 70, FRG
Weisensee, D.
Johann Wolfgang Goethe-Universität, Abt. Nephrologie, Theodor-Stern-Kai 7, W-6000 Frankfurt/Main 70, FRG
Westphal Frösch, C., Dr.
Institut für Lasertechnologie in der Medizin an der Universität Ulm, Helmholtzstr. 12, W-7900 Ulm, FRG

1. Models in Orthology and Pathology

Development of a Cell Culture System to Detect and Analyze Pathogenicity of Autoantibodies Directed to Nuclear Antigens

M. Bachmann, and *W.E.G. Müller*

Abstract

Sera of patients with certain rheumatic diseases frequently develop antibodies to nuclear antigens. These antibodies are assumed not to be involved in active disease in patients, as their antigenic epitopes localize inside of cells in the nucleus and should not be available for autoantibodies circulating in the peripheral blood of the patient. During the past few years we have studied the intracellular localization of one of the nuclear antigens known as La protein. We were able to show that this antigen shuttles between the nucleus and the cytoplasm and even to the cell surface. This shuttling occurred, simulating in cell culture just those conditions which are known to induce active disease in patients, including UV irradiation, virus infections, and certain drug treatments. In addition to these "pathogenic" conditions, we observed that the antigenic epitope recognized by anti-La monoclonal antibodies (mabs) developed in our lab is also available on the surface of living cells during transition from the G_0 to the G_1 phase, as well as on mitotic cells. Based on these observations, we developed a simple and reproducible cell culture system allowing us to prove the hypothesis that the antinuclear antibodies are not an epiphenomenon, as generally assumed, but are directly involved in disease.

Abreviations

cLSM, confocal laser scanning microscopy
FCS, fetal calf serum
HSV, Herpes simplex virus
mab, monoclonal antibody
RNP, ribonucleoprotein particle
ANA, anti-nuclear antibody
ENA, extractable nuclear antigen

A. Bernd, J. Bereiter-Hahn,
F. Hevert, H. Holzmann (Eds.)
Cell and Tissue Culture Models
in Dermatological Research

Introduction

Sera of patients with certain autoimmune disease such as systemic lupus erythematosus or Sjögren's syndrome frequently contain autoantibodies that have a nuclear staining pattern on tissues or cell culture cells (Fig. 1). The immunofluorescence staining of such antinuclear antibodies (ANAs) varies between homogeneous and speckled nuclear staining patterns (Fig. 1A and B). Based on their nuclear staining pattern, the target antigens of ANAs were originally isolated from nuclear extracts and termed extractable nuclear antigens (ENAs) known as Sm, RNP, Ro, and La. Sera of patients with rheumatic diseases are routinely screened for the presence of ANAs, because ANAs are of diagnostic and prognostic value. On the other hand, one might expect that ANAs cannot be involved in those autoimmune reactions leading to the destruction of cells during active disease, because their target epitopes are inside the cells and therefore not available for the ANAs circulating in the peripheral blood of the patients.

However, the conclusion that ENAs localize exclusively in the nucleus is based by definition only on the exclusive nuclear staining pattern of the ANAs. Just these staining patterns do not completely fit with some biochemical approaches and some staining patterns available for monoclonal antibodies with similar specificity (Fig. 1F). At least for the La protein, we demonstrated that the additional cytoplasmic staining of certain anti-La monoclonal antibodies (mabs) (e.g.), La11G; Fig. 1F results neither from a cross-reaction with a further cytoplasmic protein nor from a leakage of nuclear La protein during fixation [5]. The additional cytoplasmic staining of these anti-La mabs revealed the rough endoplasmic reticulum (Fig. 2A and B) and parts of the Golgi, especially the *trans*-Golgi network (Fig. 2C–F). The staining of these regions is in agreement with earlier observations that

a) inhibitors of N-glycosylation influenced the intracellular localization of La protein [1], and
b) La protein exists in an enzymatic active and inactive form, the latter of which was found to be sialylated [2, 3].

Glycosylation of La protein is also evident from a series of additional experiments: La protein isolated from the nucleus reacted positively with a proteoglycan kit [based on the periodate oxidation procedure (not shown)]. Moreover, the nuclear La protein reacted with the mannose-specific lectin from galanthus nivalis (GNA) (Fig. 3). This binding dissappeared when La protein was treated with endoglycosidase H or N-endoglycosidase F, while the binding was not altered during a treatment with O-glycosidase. These data hint that at least a part of the nuclear La protein belongs to the high mannose – type glycoproteins (Fig. 3). Interestingly, the mannosylated La protein is able to bind to itself but also to other mannosylated proteins, for example, to GP-120 glycoprotein of the HIV virus (Fig. 3). Therefore, La protein represents a self-binding lectin-like protein. The sugar residues reside in the 29-kD part of La protein, which forms during a proteolytic

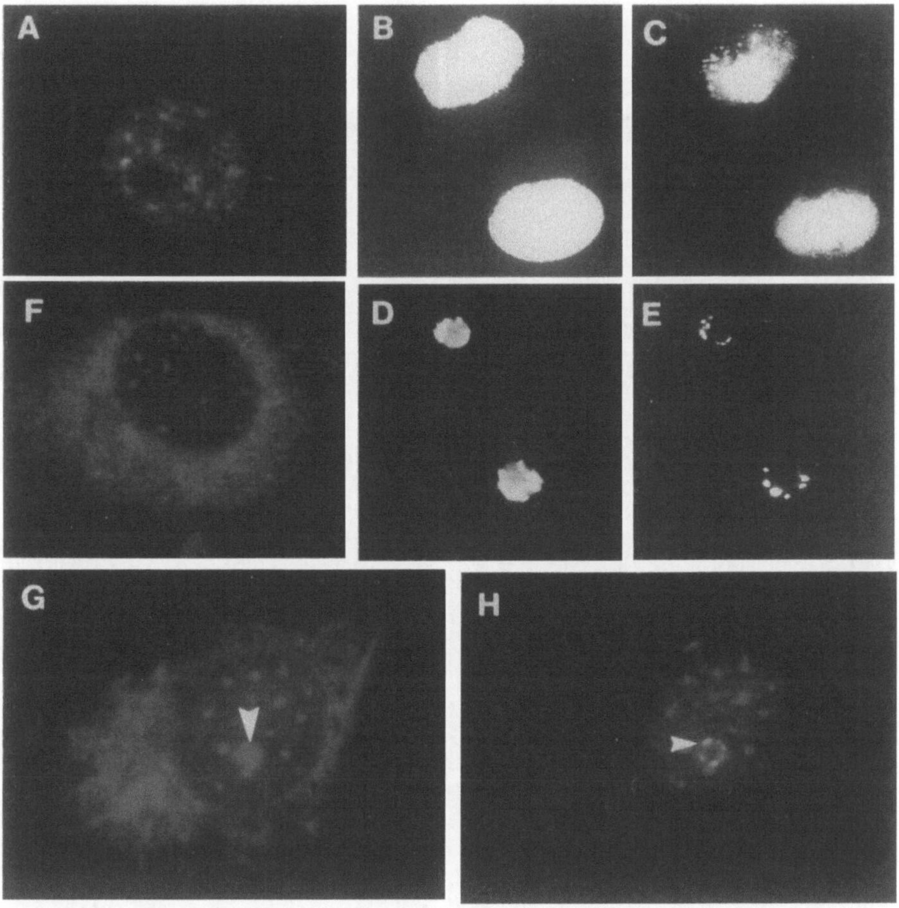

Fig. 1A–H. Staining pattern of anti-nuclear antibodies (ANAs). Sera of autoimmune patients containing ANAs usually give an exclusively nuclear staining patterns (**A, B**). These patterns are termed the "nuclear speckled type" (**A**) or the homogeneous (**B**) staining pattern. In **B–D** the staining does not show an antibody; rather, it represents the autoantigen La itself. In this experiment the FITC-labeled protein was imported into living mitotic cells. During telophase La protein entered into the reformed nuclei, resulting in a homogeneous nuclear staining (**B**). Using the image processor unit of the LSM we were able to "remove" the homogeneous staining, altering the ratio contrast/brightness. As evident from **B–E**, La protein assembled in addition to the homogeneously stained nucleoplasm with nuclear speckles. Some of the labeled nuclear speckles surrounded the nucleolus like a ring (**E**). Similar arrangements were observed in earlier studies on fixed cells using anti-La, anti-Sm, or anti-RNP mabs (**G** and **H**). In contrast to patient antibodies, which obviously prefer the nucleus, some anti-La mabs stain the cytoplasmic compartment in addition to the nucleus (**F**). Cells were analyzed by epifluorescence microscopy (**A, F, G, H**) or confocal laser scanning microscopy (cLSM; **B–E**)

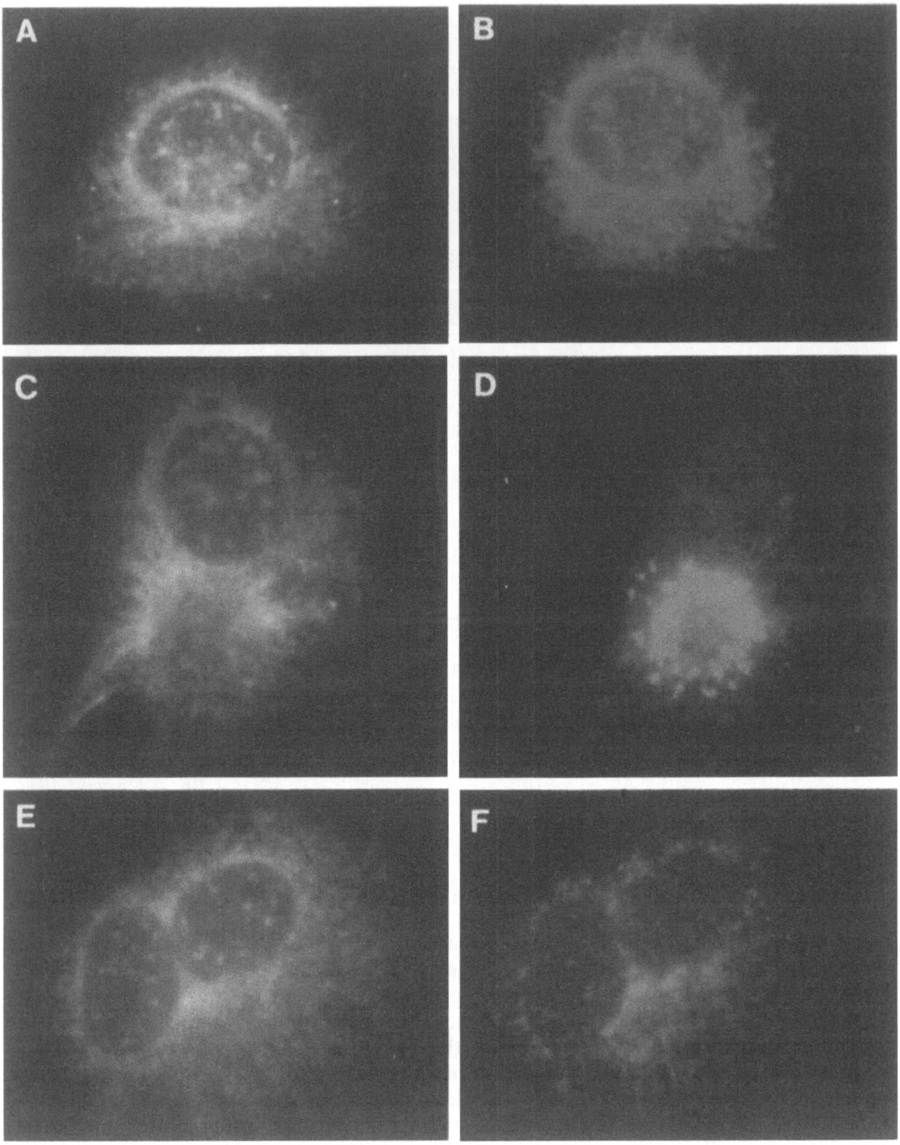

Fig. 2 A–F. Cytoplasmic localization of La protein. Anti-La mabs were directly labeled with FITC, antibodies to ER and Golgi regions with RITC. Double immunolabelings revealed that the cytoplasmic staining of the anti-La mab (**A, C, E**) costains with antibodies to ER (**B**) and Golgi regions (**D, F**). **C** and **D** Anti-La and *cis*-Golgi antibodies; **E** and **F** anti-La and *trans*-Golgi antibodies. Cells were analyzed by epifluorescence microscopy

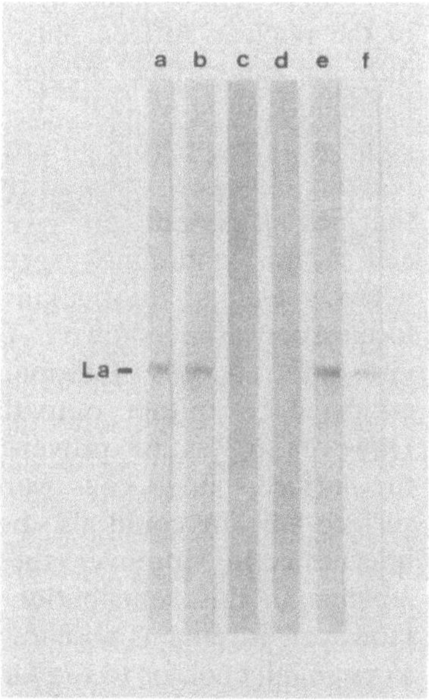

Fig. 3. Western blotting analysis of isolated La protein: characterization as a glycoprotein with lectin-like properties. La protein was isolated from nuclei of CV-1 cells and purified to homogeneity. The purified La protein reacted well with patient anti-La antibodies (*lane a*). The lectin GNA bound to the nuclear La protein, indicating terminal mannose residues (*lane b*). When La protein was pretreated with endoglycosidase H (*lane c*) or N-endoglycosidase F (*lane d*) the binding ability of GNA to La protein was destroyed, while an incubation with O-glycosidase did not alter GNA binging capability. Therefore, at least a part of the nuclear La protein is supposed to be a high mannose-type glycoprotein. Self-binding properties of La protein became evident when blotted La protein was incubated with biotinylated La protein followed by an incubation with streptavidin phosphatase (*lane e*). In addition, La protein bound to other mannosylated proteins, for example, the GP120 protein of the HIV virus. In this case La protein was incubated with GP120 followed by an incubation with anti-GP120/phosphatase complex (*lane f*)

degradation of La protein. A look at the cDNA sequence of La protein published independently by several groups indeed revealed a potential N-glycosylation site in the 29-kD region. On the other hand, the cDNA sequence lacks a typical signal peptide sequence allowing the protein to enter into the ER during translation. However, it should be mentioned that other proteins – for example interleukin 1, or the basic fibroblast growth factor, or certain extracellular lectins – also lack this typical signal peptide sequence. In spite of this they are glycosylated and secreted. In contrast to the staining pattern of patient anti-La antibodies, these data conform well to the staining pattern of our anti-La mabs.

Using these mabs we established in earlier studies a nucleocytoplasmic shuttling of La protein. This nucleocytoplasmic shuttling was inducible precisely under conditions known to cause active disease in patients. Such conditions are stress situations such as

a) UV irradiation,
b) virus infections, and
c) treatment with certain drugs.

A few earlier reports described a cell-surface labeling of human keratinocytes with ANA-positive patient sera [7]. The cell-surface labeling was obtained after UV irradiation or virus infections [3, 4, 6, 7]. However, in none of the earlier studies were the antibodies and the recognized cell surface antigen further characterized by additional biochemical approaches. Therefore, until now it remained unclear whether the observed cell-surface labeling of sera or mabs against nuclear antigens was due to a translocation of the nuclear antigen onto the cell surface or to a cross-reaction with a different cell surface antigen. A further problem is the viability of the UV-irradiated, drug-treated, or virus-infected cells. The cell-surface labelings could also result from an artificial adsorption of nuclear antigens (or cross-reacting epitopes) released from destroyed cells. Another major problem is that not all patient sera give such cell-surface stainings. Furthermore, the cell culture conditions, especially the presence of cytokines such as tumor necrosis factors, interleukins, and interferons, dramatically influence the localizations (not shown). This is important especially for studies using virus-infected cells. Baboonian et al. [6] reported a translocation of an antigenic La epitope onto the cell surface after an adenovirus infection. However, it was not proven by a double immunolabeling experiment that the surface-stained cells were indeed virus infected. Therefore, the cell-surface labeling could also be the result of cytokines produced from virus-infected cells. Moreover, there may be another reasons for the cell-surface labeling in this contribution: The only stained cells were rounded cells. However, it remains obscure whether the rounding of the cells is really due to virus infection or to the fact that the cells shown are mitotic cells. Again, this is an important point, as we recently found that La protein can be detected on the cell surface of mitotic cells (see below).

Taken together, it remains unclear if the cell-surface staining of Baboonian et al. was really due to virus infection. Nevertheless, the studies mentioned were the first indication that epitopes recognizable by ANA-positive sera can exist on the cell surface of cells in culture. At least the major interesting question remained open: Are ANAs pathogenic?

The major reasons for these unsolved problems were

a) the lack of a physiological procedure to translocalize a nuclear autoantigen to the cell surface,
b) the use of human primary keratinocytes in most of the cell-surface translocation studies, which limits the amount of available material for further biochemical approaches,
c) the lack of a simple and reproducible method to translocalize a nuclear autoantigen to the cell surface of a large amount of identically treated cells,
d) a monoclonal antibody system allowing reproducible detection of an antigenic epitope on the cell surface, and
e) the lack of a direct improvement with the antigen itself. Most of these problems have now been solved for at least one of the nuclear autoantigens, the La protein.

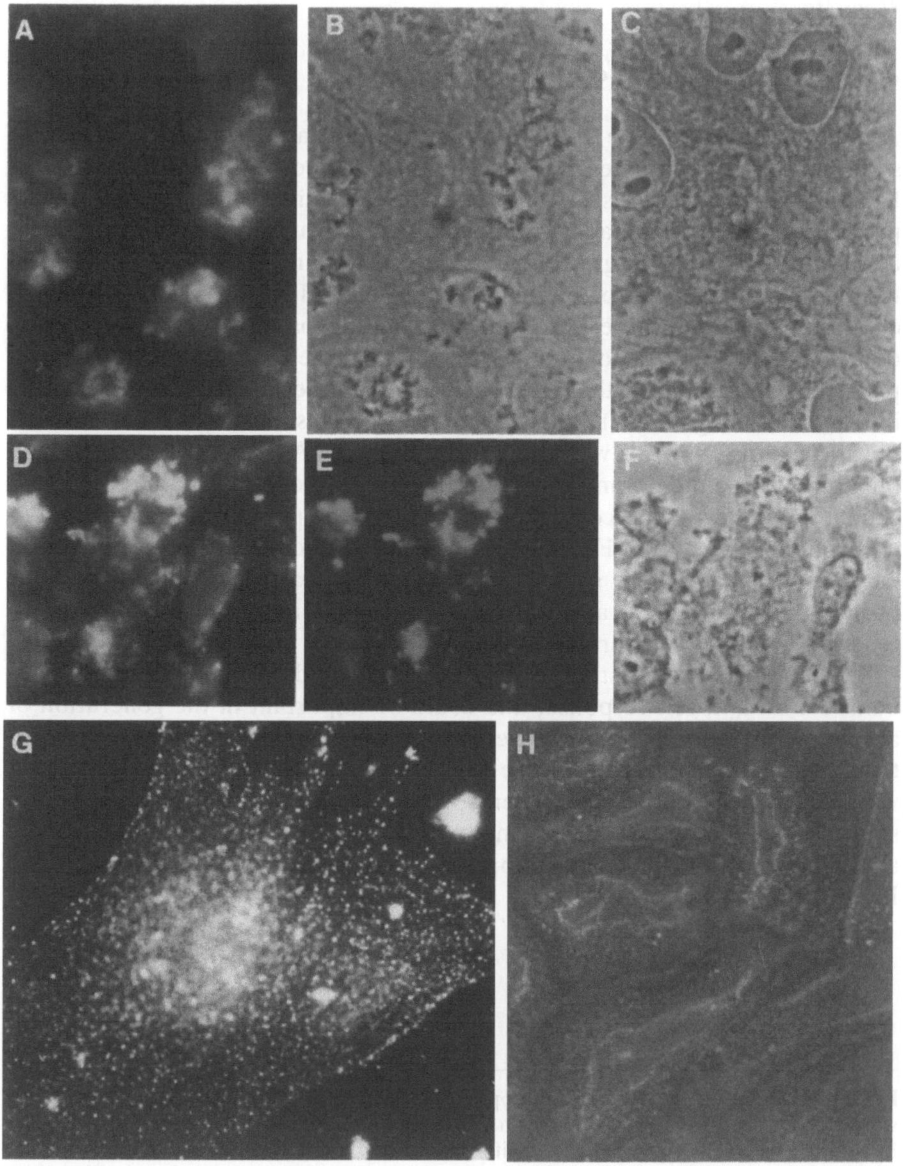

Fig. 4 A–H. Detection of a La epitope on the surface of cells treated with various pathogenic conditions. Under simulated conditions which are known to cause acute disease in patients, including UV irradiation, virus infections, and certain drugs (for example griseofulvin in La-positive patients), an anti-La epitope was detected on the cell surface: **A–F** herpes simplex virus (HSV)-infected, **G** UV-irradiated, **H** griseofulvin-treated cells. HSV type-1-infected CV-1 (**A–C**) and CHO cells (**D–F**) showed large protrusions on their surface. Using a double immunofluorescence technique, the anti-La epitope (RITC; **E**) colocalized with anti-herpesvirus antibodies (FITC; **D**) within these protrusions. The cells were analyzed with epifluorescence microscopy. **B, C, F,** corresponding phase-contrast images

Results

Development of the Cell Culture System

The development of the cell culture system is based primarily on earlier studies with UV-irradiated or herpes virus-infected cells. Both techniques, UV irradiation and virus infection, had several disadvantages. One major problem with the UV-irradiation technique was that only a minor portion of cells (about 20%–35%) gave the cell-surface staining shown in Fig. 4G. On the other hand, the herpes virus infection leads, like every virus, to asynchronous cells and thereby to varying staining patterns. However, using double immunolabeling experiments we were able to solve these technical problems and showed that the surface-stained cells were indeed virus-infected cells (Fig. 4A–F). The virus infected cells showed protuberances on the surface of CV-1- als well as CHO cells (Fig. 4A–F). As herpesvirus can only bind to the membrane of CHO cells but not penetrate it, and therefore is unable to infect CHO cells, the formation of the protrusions is due only to adsorption of the virus (Fig. 4D–F) Similar protuberances were observed in earlier studies and were shown to contain virus particles, but also cellular proteins including actin and fibronectin. Moreover, they are known to be released from the cell surface.

Such complexes consisting of self and virus could provide helper determinants for an anti-self response, and could be important in the generation of autoimmunity. Based on our observation that after UV irradiation La protein colocalized on the cell surface with growth factor receptors (Fig. 4G) [5] we decided to optimize our systems by synchronization of the cells with a serum-starvation block. Surprisingly, this treatment also resulted in the appearance of the anti-La epitopes of the anti-La mabs (LA11G7 and La1B5) Figs. 5 and 6) on the cell surface. In addition, the anti-La epitopes

Fig. 5 A–I. Translocation of an La epitope to the surface of growth-stimulated cells. Cells were starved for 2 days, followed by growth stimulation due to 1 h incubation with 10% FCS. The living cells were incubated with the anti-La mab LA11G7 for 5 min, fixed, and stained with anti-mouse Ig conjugated with FITC. The cells were analyzed either with cLSM (**A–E**) or the classical epifluorescence technique. The anti-La mab gave a punctuated staining following the cell surface (**A–E**). In addition to the punctuated staining, a homogeneous staining was obtained using the epifluorescence technique only (**G**). However, this additional staining might be an artifact due to cell surface spots being out of focus. After incubation of nonstimulated cells with the anti-La mab, no punctuated cell-surface staining was found (**H**). **I** Corresponding phase-contrast image. Using nonstimulated cells the anti-La mab entered only into dead cells (**H** and **I**). The cells next to the dead cells are not stained, supporting the idea that the cell-surface labeling shown for the serum-stimulated cells (**A–G**) cannot be an artifact due to material released from dead cells

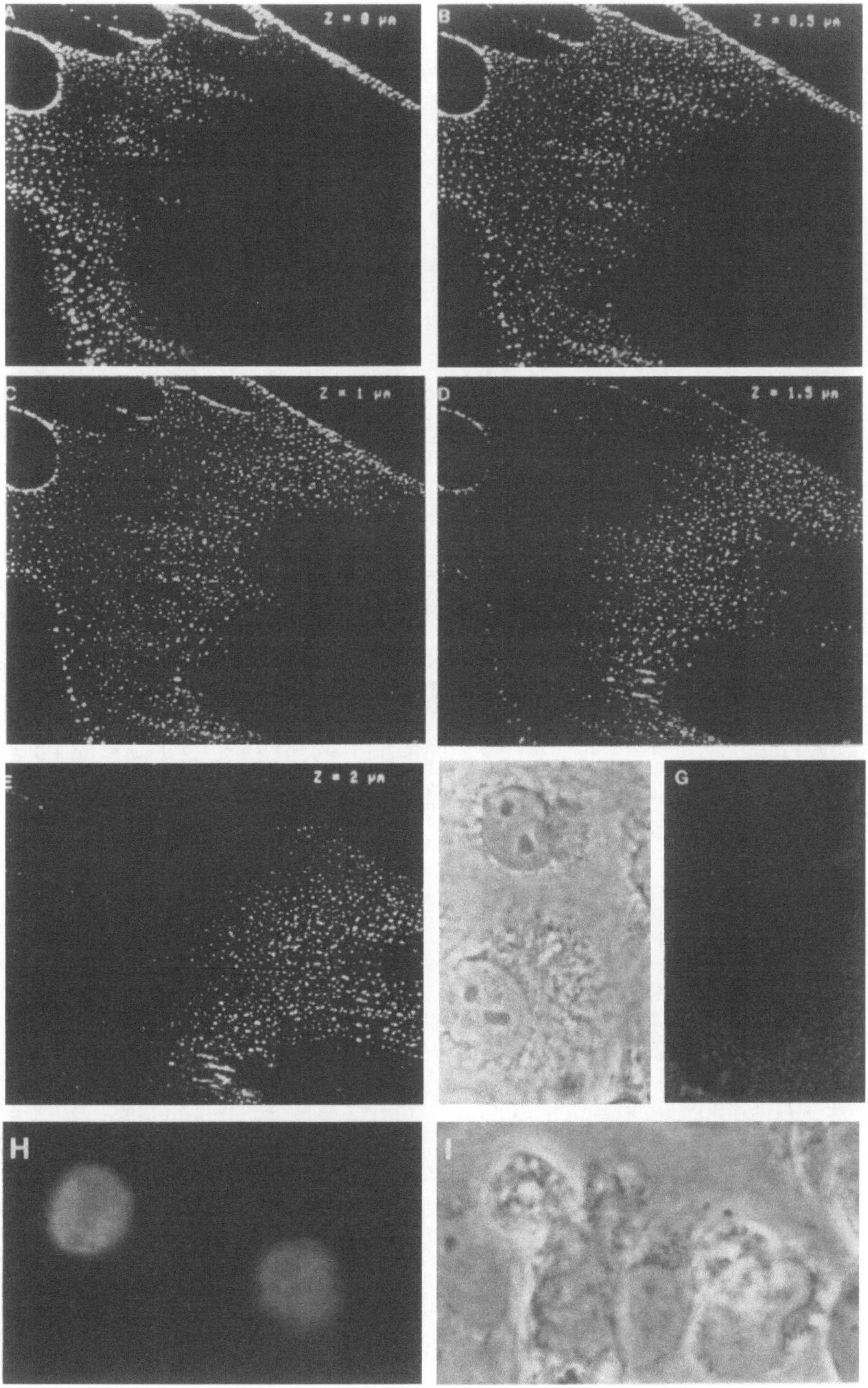
A
Z = 0 μm
B
Z = 0.5 μm
C
Z = 1 μm
D
Z = 1.5 μm
Z = 2 μm
G
H
I

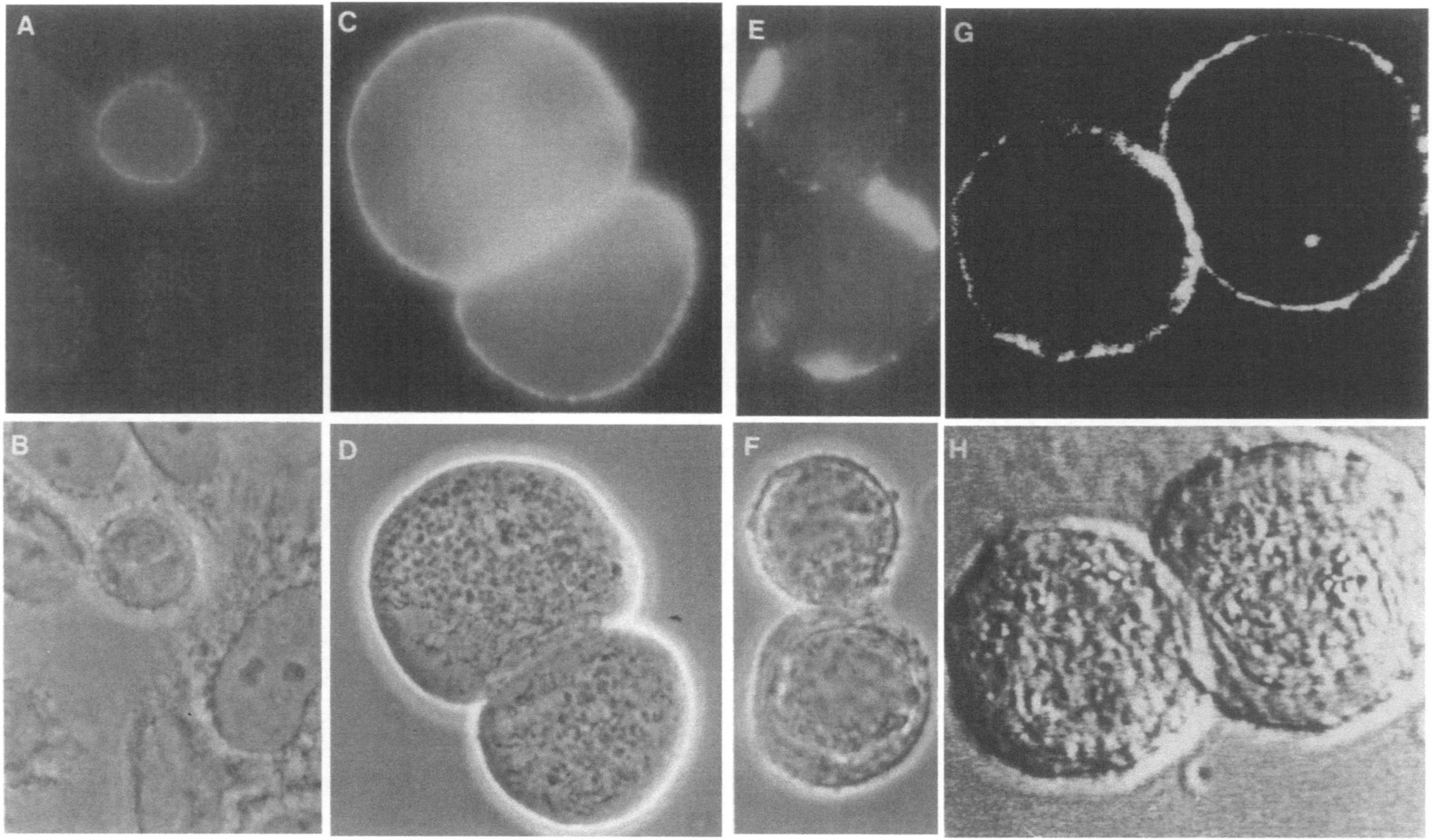

Fig. 6 A–H. Detection of La protein on the surface of mitotic cells. The surface of mitotic cells was intensively stained on coverslips of serum-stimulated cells (**A** and **B**) and cells which were synchronized into mitosis with colcemide (**C–H**). The cells were analyzed either with epifluorescence- (**C–F**) or cLSM (**G** and **H**). Prolonging the incubation time of the living cells with the anti-La mab resulted in capping effects (**E** and **F**)

appeared in the extracellular space assembled with the fibronectin network, preferentially near dividing cells. A treatment with mercury chloride – a potent inducer of autoimmunity – resulted in a displacement of the La epitopes from the extracellular matrix [3, 4].

Taken together, this simple and highly reproducible translocation procedure allowed us to determine whether (a) anti-La antibodies have pathogenic properties and (b) anti-La sera contain such pathogenic anti-La antibodies (see below). Moreover, the translocation of the antigenic epitope to the surface of mitotic cells led us to the speculation that La protein added to cell culture medium may enter into mitotic cells.

Entry of La Protein into Living Cells

To prove this hypothesis, homogeneously purified nuclear La protein was labeled with FITC and added to mitotic cells in the absence or presence of the anti-La mab. As shown in Figs. 1 (B–E) and 7, La protein entered into mitotic cells very rapidly. The entry was inhibited by anti-La mabs (not shown). In pro/metaphase cells La protein associated with or formed (presumably due to its self-binding properties) a fibrillar network (Fig. 7), while in telophase cells La protein entered into the reformed nuclei, giving a homogeneous nuclear staining pattern as described for ANAs (Fig. 1B–E). On the other hand, from such a homogeneous staining it is impossible to determine if La protein also assembled with clusters of interchromatin granules. These structures would appear on fixed cells as nuclear speckles. Using the image processor unit of the laser scanning microscope (LSM), we were able to fade out the homogeneous nuclear fluorescence, while additionally stained nuclear regions were still visible. This technique allowed us to analyze what is stained inside the nucleus in addition to the homogeneously stained nucleoplasm. With removal of this homogeneous nuclear staining by the video technique, La protein was identified in "speckles" and was inhomogeneously distributed in the nucleolus. La protein was also found in nuclear speckles surrounding the nucleolus like a ring (Fig. 1B–E). Similar ring-like arrangements were found with anti-La, anti-Sm, and anti-RNP mabs (Fig. 1G and H) [1]. To ascertain that the fluorescence was due to La protein and did not represent FITC released from degraded protein, La protein was reisolated from the mitotic cells and analyzed by immunoblotting using anti-FITC antibodies. The imported material was still intact La protein (data not shown). Moreover, the import was inhibited by unlabeled La protein.

Pathogenic Effects of Anti-La Antibodies

Finally, we looked for pathogenic effects of anti-La antibodies when bound to the surface of living cells. For this purpose, we analyzed in a first step the

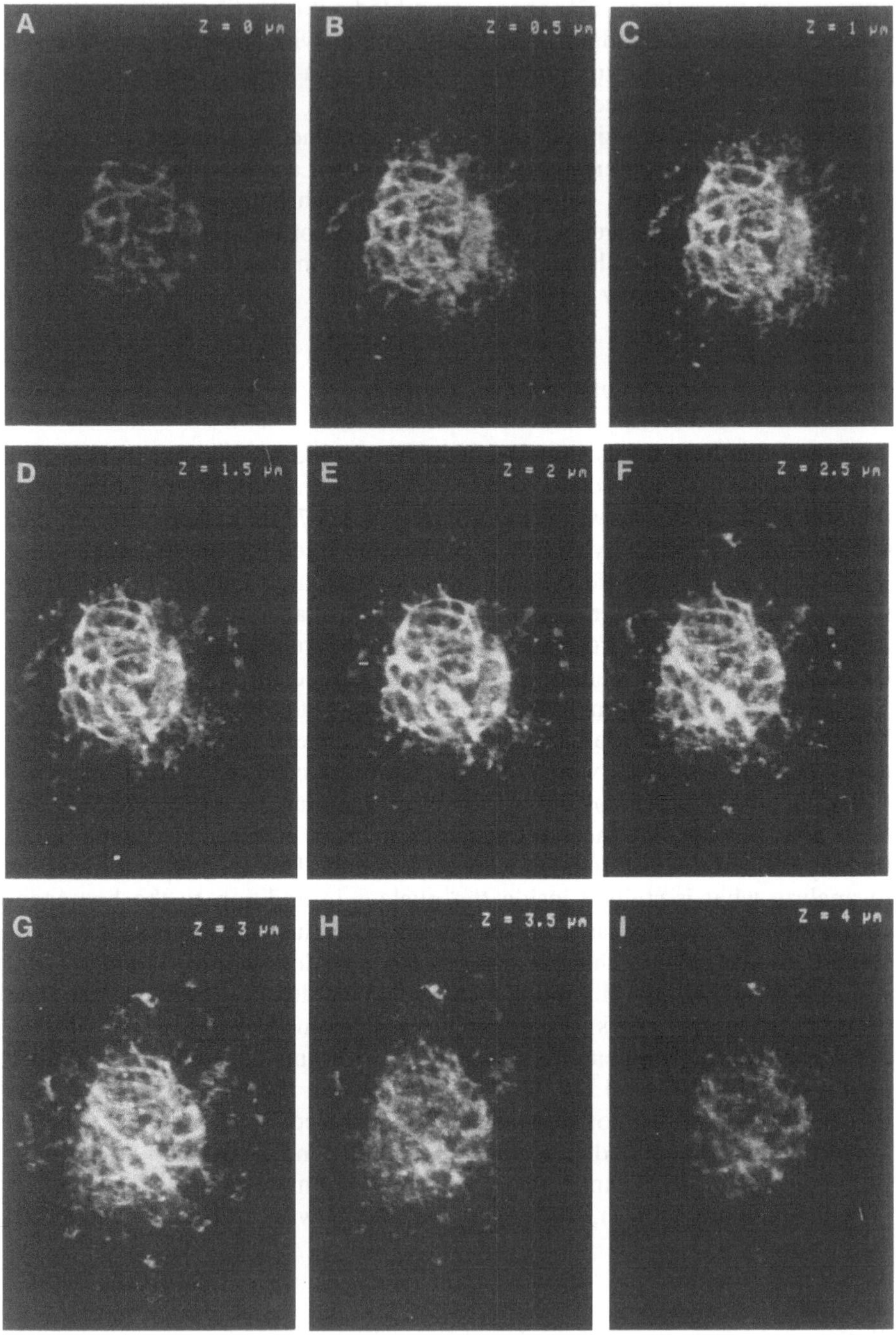

Fig. 7

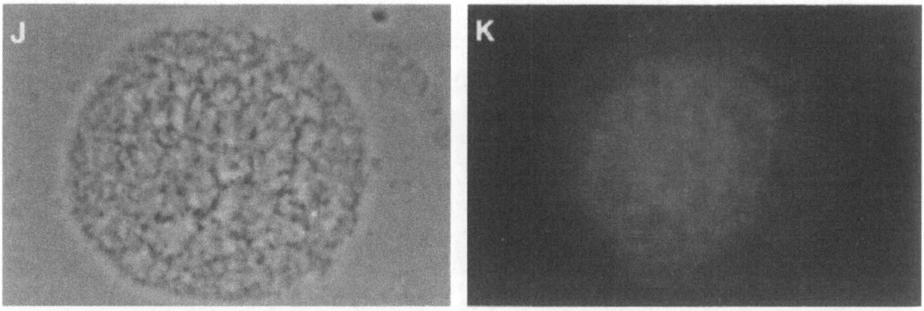

Fig. 7 A–K. Import of FITC-labeled La protein into mitotic cells. FITC-labeled La protein added to telophase or postmitotic cells entered very rapidly into the newly formed nuclei, giving a homogeneous nuclear staining pattern (see Fig. 1B). La protein imported into pro/metaphase cells very rapidly associated with or formed a fibrillar network. This network was analyzed with epifluorescence- (**J**- and **K**) or with the cLSM technique (**A–I**)

stability of the immune complexes formed on the cell surface. We observed that, once formed, the anti-La immune complexes were detectable on the cell surface for up to 64 h. An example (4 h after incubation with the anti-La mab) is shown in Fig. 8. Such immunolabeled cells were allowed to grow at low density in medium containing 10% fetal calf serum (FCS). They did not seem to undergo further cell divisions, although this observation was not made for all the analyzed cell lines. However, in most cases the mitotic cycle was at least retarded. During the first 4 h the fine punctuated pattern was altered to a coarse clumpy staining of the cell surface (compare Figs. 5 and 8). In some cases a minor portion of the immune complexes seem to be taken up by the cells and even entered into the nucleus (Fig. 9). However, further studies are required to decide if these cells were still viable when the antibodies entered into their nuclei. It is unimportant for evaluating the role of La protein in disease whether the binding of an anti-La antibody directly destroys a cell or whether the immune complexes formed on the cell surface induce the destruction via the immune system of the patient. In both cases the antibodies would be pathogenic.

In a further approach we checked and found that our cell culture system also worked with patient anti-La antibodies (Fig. 10). For this purpose the patient anti-La antibodies had been absorbed to an affinity column constructed by covalently linking recombinant La protein to CNBr-Sepharose 4B. The linked recombinant La protein had been isolated from the *Escherichia coli* extract by immunoaffinity chromatography using the anti-La mab (La11G7). Serum samples from different patients, but also different serum samples from the same patient, gave different results: For example, a series of serum samples from the patient (Ma) were obtained during a time period of about 10 years, including symptom-free but also active disease periods. Some of these serum samples gave a cell-surface labeling while others did not, although all samples were La positive and absorbed to La

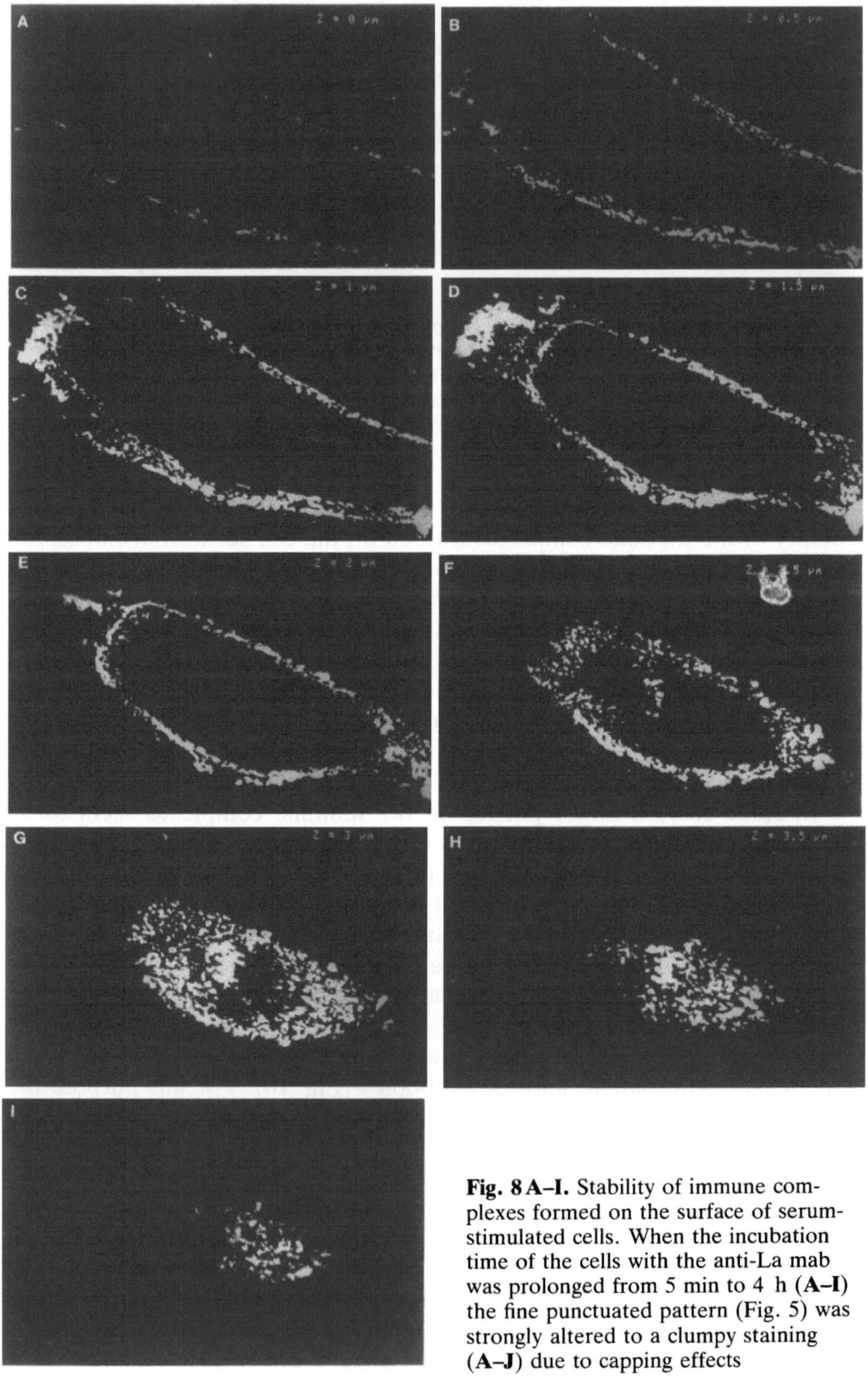

Fig. 8 A–I. Stability of immune complexes formed on the surface of serum-stimulated cells. When the incubation time of the cells with the anti-La mab was prolonged from 5 min to 4 h (**A–I**) the fine punctuated pattern (Fig. 5) was strongly altered to a clumpy staining (**A–J**) due to capping effects

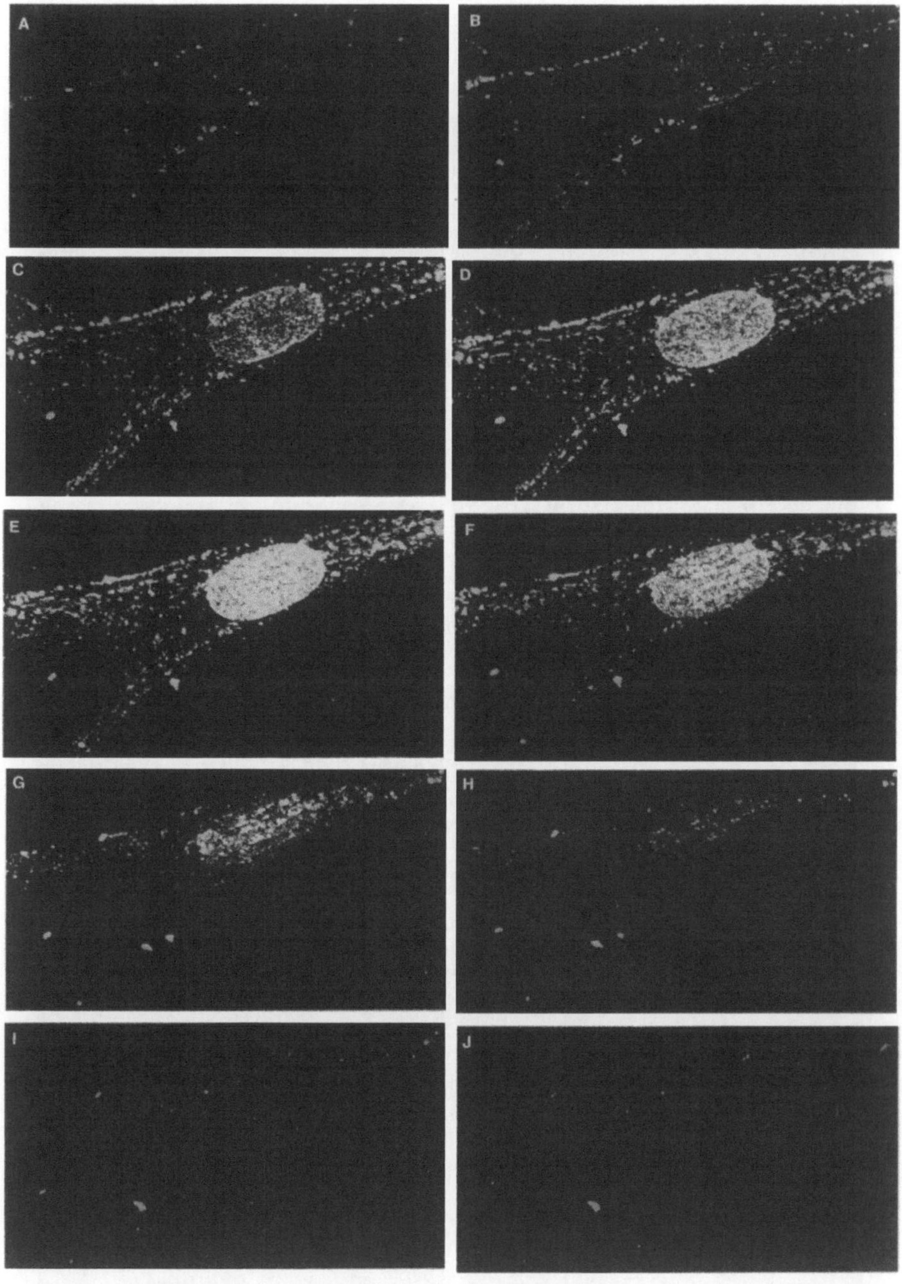

Fig. 9 A–J. Uptake of immune complexes into nuclei. In a few cases immune complexes were detected in the nucleus of living stained cells 24 h after incubation with the anti-La mab. The cells were analyzed by cLSM

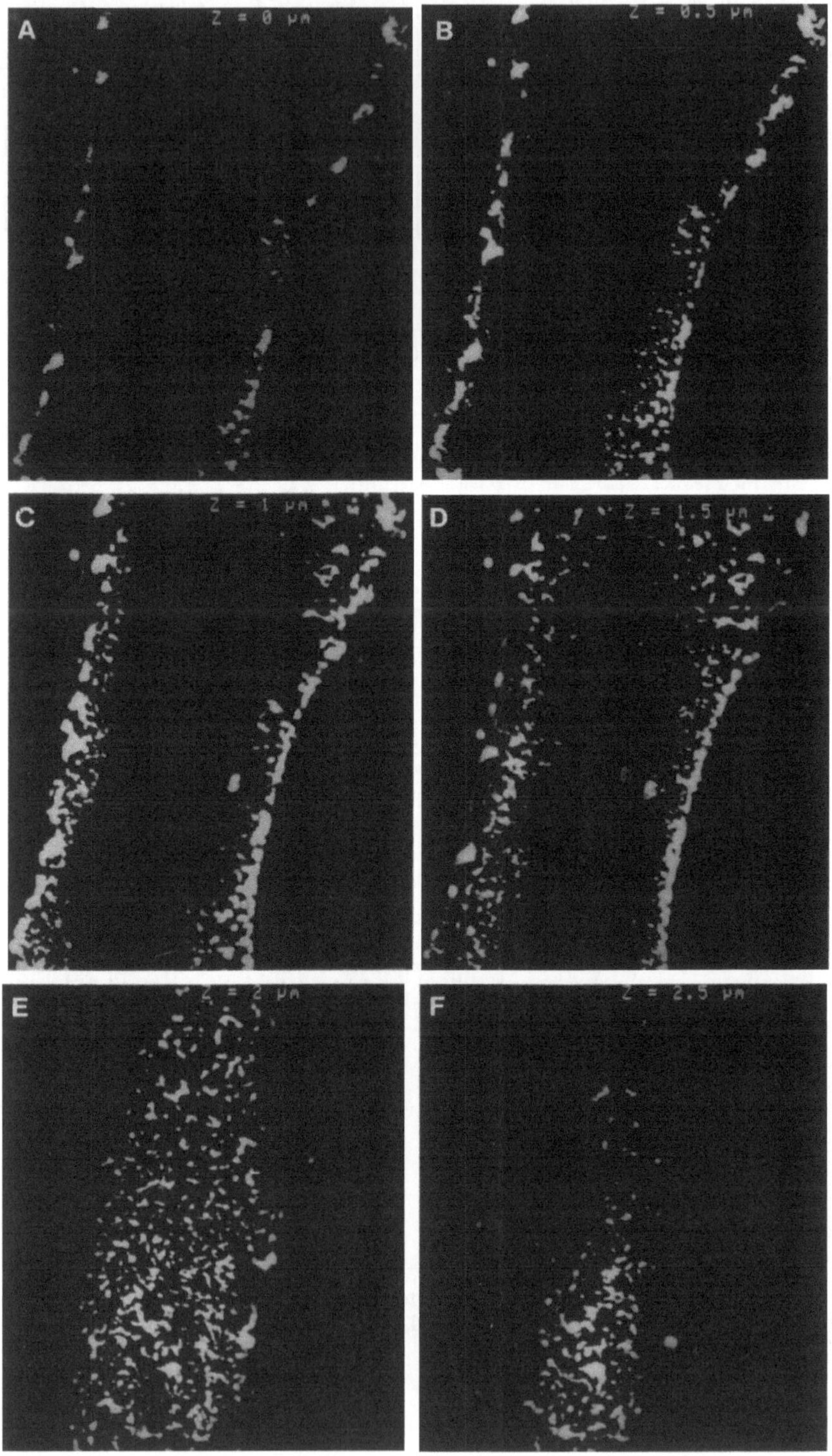
A
Z = 0 μm
B
Z = 0.5 μm
C
Z = 1 μm
D
Z = 1.5 μm
E
Z = 2 μm
F
Z = 2.5 μm

protein. In a double-blind study we found that the cell-surface staining correlated well with acute disease. Usually, the antibody titer was low during acute phases of disease (1:160), but the cell-surface staining was positive, while samples with high titers of antibodies (1:1200) obtained during symptom-free periods resulted only in an enhanced background staining. This observation implies that not all anti-La antibodies might be pathogenic to cells, but before and during disease a cell surface recognizing anti-La antibody might appear. On the other hand, our results do not exclude the possibility that pathogenic conditions migh alter the presented epitopes. For example, griseofulvin is known to evoke acute disease in La-positive patients. As shown in Fig. 4H, griseofulvin treatment induced the cell-surface staining and strongly enhanced the amount of La protein deposited to the extracellular matrix. Similar alterations were also observed for mercury chloride [5], which is known to cause autoantibody production in mice. Such conditions could convert nonpathogenic to pathogenic antibodies.

Conclusions

Taken together, our cell culture system allows us to test if sera contain pathogenic ANAs and why these ANAs are or become pathogenic. In addition, our cell culture system allows us to test whether a drug induces or protects against autoimmune reactions: Pretreatment of cultured cells with chloroquine – a drug which is used to treat autoimmune patients – followed by a treatment with griseofulvin protected the cells against antibody binding (not shown).

A further interesting aspect of our studies is the bidirectional shuttling of La protein. The reason for the cell surface translocalization of the La protein and the uptake mechanism of La protein into mitotic cells remains to be elucidated. The facts that La protein

a) is a transcription termination factor which assembles with small RNAs forming ribonucleoprotein particles,
b) colocalizes with cell-surface (growth) receptors,
c) is presented on the cell surface during mitosis and transition from the G_0 to the G_1 phase, and
d) is partially deposited to the extracellular matrix near dividing cells, combined with the knowledge that (a) viruses are nothing else than nucleoprotein particles containing polymerase(s) and transcription factors, (b) viruses can enter into cells via cell-surface receptors (for example, the T4 receptor for the HIV virus or the FGF receptor for herpesvirus), (c) some viral proteins share common regions with autoantigens that are

Fig. 10 A–F. Cell-surface staining using patient anti-La antibodies. Selected immunoadsorbed anti-La antibodies from patients with acute disease gave a staining pattern (**A–F**) similar to that of the anti-La mabs (Fig. 8). The cells were analyzed by cLSM known to exist in ribonucleoprotein particles RNPs), and (d) there are structural similarities of natural RNPs and retroviruses, give rise to the following speculation: After destruction and during regeneration of tissue, cells are activated to enter into mitosis. These premitotic cells deposit certain RNPs to the extracellular matrix. These RNPs may have the function of informing the neighboring cells of the mitotic event. These cells send RNPs back to the mitotic/postmitotic cells, leading – depending on the space remaining – either to a further mitosis cycle or to differentiation and integration in the tissue. The hypothesized natural RNPs should be the evolutionary precursors of at least the RNA viruses and should have similar cellular functions and properties, for example, activation of cellular oncogenes.

Acknowledgements. This investigation was supported by a grant from the *Deutsche Forschungsgemeinschaft* (Ba 1145/1-1). M. Bachmann is the recipient of a C3 Hermann and Lilly Schilling professorship granted by the *Stifterverband für die Deutsche Wissenschaft.* We thank Prof. Dr. W. van Venrooij and Dr. G. Pruijn (University of Nijmegen, The Netherlands) for their gift of a recombinant La protein and patient anti-La antibodies, Prof. Dr. K.-H. Meyer zum Büschenfelde and Dr. W. J. Mayet for the patient (Ma) anti-La antibodies and their support during the past 10 years, and Prof. Dr. D. Falke for his support during herpesvirus studies. Special thanks are due to graduate students K. Pfeifer, S. H. Chang, B. Diehl-Seifert, T. Zaubitzer, C. Selenka, M. Pretzlaff, and T. Metzger.

References

1. Bachmann M, Schröder HC, Falke D, Müller WEG (1988) Alteration of the intracellular localization of the La protein compared with the localization of U snRNPs. Cell Biol Int Rep 12:765–789
2. Bachmann M, Pfeifer K, Schröder HC, Müller WEG (1990) Charakterization of the autoantigen La as a nucleic acid-dependent ATPase/dATPase with melting properties. Cell 60:85–93
3. Bachmann M, Falke D, Müller WEG (1990) Is La protein involved in autoimmunization and inflammatory events during disease? Characterization of La protein as an unwinding enzyme. Mol Biol Rep 14:49–50
4. Bachmann M, Chang SH, Slor H, Kukulies J, Müller WEG (1990) Shuttling of the autoantigen La between nucleus and cell surface after UV irradiation of human keratinocytes. Exp Cell Res 191:171–180
5. Bachmann M, Chang SH, Bernd A, Mayer WJ, Meyer zum Büschenfelde KH, Müller WEG (1991) Translocation of the nuclear autoantigen La to cell surface: assembly and disassembly with the extracellular matrix. Autoimmunity 9:99–107
6. Baboonian C, Venables PJW, Booth J, Williams DG, Roffe LM, Maini RN (1989) Virus infection induces redistribution and membrane localization of the nuclear antigen La(SS-B): a possible mechanism for autoimmunity. Clin Exp Immunol 78:454–459
7. LeFeber WP, Norris DA, Ryan SR, Huff JC, Lee LA, Kubo M, Boyce ST, Kotzin BL, Weston WL (1984) UV-light induces binding of antibodies to selected nuclear antigens on cultured human keratinocytes. J Clin Invest 74:1545–1551

Cell Cultures of Fetal Human Skin Are Suitable Models for the Investigation of Mitogenic Activity in Psoriasis Sera

A. Bernd, A. Ramirez-Bosca, K. Hohlmaier, K. Dold, M. Nölle, and *H. Holzmann*

Abstract

The mitogenic activity of psoriasis sera was tested in vitro using different types of skin cells. Incubation with psoriasis sera increased the incorporation of thymidine significantly in embryonic skin fibroblasts and embryonic keratinocytes but not in adult cells. Maximum stimulation of the embryonic cells was observed with 20% of psoriasis serum in the cell culture medium. Investigations with synchronized embryonic keratinocytes indicate that the mitogenic effect results from the transfer of the cells from G_0 to G_1 phase. Our results support the existence of a mitogenic serum factor in psoriatic patients.

Introduction

Psoriasis is a chronic dermatitis which is characterized by outbreaks of papules and erythematosquamous patches. The histopathology is well known. The epidermis is acanthotic with elongated and sometimes connected rete pegs. The papillae are correspondingly elongated with slight clubbing of the tips. The stratum corneum is thickened with numerous zones of parakeratosis and small clumps of degenerated leukocytes. The stratum granulosum is absent, and the suprapapillary areas of the rete are narrow (Fig. 1).

The pathogenesis of the disease has not yet been definitively solved and a causal therapy is not possible at present. There is no doubt today that genetic processes play an important role [19]. In addition, several trigger factors are known. First manifestations of psoriasis have been observed after injury, after bacterial and/or viral infection, after application of chemical substances or drugs (e.g., lithium, chloroquine, systemic corticoids, beta blockers, ethyl alcohol, phenylbutazone, indomethacin). Psoriasis also is influenced by endocrine and metabolic factors (progesterone, hypocalcemia), and by psychological factors such as stress (for ref. see [17]).

A changed profileration behavior in connection with psoriasis has been verified for keratinocytes, fibroblasts, endothelial cells, and cells of the skeleton [5, 8, 16]. Dermal fibroblasts are supposed to play an essential role

A. Bernd, J. Bereiter-Hahn,
F. Hevert, H. Holzmann (Eds.)
Cell and Tissue Culture Models
in Dermatological Research

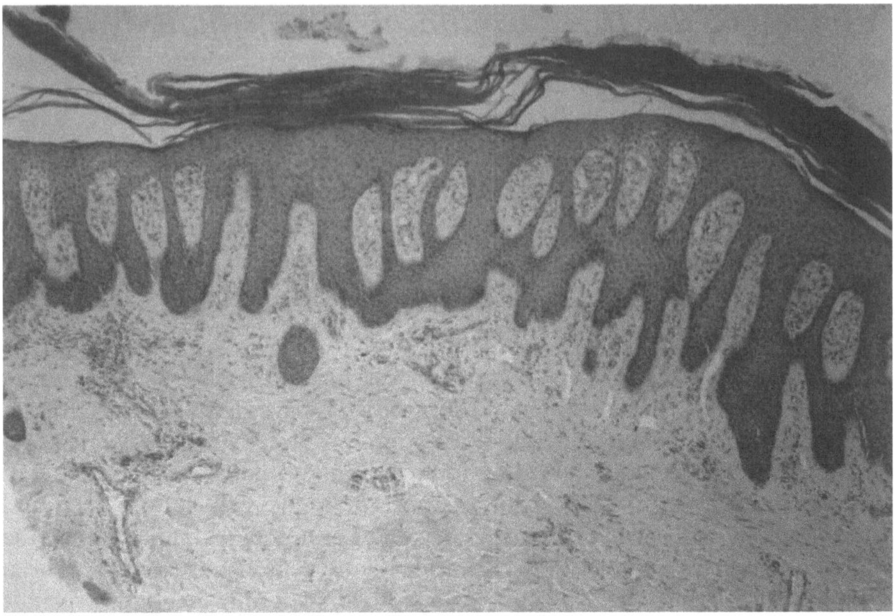

Fig. 1. Histology of psoriasis vulgaris with acanthosis, parakeratosis, elongated rete pegs, and papillomatosis. [From 9]

in the induction of epidermal cell hyperproliferation [18]. The exanthematic character of psoriasis symptoms (Fig. 2) indicates an endogenous humoral control mechanism. Thus the involvement of mitogenic factors seems reasonable. Therefore, we tried to model specific characteristics of the disease, attempting to mimick experimentally the "denominator" of psoriasis as proposed by Gans [6]. No symptoms similar to psoriasis occur in animals and none are inducible. We therefore used cell cultures of human origin and followed their reactions to serum from patients with psoriasis. Psoriasis sera increase particularly the proliferation rate of human embryonic skin cells [10, 11].

Material and Methods

Fetal calf serum was obtained from Boehringer (Mannheim, FRG), culture medium, trypsin-EDTA solution (0.125%/0.1%), and penicillin/streptomycin (10000 U/10000 μg/ml) from Biochrom KG (Berlin, FRG), hydrocortisone, cholera toxin, adenine and epidermal growth factor from Sigma (Deisenhofen, FRG), and ^{3}H-thymidine (74 GBq/mmol) from New England Nuclear (Dreieich, FRG). All remaining salts and solutions were obtained

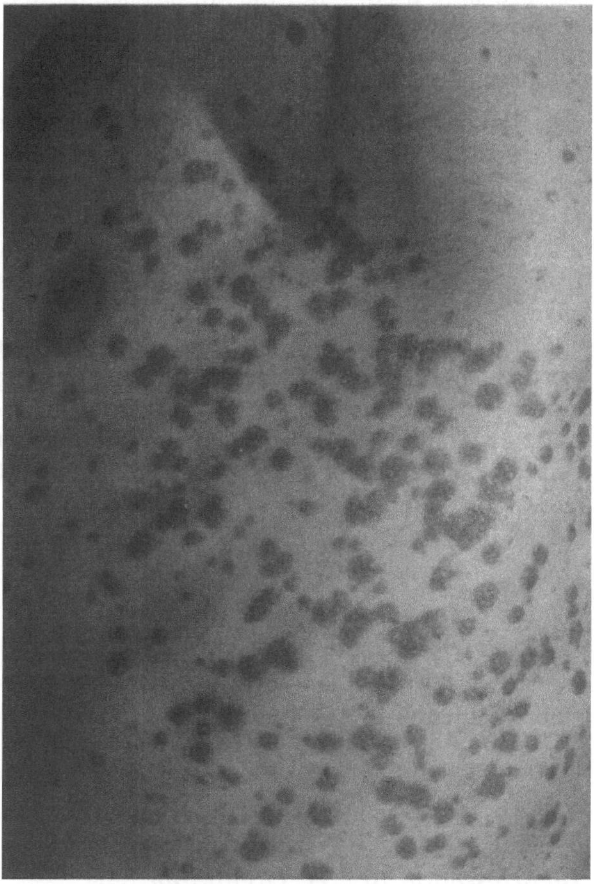

Fig. 2. Psoriasis guttata with an exanthematic dissemination of papules. [From 9]

in their highest purities from E. Merck (Darmstadt, FRG). Radiosterilized (cobalt 60) plastic bottles from Nunc (Wiesbaden, FRG) were used as culture vessels.

Cell Culture

Adult and fetal human skin fibroblasts were isolated from small skin pieces. They were cultivated in collagen-covered plastic culture flasks in RPMI 1640 culture medium, then buffered with sodium bicarbonate containing 20% fetal calf serum, penicillin (1%), and streptomycin (1%) at 37 °C in a 5% CO_2 atmosphere. The medium was changed twice a week. After approximately 14 days, the skin fibroblasts which emigrated from the skin were passaged for the first time. For the followings tests only cells of the 5th–10th passage were used. After the 4th passage the fetal calf serum in the culture medium was reduced from 20% to 10%. The doubling time of the logarithmically growing cells was approximately 48 h.

In the same way, embryonic human keratinocytes were isolated. Small pieces of skin were put into culture vessels and were incubated in Dulbecco/Ham's F12 (3:1) medium supplemented with 5% heat-inactivated fetal calf serum, hydrocortisone (0.4 μg/ml), cholera toxin ($10^{-10}M$), adenine ($1.8 \times 10^{-4}M$), and epidermal growth factor (10ng/ml) at 37 °C in a CO_2 atmosphere of 5%. After about 14 days the keratinocytes which emigrated from the skin pieces were passaged for the first time. The medium was changed twice a week.

HaCaT cells (kindly provided by Prof. Fusening from The German Cancer Research Center in Heidelberg), from a spontaneously transformed human keratinocyte cell line [3, 7], were cultivated in sodium bicarbonate-buffered Dulbecco's culture medium supplemented with 5% fetal calf serum, penicillin, and streptomycin at 37 °C in a 10% CO_2 atmosphere. The culture medium was changed twice a week.

Synchronization of Cell Cultures

For synchronization confluent cell layers were passaged; after 24 h the serum content was lowered to 0.4%. After an incubation period of 4 days at 37 °C, the cells were stimulated by increasing the serum content to 10% for fibroblasts and to 5% for keratinocytes.

Incorporation of ^{3}-H-Thymidine

In order to determine the ^{3}H-thymidine incorporation rate, the cells were sown at a density of 1×10^4 cells/cm^2 into multidishes (24 wells, culture area = 1.9 cm^2/well) as previously described [1]. After 24 h, the medium was renewed and the cultures were incubated for 48 h at 37 °C with the test sera. During the last hour of the incubation time, 37 kBq ^{3}H-thymidine (specific act. 74 GBq/mmol) per well was added. The cells were washed with PBS and harvested by trypsinization (20 min, 37 °C), and the resulting cell suspension was treated for 20 min with equal portions of 10% ice-cooled trichloroacetic acid. The acid precipitate was then isolated by means of filtration over a Whatman fiberglass filter (2.4 GF-C, Whatman, Maidstone, UK) and washed with 5% trichloroacetic acid. Each filter disc was then treated for 2 h at 70 °C with 500 μl tissue solubilizer and the radioactivity was then measured with 5 ml scintillator per disc in a liquid scintillation counter. The protein concentration of the cultures was determined according to Lowry et al. [13]. The measured radioactivity was related to the amount of protein per well.

Human Sera

Sera from psoriasis patients before therapeutic treatment (acute psoriasis guttata, chronic plaque-shaped psoriasis, and chronic active psoriasis vulgaris), sera from persons who were free of psoriasis as proven by clinical investigation and anamnesis, and sera from a human serum pool tye AB were tested. Psoriasis-associated HLA-markers were not found in the control collective. All serum samples were inactivated for 30 min at 56 °C and preserved at –20 °C until testing.

Results

First the effect of each serum of a psoriasis collective and of a control collective on the growth of different cell types and permanent cell lines was tested. The rates of thymidine incorporation are given as percentages of the controls (cells grown with 10% human AB pool serum). Psoriasis sera exerted a significant mitogenic effect in fetal skin fibroblasts only (Fig. 3).

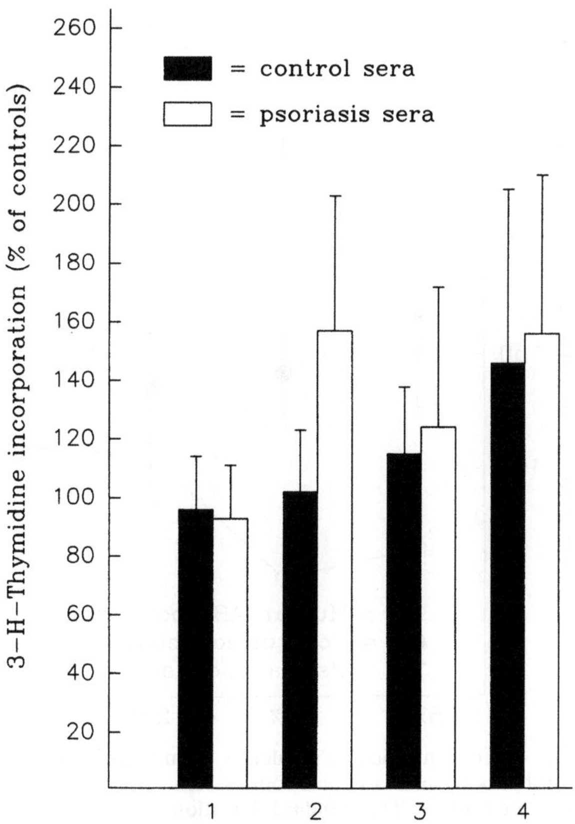

Fig. 3. Effect of psoriasis and control sera (10%) on the DNA synthesis of different cultures of human skin cells. The results of one cell line in each case are shown. The cells were sown with 10% fetal calf serum. After 24 h the medium was renewed and the fetal calf serum was replaced with test sera. After 48 h the incorporation of ^{3}H-thymidine in DNA was measured. The incorporations of ^{3}H-thymidine are presented as percent of the controls treated with AB pool serum ($n = 6$). The standard deviations are indicated. Statistics of the effect of psoriasis sera and controls: Fetal fibroblasts, different with $p < 0.05$; adult normal fibroblasts, not different; adult psoriatic fibroblasts, not different; HaCaT cell line, not different. *1*, Human embroynic skin fibroblasts; *2*, fibroblasts from adult human skin; *3*, human skin fibroblasts, adult patient with psoriasis; *4*, human keratinocytes (HaCaT-line)

No differences were induced in fibroblasts from adults. Also with skin fibroblasts of an adult patient with psoriasis, no differences between the psoriasis and control sera were found. However, the cell growth was stimulated more by the sera of both collectives than by the purchased AB pool serum. As a example for established cell lines, the results with HaCaT cells are given. Here, too, there was no difference between the two sera concerning the mitogenic effect (Fig. 3).

In a further experiment the psoriasis and the control sera each were pooled and the serum content of the culture medium was varied. The results are given in Fig. 4. The rates of thymidine incorporation are indicated in disintegrations per minute (dpm)/mg protein. At 5% serum no significant difference between the psoriasis and the control sera was found. When the serum content was increased to 10%, a significant difference between the effect of psoriasis and control and AB pool sera was observable. This difference was most distinct with the use of 20% (Fig. 4).

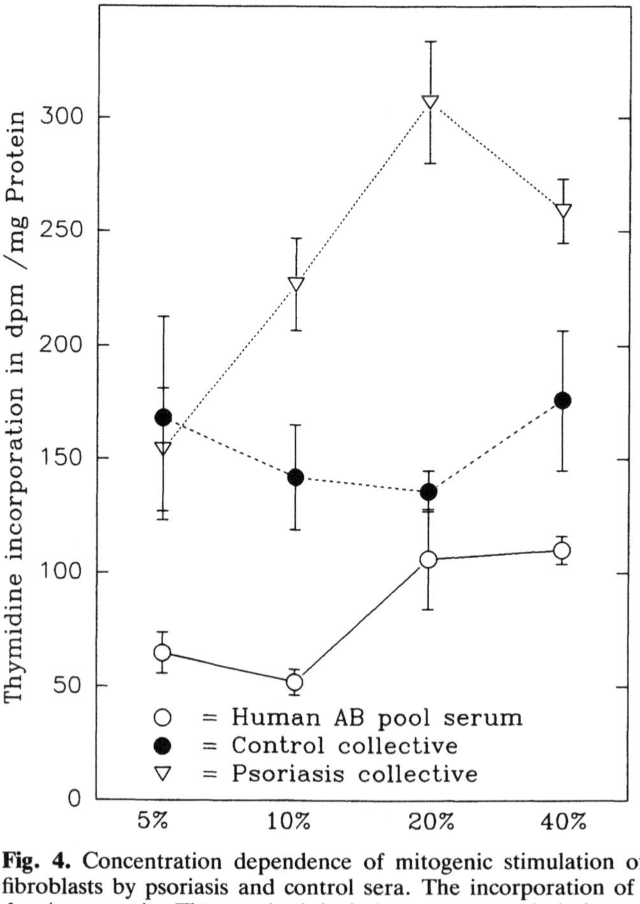

Fig. 4. Concentration dependence of mitogenic stimulation of human embryonic skin fibroblasts by psoriasis and control sera. The incorporation of ^{3}H-thymidine is given in dpm/mg protein. The standard deviations represent six independent experiments

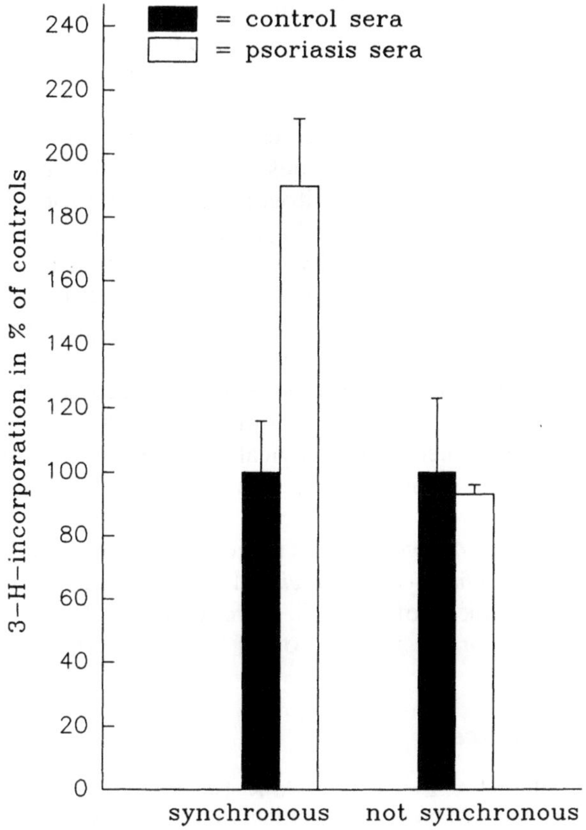

Fig. 5. Effect of psoriasis serum on the DNA synthesis of human keratinocytes in synchronized and nonsynchronized cultures. The cells were incubated with human AB pool serum, pooled serum of the normal collective, and pooled serum of the psoriasis collective at 37 °C, as described under Methods. Then the incorporation of ^{3}H-thymidine was measured and referred to the protein concentration of the cultures. The incorporations of ^{3}H-thymidine are indicated proportional to the control with 5% AB pool serum. The standard deviations of six experiments are given.

In order to test the effect of the psoriasis sera on the growth of fetal keratinocytes, the cells were arrested in the G_0/G_1 phase with a serum content of 0.4%. Afterwards, the cells were stimulated to grow synchronously with the respective test sera. With this experimental design a clear mitogenic effect caused by psoriasis sera was shown (Fig. 5).

Discussion

In 1952, an endogenous humoral control mechanism for psoriatic skin manifestations was postulated by Gans [6]. In 1985, Priestley and Adams [15] found no difference between the effect of psoriatic and non-psoriatic sera. They used cultures of human skin fibroblasts from adults to detect mitogenic effects. In 1987, Majewski et al. [14] found growth-stimulating effects of psoriasis sera especially on endothelial cells and mononuclear granulocytes. This study contains a short reference to the growth-stimulating effect in connection with fetal skin fibroblasts [14]. Recently, Laubscher et al.

showed a growth-stimulating effect of psoriasis sera on cultures of fibroblasts from adults. However, a significant difference between psoriasis sera and controls was proven only when the serum content of the culture medium was 20% [12].

Meanwhile, we have tested different cultures of human skin cells and many permanent cell lines, in order to find a good indicator for mitogenic serum activity. The mitogenic effect of psoriasis sera was clearly demonstrated with cells of embryonic origin. The optimal serum concentration in the system presented by us in 20%, as with Laubscher et al. [12].

Our investigations with embryonic keratinocytes show that under the influence of psoriasis sera possibly a larger amount of cells is transferred from the G_0/G_1 phase into the proliferation cycle, similar to observations made in situ [2, 4]. We have not yet been able to find a species of cells that would be more suitable for our studies than embryonic skin cells. Apparently, their special physiological situation makes these cells most sensitive for mitogenic stimulation by psoriasis sera.

Acknowledgements. This project was supported by the *Willy Pitzer-Stiftung*, the *Basotherm-Förderkreis e.V.*, and the *Landesversicherungsanstalt Hessen*. We thank Prof. Dr. J. Bereiter-Hahn for helpful discussions and Mrs. Kleber for the preparation of the manuscript.

References

1. Bernd A, Breuer M, Dold K, Holzmann H (1990) Uptake kinetics and the effect of minoxidil on human keratinocyte cultures. Arzneimittelforschung 40:413–416
2. Beurskens T, Chang A, van Erp PEJ, van de Kerkhof PCM (1989) Epidermal proliferation and accumulation of polymorphonuclear leukocytes in the psoriasis lesion. Dermatologica 178:67–72
3. Boukamp P, Petrusevska RT, Breitkreutz D, Hornung J,Markham A, Fusenig N (1988) Normal keratinization in a spontaneously immortalized aneuploid human keratinocyte cell line. J Cell Biol 106:761–771
4. De Mare S, de Jong E, van Erp PEJ, van de Kerkhof PCM (1990) Markers for proliferation and keratinization in the margin of the active psoriatic lesion. Br J Dermatol 122:469–475
5. Fassbender HG (1987) Reaktionsweisen der Gelenkstrukturen. In: Holzmann H, Altmeyer P, Marsch WC, Vogel HG (eds) Dermatologie und Rheuma. Springer, Berlin Heidelberg New York, pp 57–71
6. Gans O (1952) Zur Pathogenese der Psoriasis vulgaris. Hautarzt 5:193–198
7. Görmar FE, Bernd A, Bereiter-Hahn J, Holzmann H (1990) A new model of epidermal differentiation: induction by mechanical stimulation. Arch Dermatol Res 282:22–32
8. Holzmann H (1987) A dermatological view of psoriatic arthropathy. In: Orfanos CE, Stadler R, Gollnick H (eds) Dermatology in five continents. Proceedings of the XVII World Congress of Dermatology, Berlin, May 24–29, 1987. Springer, Berlin Heidelberg New York, pp 80–85
9. Holzmann H, Altmeyer P, Marsch WC, Vogel HG (eds) (1987) Dermatologie und Rheuma. Springer, Berlin Heidelberg New York
10. Holzmann H, Bernd A, Hohlmaier K (1988) Proliferation of non-psoriatic human fibroblasts in vitro by serum from patients with psoriasis. Lancet 2:1031

11. Holzmann H, Bernd A, Hohlmaier K (1989) Zur Kenntnis eines mitogenen Serumfaktors bei Psoriasis. Hautarzt 40:104–105
12. Laubscher TA, Harris N, Chalton DO (1991) The mitogenic effects of sera from psoriatic subjects on normal dermal fibroblasts: an absence of correlation with the clinical activity of psoriasis. Br J Dermatol 124:163–167
13. Lowry OH, Rosenbrough NS, Furr AL, Randall RJ (1951) Protein measurement with the Folin phenol reagent. J Biol Chem 193:265–275
14. Majewski S, Tigalonowa M, Jablonska S, Polakowski I, Janczura E (1987) Serum samples from patients with active psoriasis enhance lymphocyte-induced angiogenesis and modulate endothelial cell proliferation. Arch Dermatol 123:221–225
15. Priestley GC, Adams LW (1985) Mitogenic effects of sera from normal and psoriatic subjects on human skin fibroblasts. Arch Dermatol 277:13–15
16. Ramirez-Bosca A, Martinez-Ojeda L, Valcuende-Cavero F, Castells-Rodellas A (1988) A study of local immunity in psoriasis. Br J Dermatol 119:587–595
17. Rassner G (1980) Psoriasis. In: Korting GW (ed) Dermatologie in Praxis und Klinik. Thieme, Stuttgart, pp 10.1–10.26
18. Saiag P, Coulomb B, Lebreton C, Bell E, Dubertret L (1985) Psoriatic fibroblasts induce hyperproliferation of normal keratinocytes in a skin-equivalent model in vitro. Science 230:669–672
19. Wuepper KD, Coulter SN, Haberman A (1990)Psoriasis vulgaris: a genetic approach. J Invest Dermatol 95:2S–4S

Influence of Human Sera on the Adherence of Candida albicans to Buccal and Intestinal Cells in Culture

H. Hänel, M. Buslau, F. Dey, A. Bernd, and *J. Bereiter-Hahn*

Abstract

We investigated whether serum of healthy persons or seborrhoic patients influences the adhesion of *Candida albicans* isolated from healthy or seborrhoic patients to human intestine and buccal cells of healthy or seborrhoic patients. Serum from healthy persons inhibited the adhesion to intestine cells less than serum of seborrhoic patients did. Heat inactivation did not influence the results significantly. The adhesion of the different yeast strains to buccal cells showed a large variation. One yeast strain isolated from a seborrhoic patient adhered less than that of a healthy person.

Introduction

Cell cultures are used by numerous groups to investigate the adhesion of *Candida albicans,* because adhesion is the first and most important step in the process of body colonization. It has been a controversal subject for some decades. Ever since it became known that adhesion of *Candida* to mammalian epithelia is a specific process, it has been considered a target for chemotherapy and prophylaxis. Glycosides and phospholipase, besides other possible factors, provide a molecular basis for *Candida* adhesion and penetration of epithelia. The role of glycosides has been reviewed by Douglas [5] and by Calderone and Braun [3]. Phospholipase activity is assumed to promote penetration of hyphae through epithelia [16]. We were interested in the phospholipases as part of the adhesion and penetration process, especially with regard to different groups of patients. Seborrhoic patients [2, 10] and patients with some forms of psoriasis [9, 14, 15] harbor *Candida* strains which show a significantly higher liberation of phospholipases than *Candida* strains from the intestines of patients not suffering from these skin diseases. In the study presented here we investigated whether serum of healthy or of seborrhoic patients show any difference in standardized *Candida albicans* adhesion assays.

A. Bernd, J. Bereiter-Hahn,
F. Hevert, H. Holzmann (Eds.)
Cell and Tissue Culture Models
in Dermatological Research

Material and Methods

Cell Cultures

Epithelial intestine cells (ATCC CCL 6) from human embryonal ileum, obtained from the American Type Culture Collection (Rockville, Maryland, USA), were kept in Dulbecco's minimal essential medium (MEM; Biochrom, Seromed, Berlin, FRG) supplemented with 10% fetal calf serum (Boehringer Mannheim, FRG), 1% penicillin and 1% streptomycin in bicarbonate buffer (Flow Laboratory, Meckenheim, FRG). Temperature was 37° C with 10% CO_2 in the atmosphere of the incubator. Twice weekly the cells were trypsinized and diluted 1:5.

Yeasts

Candida albicans was obtained from human stool and cultured on Sabouraud agar plates (Merck, Darmstadt, FRG) containing 4% *D*-glucose. The plates were kept at 4° C and transferred once a month. Two different human intestine isolates were used, one from a patient with seborrhoic eczema (6216 S) and one from a healthy person (6878 N).

Buccal Cells

Two sets of human buccal cells were used; from six patients suffering from seborrhoic eczema and from ten healthy persons (age 21–50 years, male). Cells were scraped off the inner cheeks using cotton swabs. They were resuspended in PBS. From each person 4×10^5 – 1×10^6 cells were obtained. Cells were washed twice in PBS and counted in a Neubaur chamber. Final concentration was adjusted to 1–2 × 10^5/ml cells.

Sera

From a group of 14 patients with seborrhoic eczema (male, age 19–55 years) the serum was pooled. Control group subjects (n = 13) did not suffer from any skin diseases (male, 25–45 years). The sera were inactivated at 56° C for 30 min and sterilized by filtering through Servinex (0.22 μm) (Millipore, Eschborn, FRG). The sera which were mixed with Dulbecco's MEM were also sterile filtered using Servinex (0.22 μm) (Millipore).

Adherence Tests

Intestine cells were cultered on glass slides (Lab. Tek, Nunc, Wiesbaden, FRG) for 48 h: 1.3×10^5 cells/cm^2 were seeded to obtain a confluent monolayer after 48 h, and 2 × 10^7 yeast cells/cm^3 were then added. Incubation was carried out in a moist chamber at 37° C ± 2° C for 30, 60 or 90 min. During this time the cultures were shaken at a frequency of 80 shakes/min. After incubation, the medium was decanted and the cells were washed twice using PBS. Then the cultures were fixed with methanol acetone (1:1) at 4° C for 15 min, washed with PBS, and stained using Fungiqual (Ciba Corning, Fernwald, FRG). The number of years cells was counted using fluorescence microscopy (excitation: 400 nm). Yeasts with daughter cells and germ tubes were counted as one cell. Yeast cells adhering to glass were not counted. One microscopic field measured 0.0113 mm^2. The 25 fields for counting were selected arbitrarily. Mean values were compared using the Wilcoxson test for two independent samples [19]. Multiple regression analysis was carried out using a computer program (Anonymus, Hoechst AG, Frankfurt, FRG).

Buccal Cell Suspensions

One milliliter of buccal cell suspension in PBS was mixed with 1 ml yeast cell suspension (8 × 10^8/ml) in Eppendorf cups. They were kept in a waterbath at 37° C for 1 h at a shaking frequency of 80/min. A sample without yeasts was checked for natural yeast contamination. After incubation the tubes were quickly transferred to 4° C and fixed by addition of glutaraldehyde (4% final concentration) for 15 min. Nonadherend yeast cells were separated by the Percoll method [6]. Cells were transferred to slides using a Zytofuge (Hetterich Universal, FRG) and stained after Pappenheim [18]. Counting was done as mentioned above except for the microscope (brightfield technique).

Results

Intestine Cells

The adhesion of *Candida* cells to human intestinal cultures decreased with increasing concentration of human serum (Table 1). The difference was most significant between 5% and 10%. The influence of inactivated serum differed only slightly from that of normal serum. The difference was significant at 5% serum concentrations only (a = 0.5%; Wilxocson test).

The influence of pooled serum of seborrhoic patients on the adherence was as follows: In medium not containing serum (control) 115 yeast cells/1000 epithelial cells (day 1) and 78 yeast cells/1000 epithelial cells (day 2) were counted. Serum concentrations were adjusted to 10% in all experiments. The data of two experimental sets are presented in Fig. 1. Serum from healthy patients decreased the adhesion (both normal and inactivated;

Table 1. Adherence of *Candida albicans* 6878 to human intestine cells after 30 min incubation with different amounts of human serum (adherent yeast/1000 epithelial cells)

Serum %	Inactivated		Normal	
	mean	sd	mean	sd
0	–	–	116	11
5	121	16	176	23
10	75	3	67	27
20	57	21	63	9
30	56	11	62	3

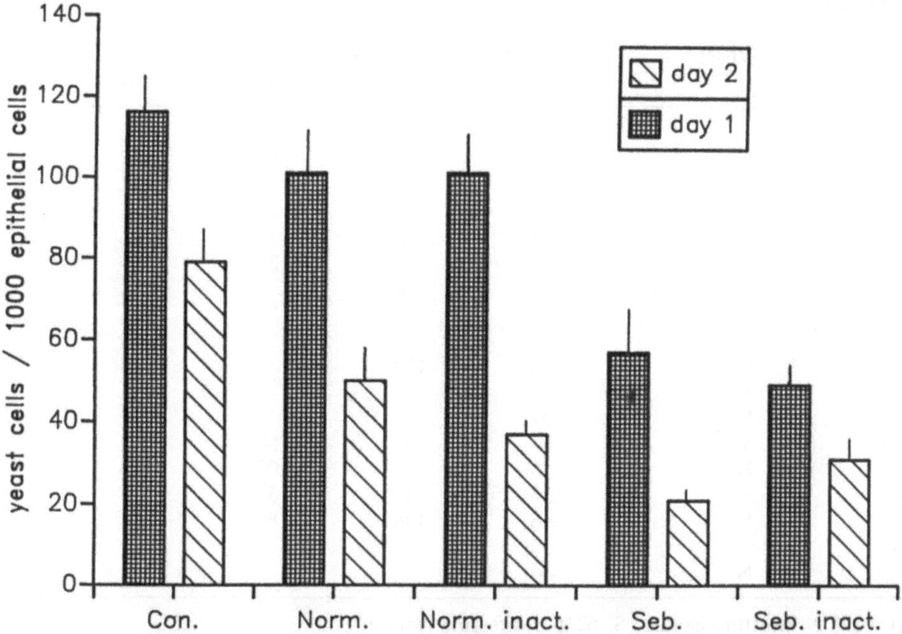

Fig. 1. Influence of human serum from seborrhoic patients (*Seb*) and healthy persons (*Norm*) on the adherence of *Candida albicans* to an intestinal cell monolayer. Incubation with 10% serum (fresh and inactivated), means from three parallel tests and standard deviations of the mean. Control group did not contain any serum. Two identical sets of experiments were carried out on different days (day 1 and day 2). *Con*, Control; *Norm*, healthy; *Seb.*, seborrhoic; *inact.*, inactivated serum

a = 5%); sera from seborrhoic patients had an even stronger inhibitory effect.

Inactivation of serum from seborrhoic patients did not further increase the inhibition. In the experiments carried out on day 2 the adherence was generally below the level of day one (Fig. 1), but the decrease due to the serum addition was obvious on both days.

Experiments with buccal Cells

Results are shown in Table 2. The yeast strain from the patient with seborrhoic dermatitis (6216 S) adhered in all cases less than the strain (6878 N; a < 1%) from the healthy donators. Results with cells from each single donator varied from day to day significantly. Yeast cells (both strains) adhered to buccal cells from seborrhoic as well as from healthy donators. The number of available donators decreased from 9 to 2 over 4 days.

Table 2. Adherence of *Candida albicans* to freshly exfoliated buccal cells of seborrhoic patients (S 1–S 6) and healthy donators (N 1–N 3) means of six countings given as yeast cells/1000 buccal cells

Donator	Yeast	Adherent yeast cells/100 buccal cells			
		day 1	day 2	day 3	day 4
S1	S	90	43	10	32
	N	250	189	48	121
S2	S	41	10	18	
	N	72	17	39	
S3	S	10	26	19	
	N	26	164	39	
S4	S	31	10	8	
	N	159	55	53	
S5	S	54	61		
	N	136	112		
S6	S	29			
	N	229			
N1	S	43	39	50	41
	N	118	81	102	73
N2	S	27	31	35	
	N	96	148	88	
N3	S	38			
	N	84			

N, 6878 N intestine isolate; S, 6216 S intestine isolate

Discussion

Adhesion to Human Intestine Cells

Numerous authors have investigated the influence of serum on the adhesion process [17]. Being a rather complex mixture with extremely varying components, serum influences the adhesion in a multitude of ways. Obviously, specific receptors on the surface of *Candida albicans* can be occupied by normal serum components, thus inhibiting the adhesion process. Inactivated serum which does not contain active complement proteins did not inhibit adhesion to buccal cells [17]. In our experiments the complement system seemed to be of inferior significance because the normal serum was about as active as the inactivated one. However, the involvement of binding to surface receptors is rather likely because of the saturation-type dependence of adherence inhibition. The increase from 10% to 20% serum had no significant effect. The variability of composition of sera from different donators as regards, for instance, carbohydrate content and nutrition-related changes was partly randomized by the process of pooling. The question remains why the serum of seborrhoic patients had a more pronounced inhibitory effect on the adhesion. Both serum pools were checked for IgG and IgM antibodies against *Candida albicans* (Elisa test); in both pools only IgG were present. The investigations did not give any indications as to how the serum of seborrhoic patients inhibits the adhesion. However, if the results are of general significance it would exclude a predisposition of seborrhoic patients for *Candida albicans* as assumed by Korting [12] and Buslau et al. [2]. Which components of the serum of seborrhoic patients exerts the inhibitory effect remains to be elucidated. Solving this problem could bring new light to the mechanism of yeast-epithelia interaction.

Experiments with Buccal Cells

Pretreatment of cells with trypsin prevents normal adhesion of yeasts [20]. Therefore, we used buccal cell cultures at least 2 days after trypsinization and buccal cells which had never been treated with trypsin. The high variation in results even from the same donators (Table 2) is in accordance with data from King et al. [11]. Types of food consumption influence the surface components of the buccal cells just as the serum does. The carbohydrate-related receptors and glycoproteins which influence the adhesion process may play a crucial role [4, 5]. Due to the variability, no difference between cells of seborrhoic patients and healthy donators was detected. In further experiments a more severe control of the nutritive conditions of the donators and the sampling time is required.

The controversal question of whether *Candida albicans* strains from healthy persons adhere less than those from patients with skin desease and/or deep-seated mycoses has been investigated by numerous authors [1, 13]. This focuses the question also on the phospholipase production of the strains. In opposition to Barret-Bee et al. [1], we found that an isolate with an exceptionally strong phospholipase production adhered much less to both intestinal and buccal cells than the two strains 6216 (S) and 6878 (N) which had a much lower phospholipase production in vitro (preliminary results). Therefore, we assume that the phospholipase production might not be a very important factor for the adhesion process [7, 8].

Our investigations with only two strains, one from a healthy person and one from a seborrhoic patient, do not indicate any general significant difference. However, preliminary investigations with large collections of *Candida albicans* strains from healthy and seborrhoic patients showed that the seborrhoic patients harbor strains with a statistically higher phospholipase production [10]. We will use these strain collections for further in vitro adhesion experiments.

References

1. Barrett-Bee K, Hayes Y, Wilson RG, Ryley JF (1985) A comparison of phospholipase activity, cellular adherence, and pathogenicity of yeasts. J Gen Microbiol 131:1217–1221
2. Buslau M, Hänel H, Holzmann H (1989) Zur Bedeutung von Hefen beim seborrhoischen Ekzem. Hautarzt 40:611–613
3. Calderone RA, Braun PC (1991) Adherence and receptor relations of *Candida albicans*. Microbiol Rev 55:1–20
4. Collins-Lech C, Kalbfleisch JH, Franson TR, Sohnle PG (1984) Inhibition by sugars of *Candida albicans* adherence to human buccal mucosal cells and corneocytes in vitro. Infect Immun 46:831–834
5. Douglas LJ (1986) Adhesion of Candida species to epithelial surfaces. CRC Crit Rev Microbiol 15:27–43
6. Fischer SM, Nelson KDG, Reiners JJ (1982) Separation of epidermal cells by density centrifugugation. A new technique for studies on normal and pathological differentiation. J Cutan Pathol 9:43–49
7. Ghannoum MA, Abu-Elteen KH (1990) Pathogenicity determinants of Candida. Mycoses 33:265–285
8. Hänel H, Pastowsky S (1990) Pathogenicity factors in *Candida albicans* (Abstr). 4th International Conference an Mycology 28 August–3 September 1990, Regensburg, p 234
9. Hänel H, Menzel I, Holzmann H (1988) Hohe Phospholipase A-Aktivität von *Candida albicans* aus dem Darm von Psoriatikern. Mycoses 31:451–453
10. Hänel H, Schütz KD, Buslau M, Holzmann H (1990) Untersuchungen von *Candida albicans* Isolaten aus Stuhlproben von Neurodermitikern. In: Kurzfassungen 24. wissenschaftliche Tagung der deutschsprachigen mykologischen Gesellschaft, 14–17 June 1990, Göttingen, p 9
11. King RD, Lee JC, Morris AL (1980) Adherence of *Candida albicans* and other Candida species to mucosal epithelial cells. Infect Immun 21:64–68
12. Korting HC (1988) Seborrhoisches Ekzem bei HIV-Infektion. Aktuel Dermatol 14:404–407
13. McCourtie J, Douglas LJ (1984) Relationship between cell surface composition, adherence and virulence of *Candida albicans*. Infect Immun 45:6–12
14. Menzel I, Holzmann H (1986) Überlegungen zum seborrhoischen Kopfekzem und der Psoriasis capilitii im Zusammenhang mit intestinalen Mykosen. Z Hautkr 61:451–454
15. Menzel I, Holzmann H (1988) Seborrhoisches Ekzem, Psoriasis und intestinaler Hefebefall. Ein neues pathogenetisches Konzept? Aktuel Dermatol 14:314–316
16. Pugh D, Cawson RA (1975) The cytochemical localisation of phospholipase A and lysophospholipase in *Candida albicans*. Sabouraudia 13:110–115
17. Ray TL, Digre KB, Payne CD (1984) Adherence of Candida species to human epidermal corneocytes and buccal mucosal cells: correlation with cutaneous pathogenicity. J Invest Dermatol 83:37–41
18. Romeis B, Böck P (eds) (1989) Mikroskopische Technik, 17th edn. Urban and Schwarzenberg, Munich
19. Sachs L (1984) Angewandte Statistik, 6th edn Springer, Berlin Heidelberg New York
20. Sobel JD, Myers PG, Kaye D, Levison ME (1981) Adherence of *Candida albicans* to human buccal epithelial cells. J Infect Dis 143:76–80

Modulation of 1,25-Dihydroxyvitamin D_3 Receptor Expression in HaCaT Keratinocytes

J. Reichrath, U. Hügel, G. Klaus, N.E. Fusenig, and *E. W. Rauterberg*

Introduction

The skin is a key tissue for the synthesis and function of vitamin D_3. The most potent metabolite 1,25-dihydroxyvitamin D_3 (1,25-D_3) exerts its effects via binding to an intranuclear vitamin D_3 receptor (VDR). Increasing evidence suggests an immunoregulatory function of the potent steroid hormone [12]. VDR is expressed by monocytes and activated lymphocytes. Since an involvement of skin immune cells (SIC) has been suggested in psoriasis, these cells might be a target of 1,25-D_3 in the topical treatment of psoriasis [6]. We recently demonstrated a significant increase in VDR expression in epidermal keratinocytes and skin immune cells in lesional as compared with nonlesional psoriatic skin by applying an immunohistochemical method and the monoclonal antibody 9A7 gamma to the VDR [10]. Since 1,25-D_3 was shown to promote differentiation and to block proliferation in keratinocytes, this might be the second effect of topical 1,25-D_3 treatment. In any case, the reason for the local overexpression of VDR in psoriasis is completely unknown. Immune cells of human skin or mediators of local inflammation might modulate the effects of 1,25-D_3 on keratinocytes via VDR expression. Release of cytokines was demonstrated under various local inflammatory conditions. We therefore focused on the question of whether VDR expression in keratinocytes might be modulated by lymphokines/cytokines and tested this hypothesis in the HaCaT keratinocyte cell line in vitro.

Materials and Methods

Cell Culture

HaCaT keratinocytes [1] were cultivated in Eagle's medium with 5% fetal calf serum at 37 °C in 5% CO_2. After an overnight period without fetal calf serum, the cells were incubated for 48 h with 1% fetal calf serum and the respective recombinant cytokines to be tested – IL-1 alpha, TNF alpha (Boehringer, Mannheim, FRG), interferon gamma (Rentschler, Laupheim, FRG), PDGF (Dianova, Hamburg, FRG), or recombinant C5a (Sigma,

A. Bernd, J. Bereiter-Hahn,
F. Hevert, H. Holzmann (Eds.)
Cell and Tissue Culture Models
in Dermatological Research

Munich, FRG). All mediators were applied at final concentrations of 10 nm/l.

Cell Preparation and Binding Studies

HaCaT suspensions obtained by gentle trypsinization with 0.25% trypsin/0.05% EDTA (Sigma, Munich, FRG) were homogenized in TED buffer. KTED cell extracts and saturation analyses according to Scatchard were carried out as described previously [4, 6, 7]. 1,25(OH)$_2$-[^{3}H]D$_3$ (Amersham-Buchler, Braunschweig, FRG) in ethanol was added in the presence or absence of a 100-fold excess of nonradioactive calciferols (Hoffmann-La Roche, Basel) to 100-μl cell extract aliquots. 1,25(OH$_2$)-[^{3}H]D$_3$ was determined using the hydroxyapatite assay.

Cell Proliferation and DNA Synthesis

[^{3}H]thymidine uptake was carried out as described elsewhere [4]. In brief, cells were incubated with 1.0 μCi/ml [^{3}H]thymidine (Amersham-Buchler, Braunschweig, FRG) at 37 °C for 3–4 h before harvesting. Before being counted ina Pharmacia counter, HaCaT cells were lysed and DNA was TCA-precipitated. Distribution in the cell cycle was determined by DNA-flow cytometry [2] (Becton-Dickinson, Heidelberg, FRG).

Immunocytochemistry

VDR were visualized by the monoclonal antibody 9A7 gamma [11] and a modification of the labeled avidin-biotin method by Milde et al. [9, 10]. Expression of the Ki-67 antigen [3] (Dianova, Hamburg: anti Ki-67) or of cytokeratins (Sigma, Munich, FRG): clone 8.60) were carried out immunocytochemically. In brief, HaCaT keratinocytes were washed in HBSS and fixed in acetone for 20 min at –20 °C. Second antibodies labeled with DTAF were from Dianova (Hamburg, FRG). Specimens were analyzed under a Zeiss microscope at 400 × final magnification. Photographs were taken on a Kodak Ektachrome 400 film.

Results

VDR in HaCaT keratinocytes were demonstrated immunocytochemically. The expression varied between the cells. However, cytokines inducing upregulation (see below) resulted in a more intense staining of VDR as revealed by MAB 9A7 gamma. Normal semiconfluent HaCaT cells revealed K_D values between 2.6 and 4.7 × 10^{-10} M by Scatchard plot and DNA cellulose binding analysis (Fig. 1). These data were completely compatible with the known values for VDR in other target cells. We observed an upregulation of VDR expression (Fig. 1) with interferon-γ as well as with IL-1α and a suppression with TNF-α by factors of 2.7, 2.3, and 0.8, respectively. The numbers of VDR are given in Fig.1. Proliferation of HaCaT keratinocytes (Fig. 2) was not modulated by IL-1α – 108% (SD 9.3) of normal controls (SD 8.4) – but was inhibited by TNF-α – 62% (SD 8.7) of normal controls (SD 8.4) – as revealed by $[^3H]$-thymidine uptake. In the cell cycle analysis

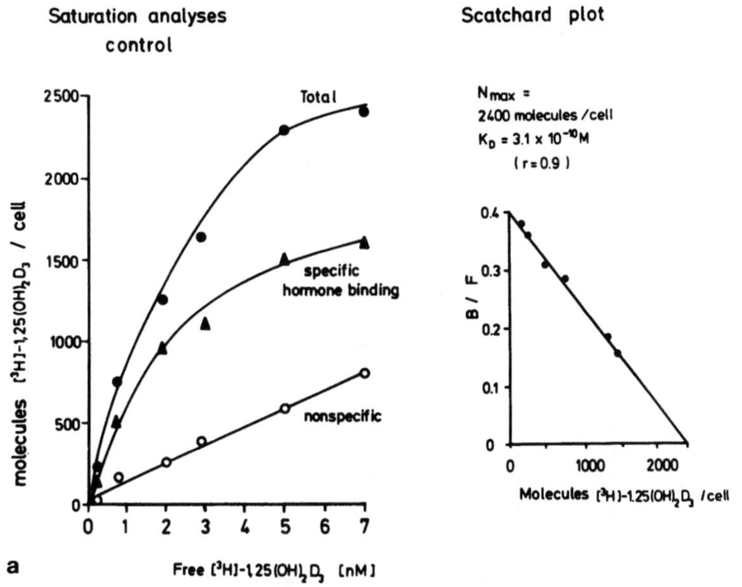

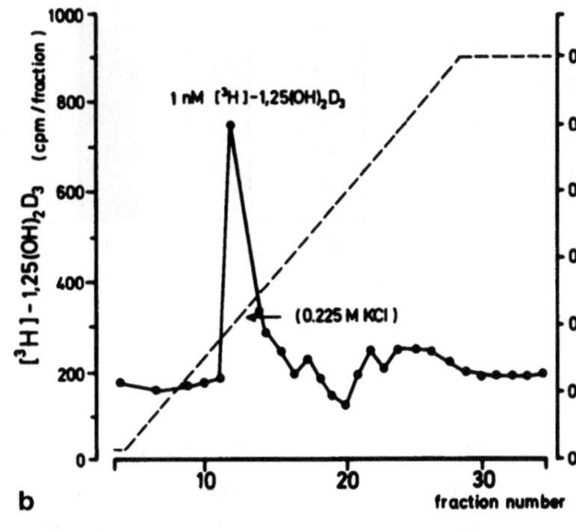

Fig. 1 a–d. VDR expression in HaCaT keratinocytes: **a** Saturation analysis of the binding of $[^3H]1,25(OH)_2D_3$ to the cytosol and nuclear fraction of HaCaT keratinocytes. The saturation plot shows total, specific, and non-specific hormone binding. **b** DNA-cellulose column shows specific elution of $[^3H]1,25(OH)_2D_3$-labeled receptor. N_{max} and K_D were revealed by Scatchard analysis, as shown for controls (**a, c**) and after stimulation (**c**) with complement C5a, PDGF, and recombinant cytokines (IL-1α, TNF-α, and inter-feron-γ). **d** N_{max} and K_D are listed for each case. The equilibrium dissociation constants K_D were calculated from the slope of the regression lines, as depicted in the Scatchard plots. N_{max} was determined by extrapolation of the lines to the abscissa (**a, c**)

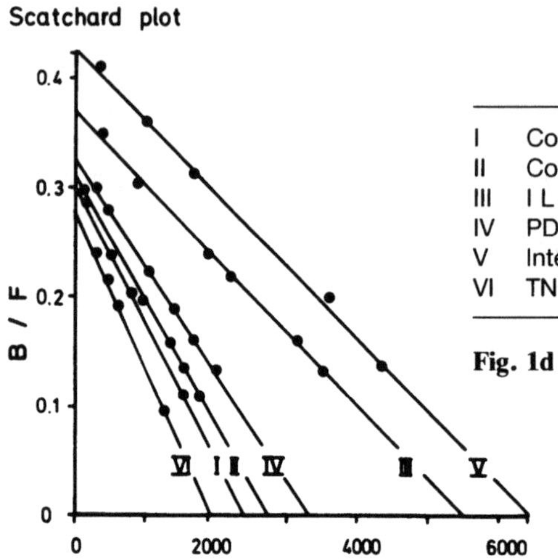

Fig. 1c

		N_{max} molecules bound/cell	K_D x 10^{-10} M
I	Control	2396	3.1
II	Complement C 5a	2790	2.6
III	I L−1α	5493	3.2
IV	PDGF	3283	3.5
V	Interferon gamma	6365	4.6
VI	TNF α	1914	3.2

Fig. 1d

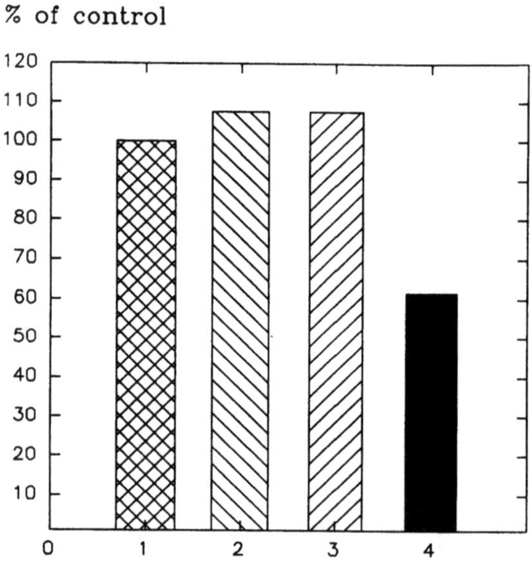

Fig. 2. [^{3}H]-thymidine uptake (1% FCS, 5% CO_2 of HaCaT cells in a representative experiment under control conditions (SD 8.4) and after stimulation for 48 h with the recombinant cytokines IL-1α (SD 9.3) and TNF-α (SD 8.7) at concentrations of 10 nm/l (1% FCS, 5% CO_2). *1,* Control medium; *2,* IL 1α; *3,* IL 1β; *4,* TNFα

(Fig. 3), IL-1α induced a slightly increased number of cells in the S and G2 + M phase (14.3% and 10.8%), while TNF-α distinctively reduced the number of cells in the two dividing cell cycle compartments. 7.2% and 5.4%). Cytokeratin 1/10 and Ki-67 expression was reduced in TNF-α- and upregulated in IL-1α-stimulated HaCaT keratinocytes (Fig. 4).

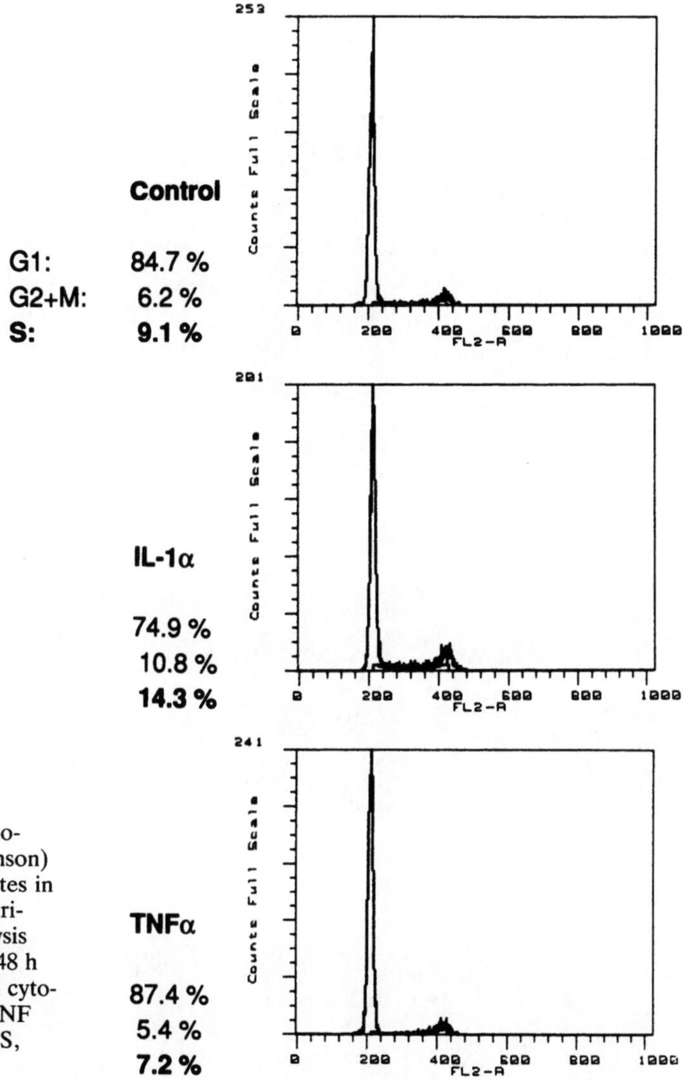

Fig. 3. DNA flow cytometry (Becton Dickinson) of HaCaT keratinocytes in a representative experiment: cell cycle analysis after stimulation for 48 h with the recombinant cytokines IL-1 alfa and TNF alfa (10 nm/l, 1% FCS, 5% CO_2)

Discussion

IL-1α and TNF-α revealed a distinct on VDR expression in HaCaT keratinocytes. This effect was specific since it was not linearly condordant with the effect of these cytokines on [^{3}H]thymidine uptake, distribution in cell cycle, or expression of the Ki-67 antigen. Our findings indicate (a) that HaCaT keratinocytes represent a promising tool for analyzing the effects of cytokines on keratinocytes, and (b) that cytokines can specifically modulate VDR expression. The results suggest that local 1,25-D_3 concentration, VDR expression, and cytokines such as TNF-α IL-1, or IFN might be of great

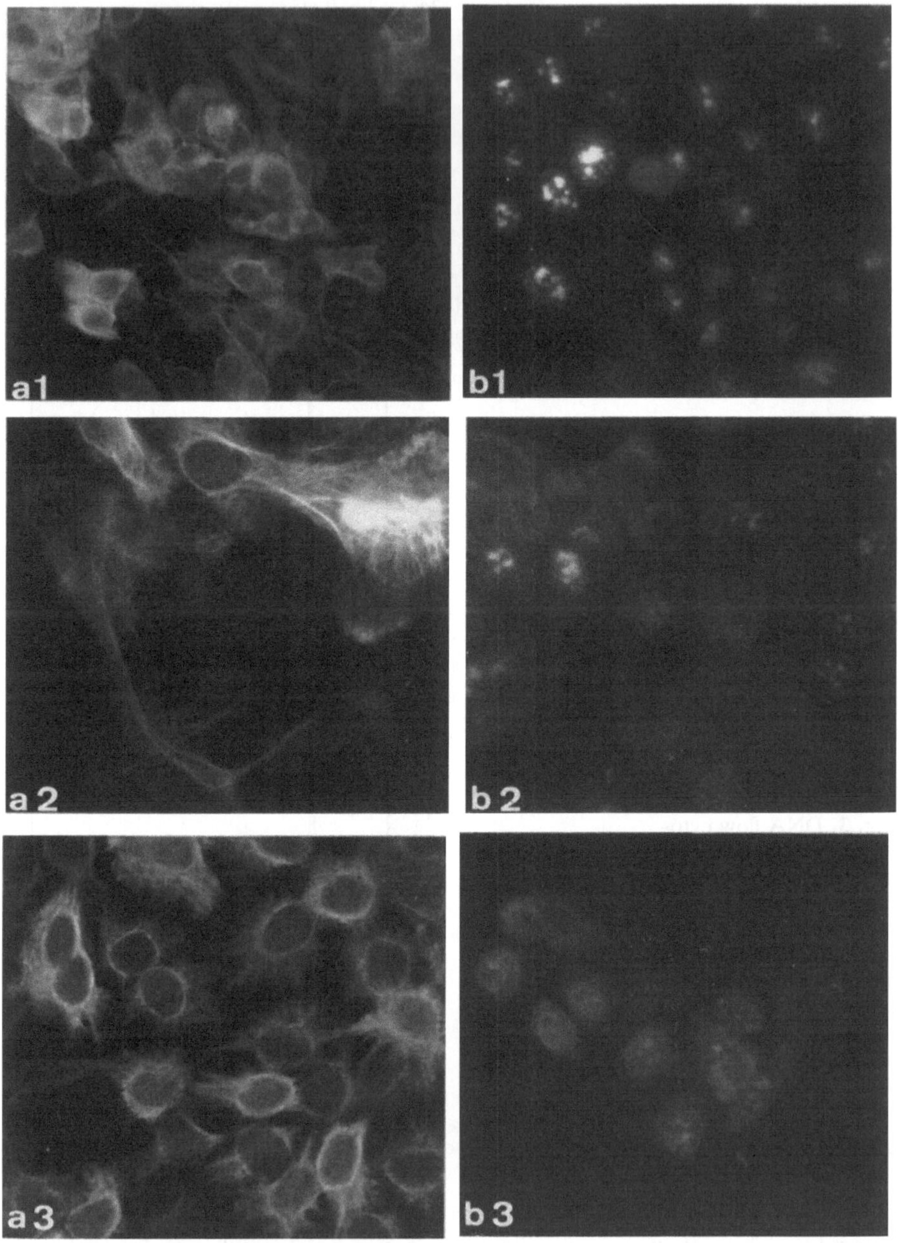

Fig. 4a, b. Immunocytochemical demonstration of (**a**) cytokeratins 1/10 (clone 8.60, Sigma, München) and (**b**) nuclear antigen Ki-67 (anti-Ki-67, Dianova, Hamburg) in HaCaT keratinocytes. Stimulation (10 nm/l, 1% FCS, 5% CO_2) with the recombinant cytokines IL-1 α (**a1, b1**) and TNF α (**a2, b2**) for 48 h and controls (**a3, b3**). Second antibodies were labeled with DTAF (Dianova, Hamburg)

importance in the connection between SIC and keratinocytes under various pathological conditions in vivo.

References

1. Bourkamp P, Petrusevska RT, Breitkreutz D, Hornung J, Markham A, Fusenig N (1988) Normal keratinization in a spontaneously immortalized uneuploid human keratinocyte cell line. J Cell Biol 106:761
2. Frenz G, Möller U, Christensen IB (1980) J Invest Dermatol 74:119
3. Gerdes J, Lemke H, Baisch H, Wacker HH, Schwab U, Stein H (1984) Cell cycle analysis of a cell proliferation associated human nuclear antigen defined by monoclonal antibody Ki-67. Immunol 133:1710
4. Görmar FE, Bernd E, Bereiter-Hahn J, Holzmann H (1990) A new model of epidermal differention induction by mechanical stimulation. Arch Dermatol Res 282:22
5. Klaus G, Merke J, Eing H, Hügel U, Milde P, Reichel H, Ritz E, Mehls O (1991) 1,25(OH) D_3 receptor regulation and 1,25 (OH_2) D_3 effects in primary cultures of growth cartilage of the rat. Calcif Tissue Int 49 (in press)
6. Kragballe K (1989) Treatment of psoriasis by the topical application of the novel cholecalciferol analogue calcipotriol (MC 903). Arch Dermatol 125:–1647
7. Merke J, Klaus G, Hügel U, Waldherr R, Ritz E (1986) No 1,25(OH) D receptors in osteoclasts of the Ca^{++}-deficient chick despite demonstrable receptors on circulating monocytes. Clin Invest 77:312
8. Merke J, Milde P, Lewicka S, Hügel U, Klaus G, Mangelsdorf DJ, Haussler MR, Rauterberg EW, Ritz E (1989): Identification and regulation of 1,25 dihydroxyvitamin D_3 receptor activity and biosynthesis of ,125-dihydroxyvitamin D_3. J Clin Invest 83:–1903
9. Milde P, Merke J, Ritz E, Haussler MR, Rauterberg EW (1989) Immunohistochemical detection of 1,25-dihydroxyvitamin D_3 receptors and estrogen receptors by monoclonal antibodies: comparison of four immunoperoxidase methods. J Histochem Cytochem. 37:–1706
10. Milde P, Hauser U, Simon T, Mall G, Ernst V, Haussler MR, Frosch P, Rauterberg EW (1992) Expression of 1,25-dihydroxyvitam D_3 receptors in normald and psoriasis skin. J Invest Dermatol (in press)
11. Pike JW, Donaldson CA, Marion SL, Haussler MR (1982) Development of hydridomas secreting antibodies to the chicken intestinal 1025-dihydroxyvitamin D_3 receptor. Proc Natl Acad Sci USA 79:7719
12. Rigby FC (1988) The immunobiology of vitamin D. Immunol Today 9:54

Aminopeptidases of Human Skin – a Survey of Enzyme Activities in Extracts from Skin and Cultured Fibroblasts

W. Hoppe, K. Flaßkamp, B. Gruber-Dunkemölle, K. Hentemann, M. Pirmann, and *E. Heymann*

Introduction

Enzymatic activities which sequentially remove single amino acids from the N-terminus of peptides are widely distributed in human tissues and cultured cells. Several aminopeptidases with different molecular and enzymatic properties have been purified from various sources [6]. Leucyl aminopeptidase and soluble alanyl aminopeptidase are cytoplasmic enzymes with a broad substrate specificity. A wide range of substrates are hydrolyzed by microsomal alanyl aminopeptidase. In contrast, a narrow substrate specificity has been reported for membrane-bound glutamyl aminopeptidase and soluble arginyl aminopeptidase. Except for arginyl aminopeptidase, these enzymes are oligomeric proteins with M_r-values of 100000–300000 and with isoelectric points in the range of pH 4–5. Many aminopeptidases are Zn-containing enzymes. Thus, chelating agents are strongly inhibitory. In addition to these more thoroughly characterized enzymes, several minor aminopeptidase activities including a proline-specific aminopeptidase [8, 12] have been reported to be present in mammalian tissues.

The function of aminopeptidases in the skin is still poorly understood. It may be assumed that they participate in the general metabolism of proteins [4] as well as in pathological processes, since increased levels of aminopeptidase activities have been found in skin inflammation, wound healing, and psoriasis [13–15]. To understand the role of aminopeptidases in these processes it is necessary to do a survey of enzymatic activities as a first step.

Materials and Methods

Cell Culture

Human skin fibroblasts were cultured in modified Eagle's minimum essential medium supplemented with 10% fetal calf serum (Boehringer, Mannheim, FRG), penicillin, streptomycin, and minocycline.

A. Bernd, J. Bereiter-Hahn,
F. Hevert, H. Holzmann (Eds.)
Cell and Tissue Culture Models
in Dermatological Research

Extraction and Fractionation Methods

Fibroblast extracts were obtained by harvesting confluent cultures with 20 m*M* Hepes-NaOH pH 7.4 + 0.1% Triton X-100 followed by centrifugation at 10000 *g* for 15 min. For cell fractionation experiments the cells were scraped from the culture dishes, suspended in 20 m*M* Hepes-NaOH + 0.25 *M* sucrose (1.2 ml/75-cm^2 dish), and homogenized by 20 strokes in a Dounce homogenizer. Cell debris and nuclei were removed by centrifugation at 1000 g for 10 min. The supernatant was then subjected to two subsequent centrifugation steps at 10000 *g* for 10 min and at 60000 *g* for 240 min. The sediments were extracted with 20 m*M* Hepes-NaOH pH 7.4 + 0.1% Triton X-100. The supernatant after centrifugation at 60000 *g* and the extracts from the sediments were assayed for enzymatic activities.

For differential extraction with digitonin (Sigma, Deisenhofen, FRG) confluent cultures of fibroblasts were exposed to isotonic 0.01% digitonin in 20 m*M* Hepes-NaOH pH 7.4 + 0.15*M* NaCl (0.4 ml/25-cm^2 dish) for 10 min at 4 °C with occasional agitation. After this extract had been removed from the culture the residual cell layer was scraped from the plate and reextracted with 0.1% Triton X-100 in 20 m*M* Hepes-NaOH pH 7.4.

Enzyme Assays

Enzyme activities were routinely assayed at 37 °C. The hydrolysis of aminoacyl naphthylamides (Bachem, Heidelberg, FRG) was determined in assays containing 50 m*M* Tris-HCl pH 8.0 and 0.2 m*M* of the naphthylamide derivative. The formation of naphthylamine was measured fluorometrically (excitation at 365 nm, emission above 400 nm) or photometrically at 546 nm after coupling with Fast Garnet GBC [18]. The effect of inhibitory agents was determined after preincubation of extracts with inhibitor for 15 min at 37 °C. The hydrolysis of Leu-NH_2 was measured in a coupled enzyme assay with glutamate dehydrogenase [1]. Glucuronidase activity was determined with *p*-nitrophenyl β-*D*-glucuronide as substrate. The assay mixture (0.2 ml) contained 50 m*M* sodium citrate pH 4.6, 75 m*M* NaCl, 0.02% sodium azide, 0.1% bovine serum albumin, and 10 m*M* substrate. The reaction was stopped by addition of 1.0 ml 0.4 *M* glycine-NaOH, pH 10.4, and the absorbance was measured at 405 nm.

Lactate dehydrogenase was assayed in 50 m*M* sodium phosphate, pH 7.5, 0.6 m*M* pyruvate, and 0.18 m*M* NADH. The decrease in absorbance was followed at 340 nm.

Other Methods

Protein was determined by a modified Folin phenol method after precipitation with trichloroacetic acid in the presence of RNA [9].

Results and Discussion

Extracts from human skin fibroblasts contain aminopeptidases capable of hydrolyzing naphthylamide derivatives of various amino acids (relative activity given in parentheses): alanine (100), leucine (34), arginine (7), glutamic acid (0.3), and proline (0.2). A similar pattern of activities was found in crude extracts from dermis and from skin (Table 1). The hydrolysis of Gly-Pro-NNap in these extracts can be ascribed to dipeptidyl peptidase IV which is highly specific for substrates with an X-Pro dipeptide at the N-terminus [10]. In contrast, the results presented in Table 1 do not allow us to identify individual aminopeptidases because their substrate specificities overlap considerably. From the data of Table 1 it is not possible to decide, for example, whether the hydrolysis of Arg-NNap is catalyzed by arginyl aminopeptidase, by one of the unspecific aminopeptidases, or by a combination of both. Thus, the determination of individual aminopeptidase activities in crude extracts requires experimental approaches that provide additional criteria for a differentiation.

A simple approach frequently employed to differentiate among proteolytic enzymes is based on their susceptibility to inhibitors. The effect of various inhibitors on aminopeptidase activities in crude extracts from human skin fibroblasts is summarized in Table 2. Chelating agents inhibited all aminopeptidase activites assayed in these extracts. As observed with many purified aminopeptidases, 1,10-phenanthroline was more efficient than EDTA [6]. Relatively high concentrations of puromycin were required for a pronounced inhibition of Ala-NNap hydrolysis. An apparent K_i value of 0.3–0.4 m*M* was determined in a separate study. Since soluble alanyl aminopeptidase is severely inhibited by puromycin at micromolar concentrations [5], it may be concluded that this enzyme does not contribute significantly to the cleavage of Ala-NNap. It has been reported that puromycin has no effect on leucyl aminopeptidase [7], whereas microsomal alanyl aminopeptidase from human kidney was inhibited about 80% by 0.3 m*M* puromycin [3]. The divalent metal ions Zn^{2+}, Cu^{2+}, and Mn^{2+} impaired the activity of aminopeptidases in a rather unspecific manner. Compared with other

Table 1. Hydrolysis of aminoacyl naphthylamides (X-NNap) by extracts from cultured human skin fibroblasts, human skin, and skin tissues

Extract	Activity (nmoles/min · mg protein) with X-NNap					
	Ala-	Leu-	Arg-	Glu-	Pro-	Gly-Pro-
Cultured fibroblasts	497	171	37.0	1.61	1.13	84.5
Skin	9.68	4.14	5.07	0.15	0.19	4.82
Epidermis, upper	0.56	–	0.12	–	–	0.40
Epidermis, lower	1.98	–	0.50	–	–	3.30
Dermis	26.2	8.20	2.92	0.14	0.28	13.1
Subcutis	3.20	–	0.56	–	–	0.72

Table 2. Effect of inhibitors on the hydrolysis of aminoacyl naphthyl amides by extracts from cultured fibroblasts

Inhibitor concentration	Relative activity with X-NNaps					
	Ala-	Leu-	Arg-	Glu-	Pro-	Gly-Pro-
None	100	100	100	100	100	100
1,10-Phenanthroline						
0.1 m*M*	88	90	82	84	58	–
0.3 m*M*	24	19	20	35	25	–
1.0 m*M*	1	1	1	0	0	91
EDTA						
1.0 m*M*	51	49	40	39	44	108
3.0 m*M*	39	20	13	30	33	79
10.0 m*M*	0	4	6	9	8	47
Puromycin						
0.3 m*M*	20	23	39	10	18	87
$ZnCl_2$						
0.01 m*M*	36	45	64	42	33	64
0.03 m*M*	14	16	39	19	11	50
0.1 m*M*	3	4	3	0	4	13
$CuSO_4$						
0.3 m*M*	35	45	50	31	28	77
1.0 m*M*	11	10	30	12	8	48
$MnCl_2$						
0.3 m*M*	55	51	77	46	54	90
pHMB						
3.0 m*M*	64	39	83	40	74	43

aminopeptidase substrates the hydrolysis of Arg-NNap was less sensitive to inhibition by these ions and also by *p*-hydroxymercuribenzoate (pHMB). The effect of this thiol-blocking agent was assayed because a partially purified prolyl aminopeptidase from kidney was completely inhibited by 2.5 m*M* pHMB [8]. The results presented in Table 2, however, indicate that in fibroblast extracts this substrate is predominantly hydrolyzed by aminopeptidase activities different from the proline-specific enzyme in kidney.

The major conclusions drawn from the inhibition experiments are supported by the distribution of enzymatic activities in cell fractions obtained from skin fibroblasts by differential centrifugation (Table 3). Lactate dehydrogenase and β-glucuronidase served as marker enzymes. The results obtained by this fractionation procedure provided direct evidence that the hydrolysis of Ala- and Leu-NNap in crude extracts from fibroblasts was due mainly to microsomal alanyl aminopeptidase. Only 5% and 6% of the total activity with these substrates were present in the high-speed supernatant that contained the bulk of lactate dehydrogenase (84%). A membrane-associated alanyl aminopeptidase was localized on the surface of fibroblasts by van den Berghe and co-workers [16]. These authors screened human cell lines and tissues for alanyl aminopeptidase: the highest specific activity was found in kidney cortex (100%), followed by normal skin fibroblasts (87%)

Table 3. Distribution of enzymatic activities in cell fractions obtained from skin fibroblasts by differential centrifugation

Cell fraction	Relative enzymatic activities in cell fractions (%) with X-NNap Ala-	Leu-	Arg-	Glu-	Pro-	Gly-Pro-	Lactate dehydrogenase	β-Glucuronidase
Sediment 1 (1000 *g*, 10 min)	72	75	59	63	24	71	10	59
Sediment 10 (10000 *g*, 10 min)	16	14	12	14	8	19	5	24
Sediment 60 (60000 *g*, 240 min)	7	5	4	8	4	8	1	5
Supernatant 60 (60000 *g*, 240 min)	5	6	24	15	65	2	84	11

and a fibrosarcoma-derived cell line (80%). The presence of alanyl aminopeptidase together with dipeptidyl peptidase IV [11] on the surface of fibroblasts points to a role of these two exopeptidases in the metabolism of extracellular matrix proteins and peptide cell-cell mediators. As reported earlier [2] these two peptidases complement one another in the degradation of casein-derived peptides. A different distribution of aminopeptidase activities was observed with Arg-NNap and Pro-NNap as substrates (Table 3). It remains to be determined whether the hydrolysis of these substrates must be assigned to microsomal alanyl aminopeptidase in the membrane fraction (sediment 1) and to soluble alanyl aminopeptidase or leucyl aminopeptidase in the soluble fraction of fibroblasts (supernatant 60). Studies on the separation of aminopeptidases from human skin, however, revealed that Pro-NNap is hydrolyzed mainly by an enzyme with an M_r of about 100000 (results not shown), whereas much larger molecular sizes have been reported for the three aminopeptidases mentioned above [6].

The treatment of cells with low concentrations of digitonin permeabilizes rather selectively the plasma membrane [17]. Cytoplasmic constituents including proteins can be extracted with digitonin in a first step, and in a second step remaining cell proteins are solubilized with a stronger detergent.

Table 4. Differential extraction of enzymatic activities from skin fibroblasts by digitonin

Extraction step	Relative enzymatic activities in extracts (%) with X-NNap Ala-	Leu-	Arg-	Pro-	Gly-Pro-	Lactate dehydrogenase	β-Glucuronidase
Extract I (0.01% digitonin)	4	7	25	63	4	85	24
Extract II (0.1% Triton X-100)	96	93	75	37	96	15	76

The distribution of aminopeptidase activities upon a differential extraction of fibroblasts (Table 4) was remarkably similar to that obtained by centrifugation (Table 3): in both experiments the major part of aminopeptidase activities acting on Ala-NNap and Leu-NNap as well as dipeptidyl peptidase IV proved to be associated with cellular membranes, while activities present in the cytoplasmic fraction contributed considerably to the total hydrolysis of Arg-NNap, and in particular of Pro-NNap. A closer characterization and identification of these activities in cultured fibroblasts will be the objective of future work.

References

1. Hafkenscheid JCM, Kohler BEM (1985) A continuous method for the determination of leucine aminopeptidase in human serum with *L*-leucinamide as substrate. J Clin Chem Clin Biochem 23:393–398
2. Heymann E, Mentlein R (1986) Complementary action of dipeptidyl peptidase IV and aminopeptidase M in the digestion of β-casein. J Dairy Res 53:229–236
3. Hiwada K, Yokoyama M, Kokubu T (1981) Comparison of membrane-bound arylamidases from human, bovine, hog, dog, rabbit and rat kidneys. Comp Biochem Physiol [B] 68:485–489
4. Hopsu-Havu VK, Fräki JE, Järvinen M (1977) Proteolytic enzymes in the skin. In: Barrett AJ (ed) Proteinases in mammalian cells and tissues. Elsevier, Amsterdam, pp 547–581
5. McClellan JB, Garner CW (1980) Purification and properties of human intestine alanine aminopeptidase. Biochim Biophys Acta 613:160–167
6. McDonald JK, Barrett AJ (1986) Exopeptidases. Academic. London (Mammalian proteases: a glossary and bibliography, vol 2)
7. McDonald JK, Reilly TJ, Ellis S (1964) The hydrolysis of amino acyl-β-naphthylamides by plasma aminopeptidases. Biochem Biophys Res Commun 16:135–140
8. Nordwig A, Mayer H (1973) The cleavage of prolyl peptides by kidney peptidases. Hoppe Seylers Z Physiol Chem 354:380–383
9. Peterson GL (1983) Determination of total protein. Methods Enzymol 91:95–119
10. Püschel G, Mentlein R, Heymann E (1982) Isolation and characterization of dipeptidyl peptidase IV from human placenta. Eur J Biochem 126:359–365
11. Saison M, Verlinden J, von Leuven F, Cassiman JJ, van den Berghe H (1983) Identification of cell surface dipeptidylpeptidase IV in human fibroblasts. Biochem J 216:177–183
12. Sarid S, Berger A, Katchalski E (1962) Proline iminopeptidase. J Biol Chem 237:2207–2212
13. Schwartze G, Hanson H (1971) Zur proteolytischen Aktivität in Psoriasisschuppen. Dermatol Monatsschr 157:315–325
14. Spector GJ (1977) Leucine and alanine aminopeptidase activities in experimentally induced intradermal granulomas and late stages of wound healing in the rat. Lab Invest 36:1–7
15. Suworow AP (1990) Zur Aktivitätsregelung der Peptidhydrolasen bei Psoriasis. Dermatol Monatsschr 176:393–397
16. Verlinden J, von Leuven F, Cassiman JJ, van den Berghe H (1981) Identification of the human fibroblasts surface glycoprotein (FSG) as aminopeptidase M. FEBS Lett 123:287–290
17. Weigel PH, Ray DA, Oka JA (1983) Quantitation of intracellular membrane-bound enzymes and receptors in digitonin-permeabilized cells. Anal Biochem 133:437–449
18. Yoshimoto T, Walter R (1977) Post-proline dipeptidyl aminopeptidase (dipeptidyl aminopeptidase IV) from lamb kidney. Biochim Biophys Acta 485:391–401

Determination of Cell Counts and Cell Proliferation by the Fluorochromes Hoechst 33342 or Hoechst 33258

R. A. Blaheta, M. Franz, M. K. H. Auth, A. Encke,
and *B. H. Markus*

Introduction

The analysis of lymphocyte activation and proliferation provides important insights into cellular immune reactions. Knowledge of these cellular reactions is of special importance, e.g., in the area of organ transplantation, as conclusions can be drawn about possible rejection of the transplant. Cell proliferation is usually measured by means of the ^{3}H-thymidine assay. The principle of this assay is to detect the uptake of ^{3}H-thymidine into newly synthesized DNA. The higher the measured radioactivity, the higher the proliferation rate of the investigated cell culture.

Although the assay is very exact, photometric and fluorometric alternatives have been tested to avoid the handling of radioactive substances and to overcome time-consuming preparations. These assays are based either on color chanes of dyes [5, 13] or on fluorochroming of the cell culture. The emergence of fluorescent complexes will be caused either by cytoplasmatic enzymes (e.g., in the case of 4-methylumbelliferylheptanoate, MUH) [6] or by binding to cellular DNA (e.g., in the case of 4,6-diamidino-2-phenylindole, DAPI; Hoechst 33342; Hoechst 33258) [7–9, 11]. The dyes are incorporated in every cell, in contrast to ^{3}H-thymidine. Therefore, all cells and not only proliferating lymphocytes can be detected with these tests. It should be mentioned that fluorometric detection of proliferating cells is also possible by monoclonal antibodies, such as Ki-67, but additional labeling with FITC (fluorescein isothiocyanate) is necessary.

In the present study we tested the efficacy of the two DNA-intercalating fluorchromes Hoechst 33258 (H33258) and Hoechst 33342 (H33342) with regard to lymphocyte counts and proliferation. According to proliferation measurement requirements, lymphocyte cultures were treated with the mitogen phytohemagglutinin (PHA; m-form) or with the immunsuppressive drugs cyclosporin A (CsA) or FK 506. All results were analyzed and compared with data achieved by the classic ^{3}H-thymidine test.

A. Bernd, J. Bereiter-Hahn,
F. Hevert, H. Holzmann (Eds.)
Cell and Tissue Culture Models
in Dermatological Research

Materials and Methods

Lymphocyte Preparation

Peripheral blood lymphocytes (PBLs) were isolated from blood of healthy donors as described elsewehre [2]. In short: Lymphocytes were separated by the Ficoll gradient, washed in Hanks' medium (Gibco, Berlin, Germany), and finally resuspended in lymphocyte medium (RPMI 1640, Gibco; 1% gentamicin, ICN/Flow, Meckenheim, Germany; 1% *L*-glutamine, ICN/Flow; 2% 1*M* Hepes, Biochrom, Berlin, Germany; 10% AB-serum).

To calibrate the fluorescence intensity related to cell numbers, various concentrations of lymphotcytes were prepared (1×10^4–1×10^7 cells/ml medium).

For lymphocyte quantification and proliferation measurements, 0.5×10^6 cells/ml were stimulated with PHA, m-form (1%; Gibco). Some of the stimulated lymphocytes were treated additionally with the immunsuppressive drugs CsA or FK 506 in various concentrations (50–5000 ng/ml or 5–500 ng/ml, respectively).

All tests were performed in 96-well flat-bottomed microtiter plates (Becton Dickinson, Heidelberg, Germany) in a final volume of 0.2 ml (100 µl cell suspension + 100 µl cell medium). The figures shown in the text represent results of several identical test series.

Fluorometric Assays

The microtiter plates were centrifuged for 10 min at 300 g to remove the serum containing medium. 100 µl of Hoechst 33342 (500 ng/ml; Riedel de Haen, Frankfurt, Germany) or Hoechst 33258 (100 ng/ml; Riedel de Haen) in PBS was added to each well. Fluorescence values (FU) were then measured on a Titertek Fluoroskan II (ICN/Flow) using a 355-nm excitation and a 460-nm emission filter. A general view of the course of the fluorometric assays is given in Fig. 1. Because Hoechst 33258 is not able to pass through cellular membranes, ultrasound or papain (Sigma, Deisenhofen, Germany) was used to improve the permeability of the lymphocyte membrane [2].

^{3}H-thymidine Assay

Cells were incubated with 0.1 µCi methyl ^{3}H-thymidine (NEN, Dreieich, Germany) for 20 h. Lymphocytes were harvested on filter papers (Schleicher & Schuell, Dassel, Germany), and scintillation fluid (Zinsser, Frankfurt, Germany) was added to quantify tritium incorporation in activated cells. Results, measured with a liquid scintillation counter (Zinsser), are expressed as dpm (disintegrations per minute).

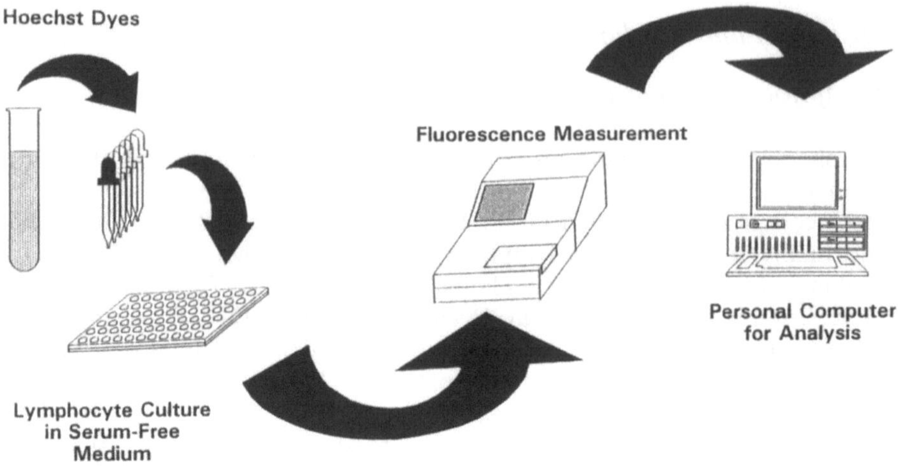

Fig. 1. The fluorometric assay (10 Minutes for Complete Procedure)

Results

Calibration of Fluorescence Values

Differences of the linear relation between the fluorescence intensity and cell number were observed, depending on the two tested fluorochromes. Measured fluorescence of H33342 was linear from 2.5×10^3 to 0.5×10^6 lymphocytes/well (SD = 1.7%, $n = 6$). The application of H33258 (after papain pretreatment of the cell membrane) led to a nonlinear relationship in the upper range of cell concentration (SD = 2.4%, $n = 6$; Fig. 2).

Cell Quantification

Fluorescence values of PHA-stimulated and unstimulated lymphocytes were measured between day 1 and day 10, applying (a) H33342, (b) H33258 + ultrasound pretreatment, or (c) H33258 + papain pretreatment. An enhancement of FU in comparison to the unstimulated controls was measured, depending on the incubation time of PHA – (a): SD = 2.5%, (b): SD = 2.3%, (c): SD = 2.2%; ($n = 4$ in all cases). Highest FU were detected when H33342 was added to the cell cultures. Combination (b) showed the poorest results (Fig. 3).

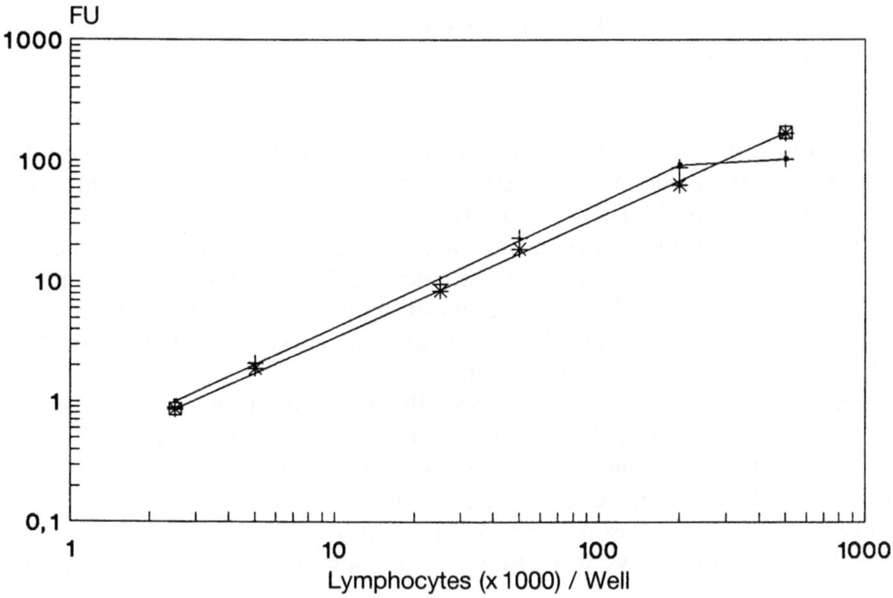

Fig. 2. Calibration curves of H33342 (*) and H33258 (+): SD < 2%

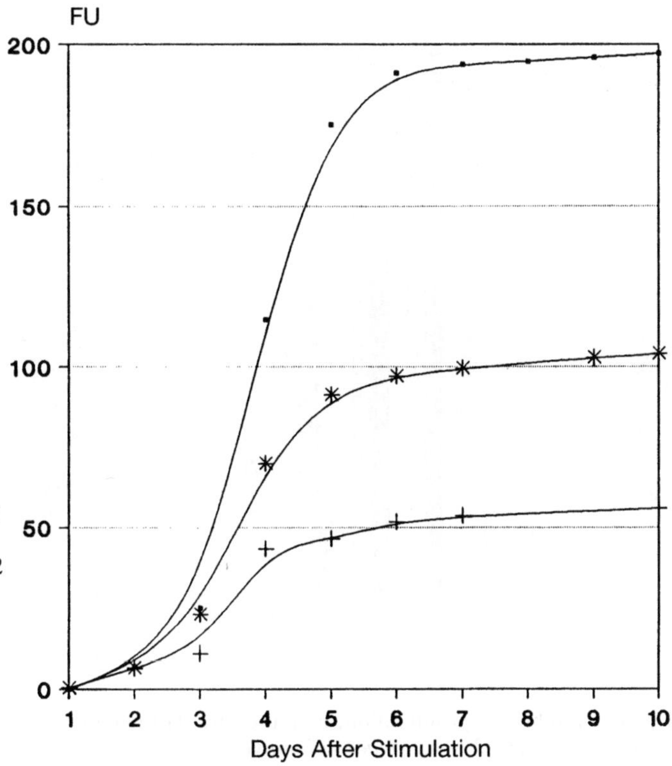

Fig. 3. Comparison of the efficacy of the fluorochromes H33342 (• • •), H33258 + papain pretreatment (*), and H33258 + ultrasound pretreatment (+), given to PHA-stimulated lymphocytes

Proliferation Assay by ^{3}H-Thymidine and Comparison with the Fluorescence Assays

Measurements revealed in increase of DNA synthesis 2 days after PHA stimulation. A maximum of ^{3}H-thymidine-incorporation was reached on day 4, followed by a decline up to day 10. The SD of all measurements were 9.6% (n = 3; Fig. 4, striped columns).

The results of the fluorochrome assay (Fig. 3) refer to the whole cell number but proliferation rates can be calculated easily by $FU_{day\ n+1} - FU_{day\ n}$ (Delta FU, Fig. 4). The statements of cellular proliferation obtained by this calculation via the H33342 assay were in good agreement with those obtained by the thymidine assay (Fig. 4).

The influence of CsA or FK 506 on cellular proliferation or increase in cell numbers was investigated with ^{3}H-thymidine and H33342, respectively (day 5 after PHA stimulation). The results of these assays were nearly identical. Radioactivity or fluorescence values of CsA- or FK 506-treated cultures were lower in comparison to the controls (CsA- or FK 506-free cultures; Fig. 5a, b). Both assays were therefore equally able to detect the drug –

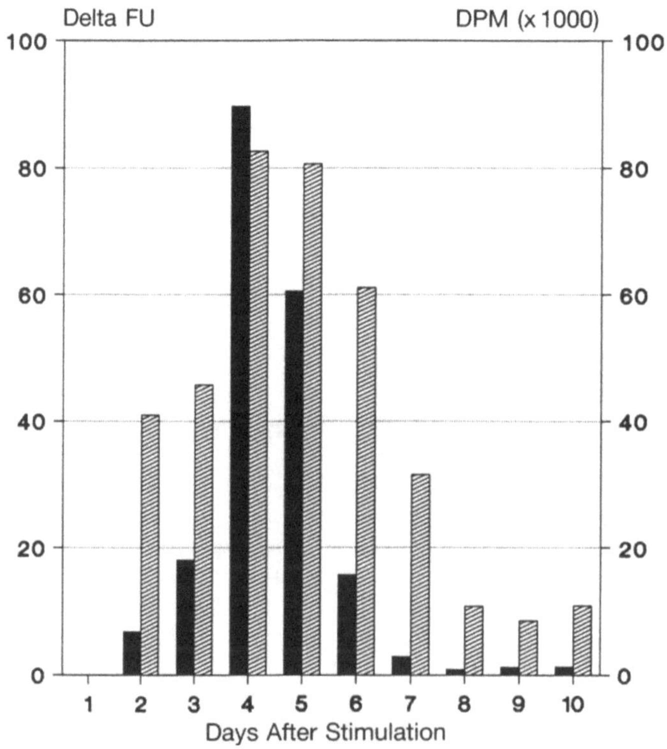

Fig. 4. Lymphocyte proliferation, measured by ^{3}H-thymidine (*striped columns*) or calculated by the H33342 assay (*black columns*)

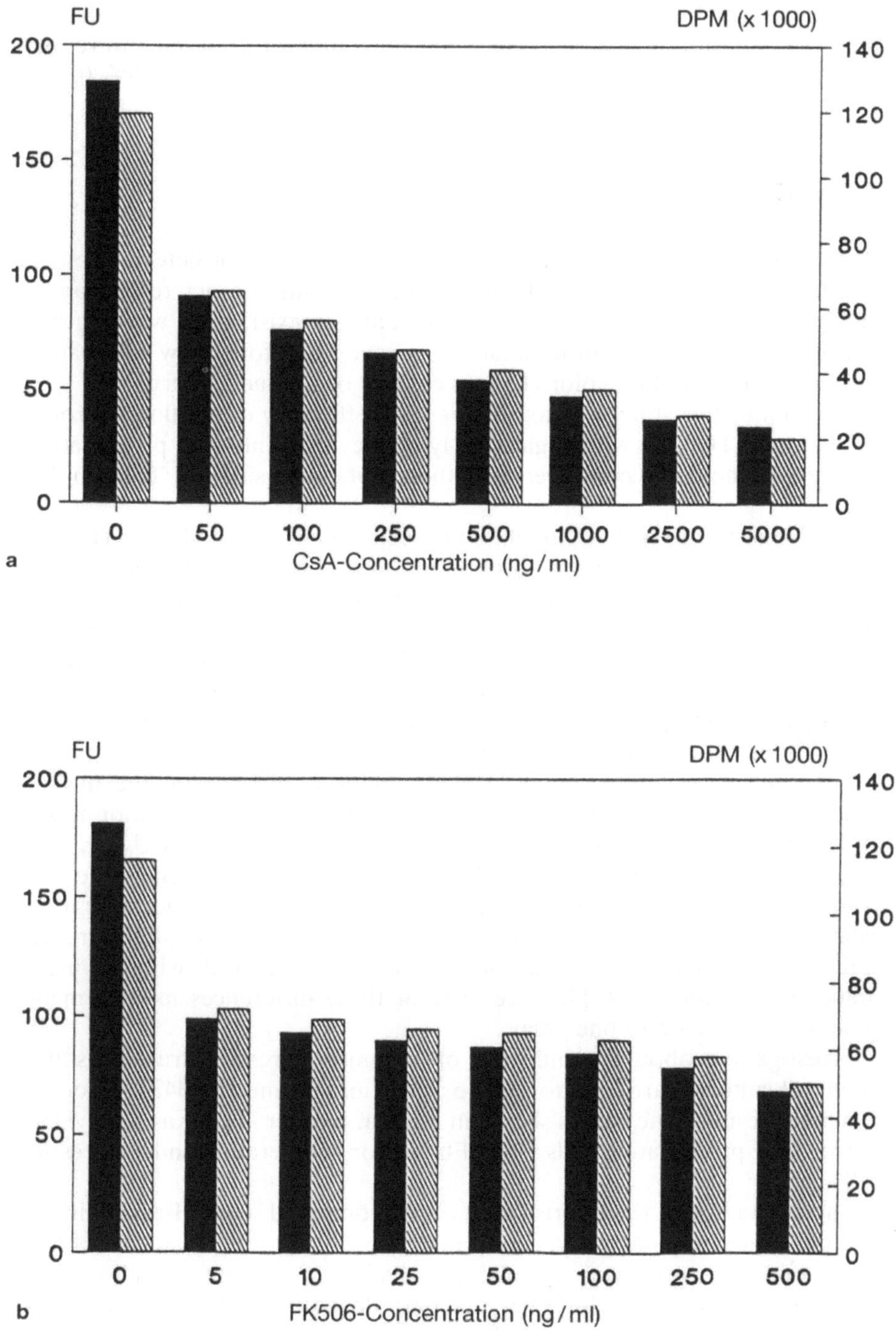

Fig. 5 a, b. Influence of CsA (**a**) and FK 506 (**b**) on lymphocyte proliferation, measured by ^{3}H-thymidine (*striped columns*) or H33342 (*black columns*)

dependent inhibition of cell proliferation/increase in cell number, but remarkable differences were seen when the standard deviations were analyzed: $SD_{Thymidine}$ = 9.1% (CsA), 9.0% (FK 506); SD_{H33342} = 2.8% (CsA), 2.7% (FK 506); n = 4 in all cases.

Discussion

Radioactive marking of chemicals is a common method of detecing cellular events. The ^{3}H-thymidine test in particular promises exact results and is highly reproducible. Nevertheless, disadvantages exist, e.g., working with radioactivity and long duration of the assay. Therefore, new alternatives were created, based on color changes of dyes or on fluorometry.

The aim of this study was to analyzed the efficiency of the fluorochromes H33342 and H33258 with regard to lymphocyte counts and proliferation. The results should be compared with those obtained using the ^{3}H-thymidine assay.

H33342 seems to be more qualified for cell counting than H33258 (Fig. 3). The use of H33342 requires no prior membrane permeabilization step. More important is the fact that FU of H33258 show nonlinearity when high concentrations of lymphocytes are measured (Fig. 2). The same phenomenon can be observed in the interaction of H33258 with highly concentrated isolated DNA [2]. Araki et al. [1] and Lipman [10] explain this by a quenching effect exerted by nuclear histones. At present it is unclear why H33342 does not show such behavior.

The calibration curve reveals a close relationship between the fluorescence values and the actual cell number of a culture. The calculation is also applicable to proliferation assays because $FU_{Proliferation} = FU_{day\ n+1} - FU_{day\ n}$ (Delta FU). Information about the cell activity, especially the time point of maximal cell proliferation, measured by the ^{3}H-thymidine assay or calculated by the H33342 assay is nearly identical (Fig. 4). Small differences could be explained by different interactions of the chemicals with cell membranes and/or the DNA [12]. Because of these differences measurements should be restricted to one assay.

Investigations about the influence of immunosuppressive drugs on stimulated cell cultures carried out with the ^{3}H-thymidine and H33342 method led to similar results. The values shown in Fig. 5a, b differ insofar as dpm values mirror only proliferating cells while FU mirror proliferating and also resting cells.

The advantage of the fluorescence assay, compared with ^{3}H-thymidine, is twofold: The complete assay, set up in a 96-well microtiter plate, takes only a few minutes. In contrast, preparation and measurement of radioactivity will take several hours. Moreover, standard deviations of FU are very small. Our experiments reveal a mean SD of approximately 2% (H33342), but one of 9%–10% when ^{3}H-thymidine is used. The possibility of getting information about cell numbers expands the application of fluorometry. Not only cell quantities and proliferation, but also cell adhesion can be measured (detection of the number of adherent cells), the latter without the use of ^{51}Cr [3, 4].

References

1. Araki T, Yamamoto A, Yamada M (1987) Accurate determination of DNA content in single cell nuclei stained with Hoechst 33258 fluorochrome at high salt concentration. Histochemistry 87:331
2. Blaheta RA, Franz M, Auth MKH, Wenisch HJC, Markus BH (1991) Basic principles of three rapid fluorochrome assays for DNA quantification and cell counting. Transplant Proc 23:461–463
3. Brenan M, Parish CR (1984) Intracellular fluorescent labelling of cells for analysis of lymphocyte migration. J Immunol Methods 74:31
4. Cranston D, Wood KJ, Morris PJ (1989) A comparison of ^{51}chromium and Hoechst 33342 for monitoring lymphocyte migration in vivo. Transplant Proc 21:192
5. Denizot F, Lang R (1986) Rapid colorimetric assay for cell growth and survival. Modifications of the tetrazolium dye procedure giving improved sensitivity and reliability. J Immunol Methods 89:271
6. Dotsika EN, Sanderson CJ (1987) A fluorometric assay for determining cell growth in lymphocyte proliferation and lymphokine assays. J Immunol Methods 105:55
7. Jennings BR, Ridler PJ (1983) Interaction of chromosomal stains with DNA: an electrofluorescence study. Biophys Struct Mech 10:71
8. Kapuscinski J, Skoczylas B (1977) Simple and rapid fluorometric method for DNA microassay. Anal Biochem 83:252
9. Kim YJ, Sah RLY, Doong JYH, Grodzinsky AJ (1988) Fluorometric-assay of DNA in cartilage explants using Hoechst 33258. Anal Biochem 174:168
10. Lipman JM (1989) Fluorophotometric quantitation of DNA in articular cartilage utilizing Hoechst 33258. Anal Biochem 176:128
11. Lydon MJ, Keeler KD, Thomas DB (1980) Vital DNA staining and cell sorting by flow microfluorometry. J Cell Physiol 102:175
12. McCaffrey TA, Agarwal LA, Weksler BB (1988) A rapid fluorometric DNA assay for the measurement of cell density and proliferation in vitro. In Vitro 1:247
13. Moosmann T (1983) Rapid colorimetric assay for cell growth and survival: application to proliferation and cytotoxicity assays. J Immunol Methods 65:55

Volume Determination of Adhering Cells in Culture by Means of Acoustic Interferometry

S. Hegner, and *J. Bereiter-Hahn*

Abstract

Using interferences of ultrasound (1 GHz) reflected from the supporting substratum and the medium-facing side of well-spread cells, the volume of these cells was determined. The method makes it possible to follow volume changes which (for instance) are evoked by the addition of 1 m*M* adenosine. Madine Darby canine kidney cells (MDCK cells) show an initial phase of shrinkage which is followed by regulatory volume increase, sometimes exceeding the original volume. The reaction of 3T3 cells to this substance was very inhomogeneous. Compared with a similar method used with visible light, the large depth in focus and the additional information on mechanical parameters makes this method favorable for studies of cellular volume changes.

Introduction

The water content of human skin is an important factor in normal physiology, in aging, and in some diseases. Urea was found to be one of the most important substances enhancing skin humidity. How this water content is regulated and which biochemical mechanisms are involved in its control are still under debate.

Considering the ability of urea to induce swelling of cells under isotonic conditions, an involvement of cell volume in the regulation of skin humidity is highly probable. Evocation of cell volume changes initiates a series of further reactions; primarily, the ion transport activities become altered, which may result in regulatory volume changes [14, 16]. Furthermore, secretory processes can be activated [19] and the metabolic activities may change [7]. Thus, a series of interactions between the cells with their extracellular environment might be triggered by "simple" volume shifts. If several cell types are involved in organs, such as the skin, the situation is even more complex because of different modes of reactions of fibroblasts and epithelial cells and of quiescent and proliferating cells [15]. The volume regulatory responses of epithelia include polarized ion fluxes, different at the apical and the basal membrane areas [1, 13].

Investigations on the cellular and biochemical basis of volume regulatory responses and of other reactions induced by volume changes are limited by

A. Bernd, J. Bereiter-Hahn,
F. Hevert, H. Holzmann (Eds.)
Cell and Tissue Culture Models
in Dermatological Research

the methods available for volume determinations. The most convenient methods are based on the signal modulation in Coulter counters or on light scattering. These methods, however, are limited to cell suspensions. Fibroblast – like cells as well epithelia require for their "normal metabolic activities" an appropriate extracellular substrate to attach to. In suspension they may immediately begin new activities [8]. In addition, the organization of the cytoskeleton is altered by loss of adhesion to a solid substratum [6, 11], and we have good reason to assume that an intact cytoskeleton participates in the maintenance of the cell volume [4, 12, 13]. Therefore, only those procedures can be used which allow volume determination of well-spread, irregularily shaped cells in culture. One such method would be optical sectioning of the cells using confocal laser-scanning microscopy and computing the sum of the volumes of the total number of sections. This method works very well with fluorescence images of cells with fluorochromed plasma membrane or cortical cytoplasm, it is difficult to apply to living, unlabeled cells.

We therefore developed an interferometric method for volume determination using light reflected from cell surfaces [5] which has now been extended to acoustic microinterferometry using an acoustic microscope operated in the GHz range. This method is noninvasive (high-frequency ultrasound is not harmful to living cells [9]) and thus it allows us to follow the volume changes of single cells.

Material and Methods

Cell Culture

The experiments were performed with the cell line 3T3, derived from mouse fibroblasts. The cells were cultured in Dulbecco's minimal essential medium modified by Eagle (Gibco) containing 5% fetal calf serum and 1% PSN-antibiotic mixture at 37° C. Every 3 days the cell layer was trypsinized; the cell suspension was diluted 1:3 with new medium.

For the experiments cells were grown on plastic coverslips of 2 cm diameter for 3 days in medium containing 1% serum. Before the experiment the medium is diluted 10:1 with PBS (pH 7.2). The coverslip is placed on a glass slide and fixed with silicone grease. The coverslip is surrounded by a circular silicone rubber, that is fixed with glue on the slide. For exchange of the media a disposable scalp veinset (Venofix, Braun Melsungen) that penetrates the silicone rubber is used.

Volume Determination

The method was initially developed for images of the interference reflection microscope RIC [2, 5, 8], and it has now been extended to images obtained with the acoustic microscope.

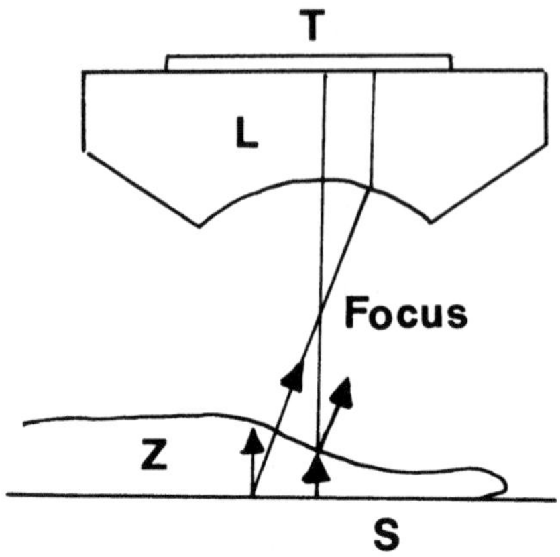

Fig. 1. Model of the acoustic path for imaging a cell. Sound reflections occur at the boundaries plastic/cell and cell/medium. Differences in impedance of the transversed media and incident angle of the waves determine the reflectivity of the interface. *T*, transducer; *L*, lens; *Z*, cells; *S*, substratum

The derivation of the cell volume from the interference pattern will be explained for the type of imaging with reflected beams, no matter whether it is a sound or a light beam: Sound waves as well as light waves are reflected at the boundaries of cell/substratum and cell/medium. The beams reflected from these two boundaries interfere with each other, thus generating an interference pattern which depends on the optical/acoustical path differences between the interfering beams. Each interference fringe connects points with the same thickness; thus, the interference pattern (Fig. 1) delineates the surface topography of the sample (cell). Assuming a constant refractive index (or acoustic impedance), the thickness difference between adjacent minima and maxima is equal to a quarter wavelength. Thus multiplication of the area circumscribed by every ring with a quarter of the wavelength gives the volume of one slice defined by a single constructive or destructive interference. The total cell volume is obtained by summing up the volumes of all these slices.

In this first approach, constant density of cytoplasm and constant sound velocity throughout the cell are assumed. In addition, the presence of a thin fluid layer between the cytoplasm and the supporting solid substratum is neglected. For well – adhering cells these simplifications have only minor influence on thickness determinations (in the range or error introduced by marking the course of the fringes).

Acoustic Microscopy and Image Analysis

A scanning acoustic microscope (SAM) "ELSAM" (Wild-Leitz, Wetzlar) operated at 1 GHz is used. This frequency corresponds to a wavelength of approximately 1.5 μm in water at room temperature and gives a lateral resolution of about 1.2 μm. The acoustic lens has a radius of curvature of 40 μm and an opening half angle of 50°. The images were recorded at the

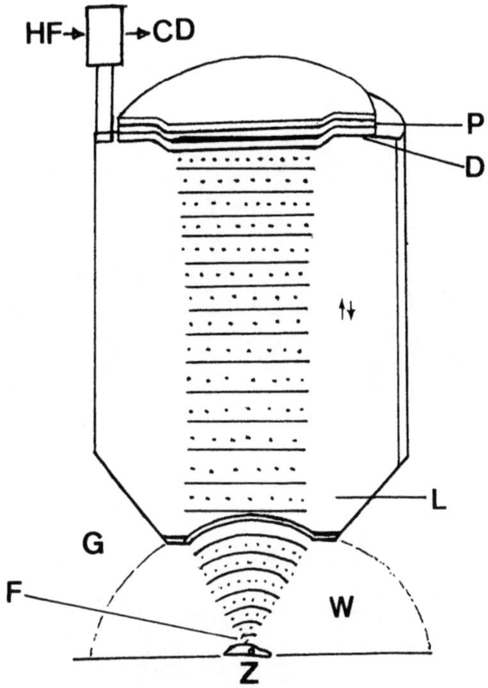

Fig. 2. Acoustic objective lens. *HF,* High-frequency generator; *CD,* control and display system; *P,* piezoelectric layer (transducer); *D,* iris diaphragm; *L,* acoustic lens made of saphire; *G,* antireflex layer; *W,* coupling fluid; *F,* focus of the lens; *Z,* cell

focus level on the surface of the substratum (plastic slide, Fig. 2) and stored on the hard disk of a host computer (ICD, Hanau, FRG) via an AT-FS 100 (Imaging Technology Inc., Massachusetts) image storage board (512 × 512 pixels, 256 gray levels). An interactive program developed by us makes it possible to sketch the course of the interference lines in the acoustic images and to measure the area covered by each interference line (representing one volume slice). This procedure was performed with a Quantimet 520 image analysis system (Cambridge Instruments).

Results

The SAM images allowing the measurement of volume changes induced by 1 m*M* adenosine are exemplified for two cell types: the fibroblastic 3T3 cells (Fig. 3) and the epithelial MDCK cells (Fig. 4). Due to large differences in the volume of untreated cells, the control values are set to 100 and the changes (as determined from the interference pattern as described above) are presented in percent (Figs. 5 and 6). The reaction of the four 3T3 cells (shown in Fig. 6) to adenosine does not follow a constant pattern. The initial response may be swelling or shrinkage or an almost constant volume; the long term response (30 min) is variable as well. Adenosine seems to destabilize the volume regulation of these 3T3 cells. The reasons for this behavior

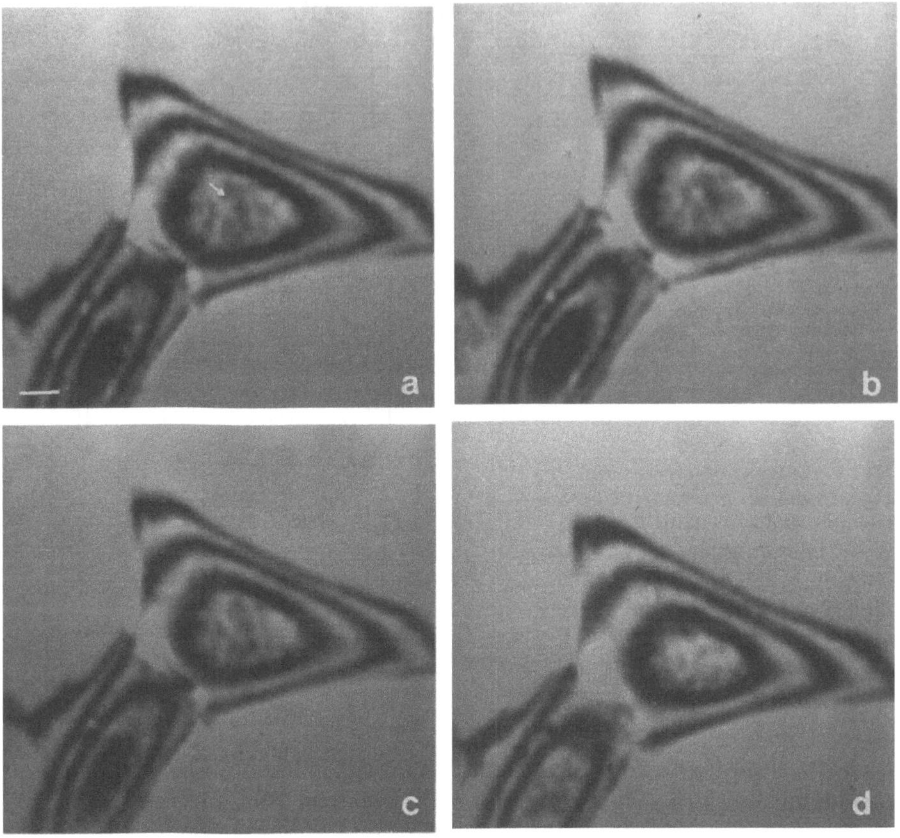

Fig. 3a–d. SAM images of 3T3 cells spread on plastic. Image **a** depicts the cell before perfusion with 1 m*M* adenosine. **b** Two minutes *after addition of adenosine;* **c** 10 min; **d** 30 min after substance application. The interference fringes delineate the surface topography of the cell. In addition to the three outer rings, the dark apex in **b** (*arrow* in **a**) indicates an increased cell thicknes compared with **d**, where the apex is white. *Bar* = 10 μm

are unknown; further experiments are required to rule out influences of the cell cycle stage.

Reactions of MDCK cells are more constant. A phase of shrinkage starting immediately on adenosine addition is followed by a slow phase of recovery, during which the original volume may be considerably exceeded (Fig. 5).

Discussion

For the determination of cell volumes the acoustic method has some advantages compared with RIC microscopy. The main advantage of the acoustic

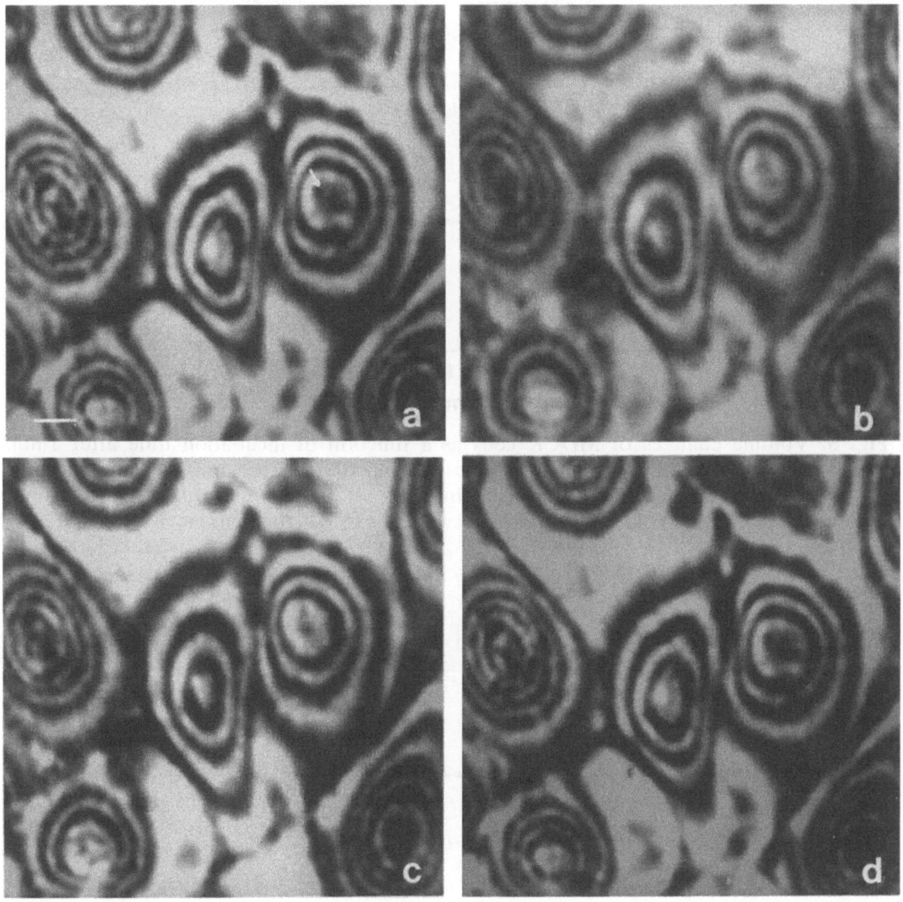

Fig. 4a–d. Volume changes of MDCk cells on addition of 1 m*M* adenosine: **a** Control; **b** 2 min after addition, the dark area at the apex (*arrow* in **a**) disapears, indicating a volume decrease (*arrow*) which reverses in **c** (10 min) and **d** (30 min). *Bar* = 10 μm

microscope is its great focal depth: Images resulting from ultrasound microscopy provide interferences of the whole upper cell surface, whereas with the RIC microscope no focusing plane is available for normal-sized and large cells that gives interferences describing the whole surface topography. The dome-shaped cell center has to be estimated according to the known slope of the lower cell regions. Due to the lower resolution of the acoustic image, interferences caused by irregular attachment of the cell contribute less to the image than light interferences. The acoustic interferences are wider and fewer in number compared with the light interferences. The increased homogeneity of the interference fringes facilitates the evaluation.

Besides these advantages, based mainly on the lower resolution of acoustic images, SAM provides information on the mechanical properties of cells.

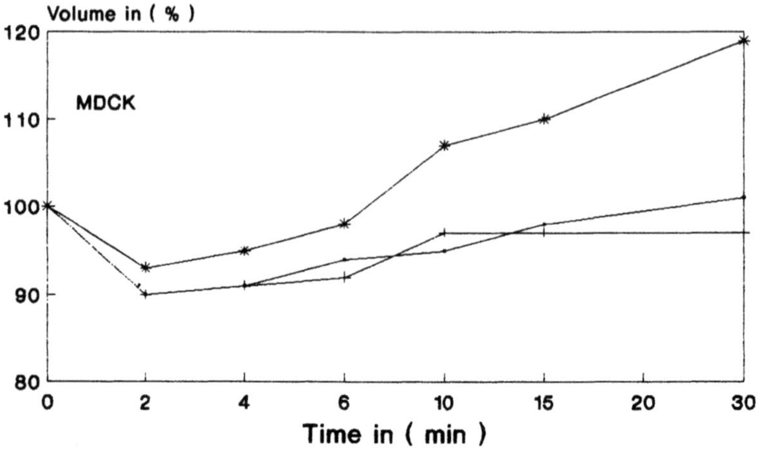

Fig. 5. Volume changes of MDCK cells as a function of incubation time after 1 m*M* adenosine was added. The original volume is set t_0 100%. Each line indicates the time course of the volume change of one cell. Initially, all cells show a shrinkage; then they recover to their initial volume. In one case swelling exceeds the amount of shrinkage. *Bar* = 10 μm

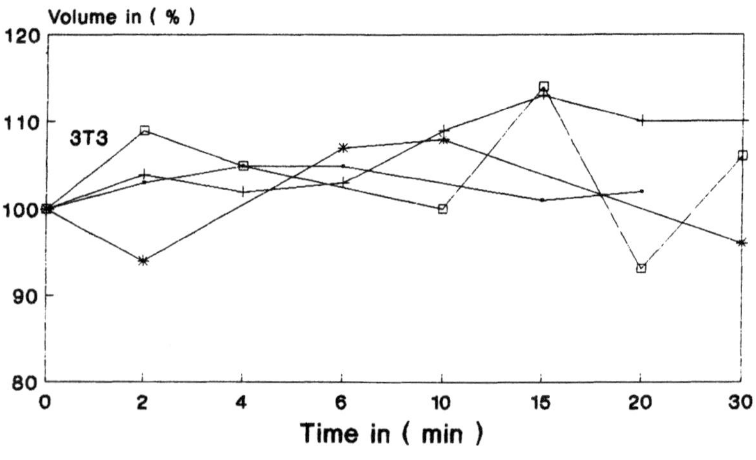

Fig. 6. Relative volume changes of 3T3 cells versus incubation time with 1 m*M* adenosine. The initial volume is set t_0 100%. No regular behavior can be observed

Changes in surface reflectivity indicate changes in tension [3, 17]. Tension of the membrane-cortex complex could well be the parameter responsible for information storage of cellular volume during volume recovery responses.

Regarding the observed volume changes on the addition of adenosine (and other substances), two questions have to be discussed: the mechanisms underlying the volume changes and their mutual significance in ameliorating skin moisture. Because of the very preliminary character of the results at present, only a theoretical outline of these questions can be provided.

The initiation of volume changes by adenosine is most probably initiated by specific receptors activating ion fluxes, the exchange of Ca^{2+} in particular [10]. The variability of the time course and in the extent of the volume changes could be caused by differences in the metabolic states influencing the number and activity of channels.

A series of factors influencing smoothness and moisture of skin must be considered. These include direct effects of changes of cellular volume and consequences of swelling, as well as increased substance release by the cells. Swelling of epidermal cells would reduce the intercellular spaces, and so reduce the amout of exchangeable water, thus retaining humidity in this region. Volume changes in fibroblasts can be supposed to alter the ionic condition of their immediate environment and concomitantly shift the water-binding capacity of the extracellular matrix compounds. In addition, volume changes of these cells could be involved in the release of extracellular matrix precursor material, and could thus control the water-binding capacity of the connective tissue layers. Observations on the release of TSH by adenohypophysis cells, induced by a swelling of the cells that is evoked by urea or glycerol treatment [19], support this hypothesis. Whether the induction of cellular product release by volume changes represents a general mechanism has to be elucidated by further studies.

Acknowledgements. Support of these studies by Basotherm (Biberbach a. d. Riß) and by the *Deutsche Forschungsgemeinschaft,* grant Be 423/13) is gratefully acknowledged.

References

1. Allen JC, Mills JW (1988) Two lines of MDCK epithelial cells with different volume and ion responses to calcium ionophore A23187. In vitro Cell Dev Biol 24:588–592
2. Bereiter-Hahn J (1985) Computer-assisted microscope interferometry by image analysis of living cells. Adv Microsc 196:27–44
3. Bereiter-Hahn J (1987) Scanning acoustic microscopy visualized cytomechanical responses to cytochalasin D. J Microsc 146:29–39
4. Bereiter-Hahn J, Strohmeier R (1987) Hydrostatic pressure in metazoan cells in culture: its involvement in locomotion and shape generation. In: Bereiter-Hahn J, Anderson OR, Reif WE (eds) Cytomechanics. Springer, Berlin Heidelberg New York, pp 261–267
5. Bereiter-Hahn J, Strohmeier R, Beck K (1983) Determination of the thickness profile of cells with the reflection – contrast microscope. Sci Tech Inf 8:125–150
6. Bereiter-Hahn J, Lück M, Miebach T, Stelzer HK, Vöth M (1990) Spreading of trypsinized cells: cytoskeletal dynamics and energy requirements. J Cell Sci 96:171–188
7. Clegg JS, Jackson SA, Fendl K (1990) Effects of reduced cell volume and water content on glycolysis in L-929 cells. J Cell Physiol 142:386–391
8. Emerman JT, Pitelka DR (1977) Maintenance and induction of morphological differentiation in dissociated mammary epithelium on floating collagen. In Vitro 13:316–328

9. Hoppe M, Bereiter-Hahn J (1985) Applications of scanning acoustic microscopy – survey and new aspects. Trans Sonics Ultrasonics 32:289–301
10. Javors MA, Liu M, Cuvelier BS, Bowden CL (1990) Characterization of the effect of the adenosine agonist cyclohexyladenosine on platelet activating factor-induced increases in $(Ca^{2+})_i$ in human platelets in vitro. Cell Calcium 11:647–653
11. Kajstura J, Bereiter-Hahn J (1989) Loss of focal contacts accompanies the density – dependent inhibition of cell growth. Cell Biol Int Rep 13:377–383
12. Kajstura J, Reiss K (1989) F-actin organization influences the osmotic reactions of animal cells. Folia Histochem Cytobiol 27:201–208
13. Kleinzelleer A, Ziyadeh FN (1990) Cell volume regulation in epithelia – with emphasis on the role of osmolytes and the cytoskeleton. Cell Vol Regul 4:59–86
14. Larsen T, Solberg S, Johansen R, Jorgensen L (1988) Effect of cooling on the intracellular concentration of Na^+, K^+ and Cl^- in cultured human endothelia cells. Scand J Clin Lab Invest 48:565–571
15. Lee SC, Price M, Prystowsky MB, Deutsch C (1988) Volume response of quiescent and interleukin 2-stimulated T-lymphocytes to hypotonicity. Am J Physiol 254:C286–C296
16. Levison C (1987) Volume regulatory activity of the Ehrlich ascites tumor cell and its relationship to ion transport. J Membr Biol 100:183–191
17. Litniewski J, Bereiter-Hahn J (1990) Measurements of cells in culture by scanning acoustic microscopy. J Microsc 158:95–107
18. Piper J, Pera F (1980) Rekonstruktion des Oberflächenreliefs von Erythrocyten mit Hilfe der Leitz-Reflexionskontrast-Einrichtung. Leitz Mitt Wiss Tech 7:230–234
19. Wang X, Sato N, Greer MA, Greer SE, McAdams S (1989) Cell swelling induced by the permeant molecules urea or glycerol induces immediate high amplitude thyrotropin and prolactio secretion by perifused adenohypophyseal cells. Biochem Biophys Res Commun 163:471–475

Effects of Uremic Compounds on Cultured Myocytes: a Model for Uremic Myocardiopathy

D. Weisensee, I. Löw-Friedrich, W. Schoeppe,
and *J. Bereiter-Hahn*

Abstract

Respiration is a central factor of metabolism in almost all tissues; thus, all pharmacological effects also concern the energy metabolism in a direct or an indirect manner. A system which allows determination of oxygen consumption and simultaneously application of the methods of light microscopy is therefore of general pharmacological interest, irrespective of the cell type under investigation. We present the application of such a test system to the study of myocardial cell responses after exposition to uremic compounds.

Congestive heart failure is the most common cause of mortality in patients of maintenance hemodialysis. However, no unambiguous evidence of specific uremic damage to the myocardium has yet been provided. Therefore, we exposed fetal mouse cardiac myocytes in culture to "uremic" conditions and investigated the responses in oxygen consumption and contractility using a microperfusion system. The uremic state was achieved by perfusion of the cultures with sera obtained from chronic hemodialysis patients, immediately before dialysis after the long interval. Cultured myocytes were spontaneously contractile with a constant beating frequency during perfusion with serum-free medium. During perfusion with high concentrations of uremic sera (5%–20%) contractions ceased completely within 9–10 min; with sera from healthy controls spontaneous contractions continued for at least 50 min. Perfusion of the cardiomyocyte cultures with urea (10–500 m*M*), creatinine (5–20 m*M*), or combinations of both compounds also inhibited cellular contractility. Application of 10 m*M* urea and of "uremic" sera (> 5%) increased cellular steady-state respiration while high concentrations of urea (50–500 m*M*) and creatinine (5–20 m*M*) provoked a concentration-dependent decrease in steady-state oxygen consumption. Obviously, changes in osmolarity did not contribute to these effects. We conclude that the sera of patients on maintenance hemodialysis contain factors which adversely affect myocyte contractility and respiration in culture. High urea and creatinine concentrations are assumed to contribute to this effect. The in vitro test system seems to be a useful assay for testing further substances suspected to be "uremic toxins" with respect to their impact on the myocardium.

A. Bernd, J. Bereiter-Hahn,
F. Hevert, H. Holzmann (Eds.)
Cell and Tissue Culture Models
in Dermatological Research

Introduction

Respiration plays a central role in the maintainance of the basic metabolism of the cell. Consequently, alterations of energy metabolism should reflect a wide variety of cellular responses to pharmacological actions. A test system was therefore developed which enables the investigation of oxygen consumption and morphology of cell cultures simultaneously. This is exemplified by mouse cardiomyocyte cultures following the effects of uremic substances on oxygen consumption and contractility.

Patients on maintenance hemodialysis frequently suffer from cardiac impairment [1]. Even in patients without manifest cardiac dysfunction, a high incidence of cardiac hypertrophy [2] and a reduced diastolic compliance of the left ventricle have been documented [3]. Chronic experimental uremia in rats evokes a cardiac hypertrophy which is accompanied by an increase in heart weight based on an elevated myocardial fiber mass; the interstitial cells become activated, and the number of mitochondria and that of structures derived from the endoplasmic reticulum are increased [4]. Normalization of renal anemia and arterial blood pressure, beta- and alpha-adrenoreceptor blockade, and parathyroidectomy fail to prevent cardiac hypertrophy in experimental uremia [4]. The mechanisms and compounds responsible for the cardiac alterations in chronic uremia are still unknown. Therefore, we developed an in vitro model for "uremic myocardiopathy": the measurement of contractility and respiration of cultured mouse cardiomyocytes under uremic conditions and tentative uremic toxins.

Material and Methods

Cell Culture

Hearts from 18-day-old fetal mice were washed with phosphate-buffered saline to remove blood. Trypsin digestion (final concentration of trypsin: 0.1%) was performed for 15 min at room temperature under slight agitation.The supernatants of the first two digestion steps were discarded; those following were collected in alpha medium (Biochrom Seromed) supplemented with 20% fetal calf serum (FCS) and antibiotics. The heart cells were collected by centrifugation at 4° C in a Sorvall SS 34 rotor at 1000 rpm for 10 min and seeded in Petri dishes (21 cm^2) with a coverslip on the bottom. The cell number was adjusted to 1×10^6 cells per dish (4 ml medium). The adherent cardiac myocytes were incubated for 2 days in alpha medium with 20% FCS [5] and for a further 2 days in Dulbecco's minimum essential medium (DMEM) with 10% FCS. Later on, the cultures were transferred to L-valine-deficient DMEM to prevent the growth of the remaining fibroblasts. The cardiac myocytes which were spontaneously contractile in culture were incubated for 5–10 days in a water-saturated atmosphere with 5% CO_2 (Fig. 1). During this period, they formed a synchron-

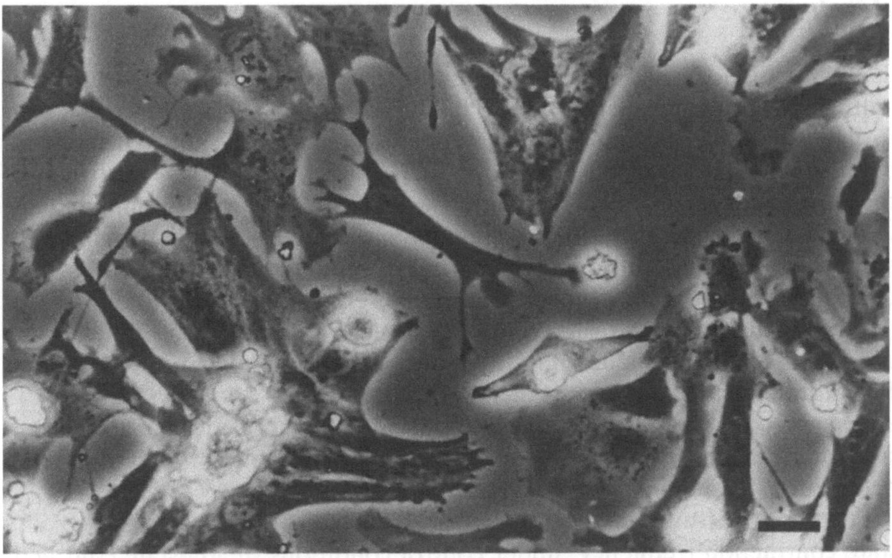

Fig. 1. Phase-contrast micrograph of spontaneously contractile cardiomyocytes cultured on a coverslip for 2 days. *Bar* – 20 μm

ously beating confluent monolayer. For the experiments the coverglasses with the cell cultures were integrated into a microperfusion system [6] based on a modified Dvorak-Stotler chamber. Perfusion was performed at a constant flow rate of 5.97 ml/h and a temperature of 37 ± 1° C.

Evaluation Procedure

Parameters of contractility – including contraction frequency, development of arrhythmia, and inotropy – were determined by computer-assisted analysis of video-tape records using a VT-100 board (Image Technologies) for real-time subtraction and a Q520 (Cambridge Instruments) image analysis system to binarize the resulting difference image [7]. For the measurement of myocyte respiration, a Clarke oxygen electrode was integrated into the perfusion system allowing permanent registration of oxygen consumption. Intracellular adenine nucleotide content was determined by extraction with 0.1% Triton X-100 on ice and subsequent quantitation by HPLC. ATPase activity was completely inhibited by 1 m*M* dispase during the extraction time (6 min) and the following measurement procedure. The data shown are representative; all experiments were repeated at least three times. The sera used for perfusion were obtained from patients with end-stage renal failure immediately before dialysis after the long interval ("uremic" sera) or from healthy individuals (control sera). The samples were pooled and heat inactivated at 56° C for 30 min.

Results

The individual cardiac myocytes in culture were spontaneously contractile with an average frequency of 86 ± 28 cycles/min. As soon as the cells formed a confluent monolayer, they beat synchronously at a constant frequency which spontaneously did not vary more than ± 10%. Perfusion of the cardiac myocytes in culture with 5%–20% of sera from uremic patients induced a complete cessation of spontaneous beating within 9–10 min; with control sera from healthy individuals the spontaneous contractions continued for at least 50 min. When 0.5% uremic sera was applied, no differences in beating frequency to the control condition was observed. While 1.0% uremic sera initially induced an inhibition of contraction for about 15 min, 1.5% uremic sera led to a prolonged inhibition for about 55 min. These effects were reversible. The same amount of time was needed for the cells to recover from the uremic influence and to display their normal contractile behavior.

Perfusion with urea alone (10–500 m*M*) led to a decrease in the steady-state contraction frequency; under control conditions steady-state contractility was not affected. Perfusion with 200 m*M* urea caused, in addition to the frequency decrease, arrhythmias. These effects cannot be attributed to an increased osmolarity, because perfusion with 200 m*M* mannitol showed normal contractile behavior (Fig. 2). When the cardiac myocytes were perfused with creatinine (5–20 m*M*), contraction frequency decreased (–23%). A combined perfusion of the cardiomyocyte cultures with 5 m*M* creatinine plus 200 m*M* urea caused arrhythmia and a decrease in beating frequency (–29%). In order to simulate the situation of a long-lasting uremic condition, the heart cells were pre-incubated for 20 h with 1% uremic sera. Following perfusion with 5 m*M* creatinine and 200 m*M* urea induced permanent arrhythmias and a loss of synchrony. Cellular steady-state respiration was enhanced with 10 m*M* urea, with high concentrations of uremic sera, and after 1% uremic sera for 20 h. Perfusion with urea (≥ 50 m*M*) or creatinine provoked a concentration-dependent decrease in steady-state oxygen consumption (Fig. 3). Inotropy, revealed by the cellular shortening during the contractions, tended to be reduced under "uremic" conditions. Determinations of intracellular adenine nucleotides in the perfused cardiac myocytes attest to a great variability of ATP content and ATP/ADP ratio without any direct correlation to the test conditions. The creatine-creatine phosphate shuttle, which represents a very effective system for rapid energy regeneration in muscle cells, might be responsible for this effect.

Discussion

Impaired cardiac performance is an important cause of morbidity and mortality in patients with chronic renal failure. Multiple factors may be involved in the development of cardiac failure in patients on maintenance hemodialy-

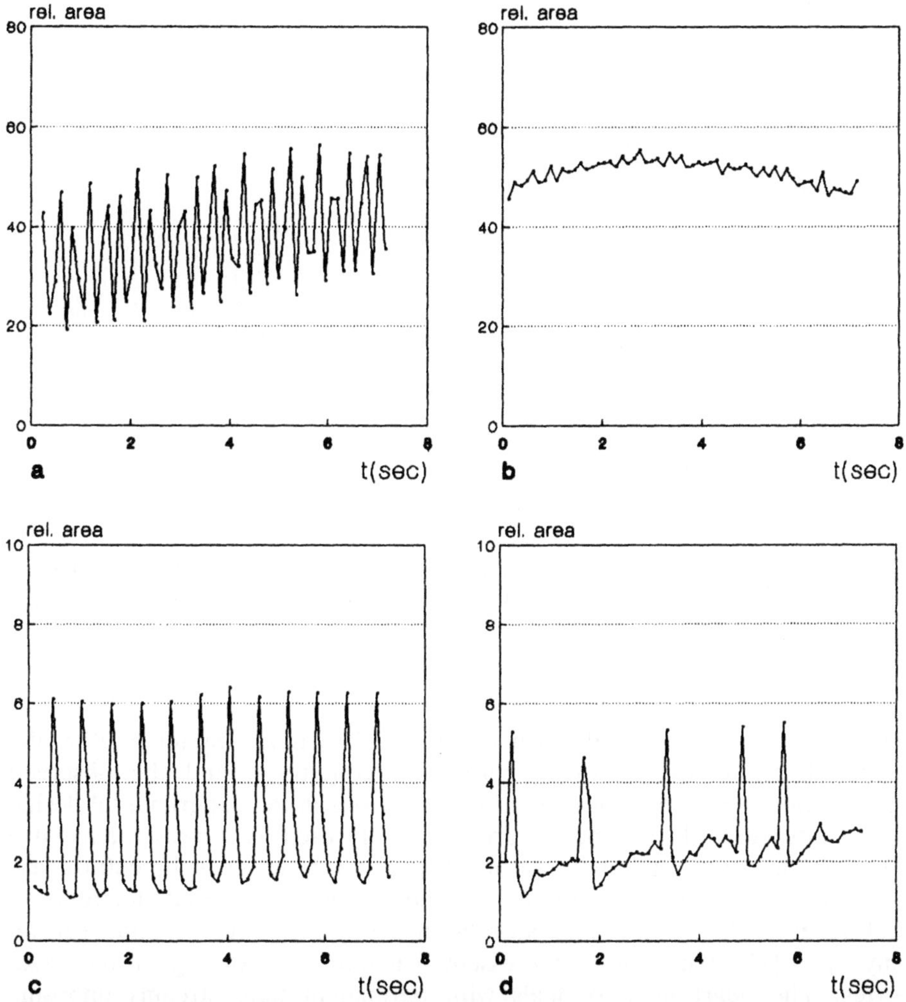

Fig. 2 a–d. Area quantification of cardiomyocyte contractions as described in Material and Methods. Each oscillation represents one synchronous contraction of the monolayer, whereas the peak height demonstrates the degree of inotropy. Values were determined in the 9th min of perfusion with 1.0% of either (**a**) control sera or (**b**) sera from uremic patients. (**c**) Perfusion with 200 m*M* mannitol as an osmotically active control substance (497 mosmol/l) for 120 min; regular and periodic contractions. (**d**) Perfusion with 200 m*M* urea (495 mosmol/l) lead to strong arrhythmias and a decrease of beating frequency during the same period ($n = 3$)

sis, e.g., arterial hypertension, anemia, fluid overload, the arteriovenous fistula, electrolyte imbalance, calcifications. In addition, the "uremic" state per se may impair myocardial performance, resulting in an "uremic myocardiopathy" [8]. In vivo investigations can hardly contribute to the solution of this problem because in the living "uremic" organism all parameters men-

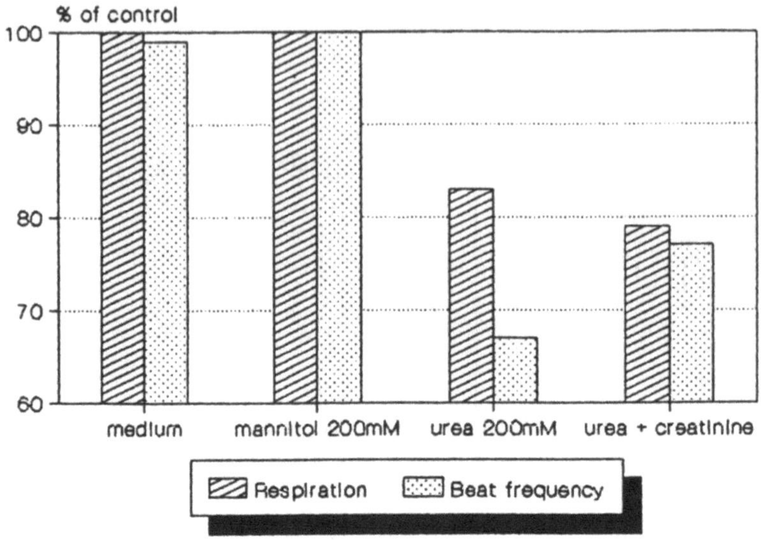

Fig. 3. Steady-state respiration and contraction frequency of perfused cells referred to the control level before exposure to the test conditions. On mannitol addition, no decrease of respiration or contraction frequency occurs ($n = 3$)

tioned above interact with each other. In vitro model systems allow the exact determination of the experimental conditions, and the observed effects can therefore be addressed to alterations in the experimental design [9]. Cultured cardiac myocytes are frequently used as a basis for in vitro studies on myocardial metabolism [10]. Fetal heart cells in culture are spontanously contractile, forming a monolayer with restituted electrophysical signal conductance and the morphological characteristics of intact cardiomyocytes [11]; they display the essential features of working cardiac myocytes. The heart in vivo deals with permanent fluid streams imposing pressure and volume load on the cells. Therefore, heart cells under a continuous perfusion condition resemble the physiological state more closely than nonperfused cells [12]. Our results show a harmful influence of complete "uremic" sera on essential parameters of myocyte performance, incidating an interference with cellular vitality. However, the mechanism of the reversible contraction inhibition by 1.0% or 1.5% "uremic" sera is not understood. If the cardiac myocytes are exposed to the clinically predominant "uremic" compounds – i.e., urea and creatinine – changes in the parameters investigated can be detected, too. Depression of myocardial force development as well as decrease of oxygen consumption is found during perfusion of these substances. The loss of rhythmicity and synchrony influenced by the membrane-permeable molecule urea might be the result of disturbances in ion transport control mechanisms of the cytoplasm and/or the glycocalyx, thus modifying the excitation-contraction coupling mechanism of the heart cells [13]. These effects are different from those of complete sera; if combined, both chemicals act additively. Patients depending on hemodialysis undergo their dialysis treatment at intervals of 2 or 3 days. Therefore, the composition of plasma components alternates to a very high degree. These permanent fluctuations of uremic metabolites circulating in the patient's serum might further promote the pathogenesis of "uremic myocardiopathy". The presence of circulating toxins which are inhibitory to myocardial contractility [14] and impair oxygen consumption in the isolated perfused guinea pig heart [15] has been postulated and is in good agreement with our in vitro results.

In vitro models are always limited to a certain chain of reactions. The construction of the model has to be thoroughly related to the in situ event which is going to be modeled. As exemplified by the cardiomyocyte cultures, acute as well as chronic actions can be studied. The combination of respiration measurements with the analysis of the contents of the perfusate and with the broad range of microscopic methods available (reaching from general descriptions of morphology to advanced fluorescence techniques and image analysis) allows a multiparameter analysis of cellular reactions, which is required to gain some understanding of orthological and pathological events.

References

1. Wizemann V, Kramer W, Bechtold A (1984) Ventrikelfunktion bei chronisch dialysepflichtiger Niereninsuffizienz. Nieren-Hochdruckkr 13:405–408
2. Ikram H, Lynn KL, Bailey RR, Little PJ (1983) Cardiovascular changes in chronic hemodialysis patients. Kidney Int 24:371–376
3. Kramer W, Thormann J, Kindler M et al (1985) Urämische Herzkrankheit. I. Inzidenz und klinische Wertigkeit nicht-invasiver kardialer Befunde bei dialysepflichtiger Niereninsuffizienz. Med Welt 36:1228–1233
4. Rambausek M, Mann J, Mall G et al (1986) Cardiac findings in experimental uremia. Contrib Nephrol 52:125–133
5. Löw I, Friedrich T, Schoeppe W (1989) Synthesis of shock proteins in cultured fetal mouse myocardial cells. Exp Cell Res 180:451–459
6. Schlage WK, Bereiter-Hahn J (1983) A microscope perfusion respirometer for continuous respiration measurement of cultured cells during microscopic observation. Microsc Acta 87:19–34
7. Riehle M, Bereiter-Hahn J (1991) Amplitude and frequency determination of cardiomyocytes in culture by means of digital image analysis. Eur J Cell Biol 54 [Suppl 32]:81
8. Drueke T, LePailleur C, Zingraff J, et al (1980) Uremic cardiomyopathy and pericarditis. Adv Nephrol 9:33–70
9. Löw-Friedrich I, von Bredow F, Schoeppe W (1991) A cell culture assay for the detection of cardiotoxicity. J Pharmacol Methods 25:133–145
10. Piper HM, Hütter JF, Spieckermann PG (1984) Relation between enzyme release and metabolic changes in irreversible anoxic injury of myocardial cells. Life Sci 35:127–134
11. Kandolf R, Canu A, Hofschneider PH (1985) Coxsackie B3 virus can replicate in cultured human foetal heart cells and is inhibited by interferon. J Mol Cell Cardiol 17:167–181

12. Bhatti S, Zimmer G, Bereiter-Hahn J (1989) Enzyme release from chick myocytes during hypoxia and reoxygenation: dependence on pH. J Mol Cell Cardiol 21:995–1008
13. Abaurre PF, Vassallo DV, Mill JG, Leite CM (1988) Depression of myocardial force development after urea treatment. Braz J Med Biol Res 21:1049–1052
14. Scheuer J, Stezoski SW (1973) The effects of uremic compounds on cardiac function and metabolism. J Mol Cell Cardiol 5:287–295
15. Kersting F, Brass H, Heintz R (1978) Uremic cardiomyopathy: studies on cardiac function in the guinea pig. Clin Nephrol 10:109–113

Cultivation of Trophoblast Tissue of Human Intrauterine and Ectopic Pregnancies and Treatment with Methotrexate

D. Schäfer, J.-P. Pfuhl, R. Baumann, S. Neubert, H.-G. Bender, and *H. Naujoks*

Introduction

During the past two decades there has been an obvious increase in extrauterine pregnancies (EUP) worldwide. Several factors seem to be responsible for this development, e.g., the increase in antibiotic treatment of salpingitis, the higher frequency of Fallopian tube surgery, and the use of intrauterine devices.

Some years ago, salpingectomy was the standard therapy for EUP, but recently more conservative methods have been used. For a few years now we have known about local treatments, such as injection of various substances like methotrexate (MTX) or prostaglandins directly into the EUP [1, 2]. A nonsurgical approach is also possible, i.e., systemic treatment with substances like MTX [6]. Whereas many case reports show the clinical value of MTX, prostaglandins or other substances with regard to this topic, only a few authors have been concerned about the effects of these drugs on trophoblastic tissue in vitro [3].

This prompted us to start a study in 1987 comparing trophoblast tissue cultures of ectopic and intrauterine (IUP) pregnancies. The aim was to gain more information about the cause and the genesis of EUP and to reveal the differences and similarities between EUP and IUP. We investigated the time period in which trophoblastic tissue from ectopic sites was capable of producing HCG in culture, and we established a model for testing the effect of antitrophoblastic agents in vitro [4]. Here we describe the effects of MTX on trophoblast tissue cultures derived from ectopic and intrauterine pregnancies using the above-mentioned culture system.

Materials and Methods

Evaluation of EUP was carried out by vaginal and abdominal ultrasound and determination of serum HCG levels. Gestational age ranged from 5 to 8 weeks after menstruation. Tissue was obtained directly after surgery. Contaminating material was removed carefully. We set up both long-term (LTC) and short-term (STC) cultures whenever a reasonable amount of tissue was available. Otherwise, we set up only LTC. For LTC the tissue obtained was

A. Bernd, J. Bereiter-Hahn,
F. Hevert, H. Holzmann (Eds.)
Cell and Tissue Culture Models
in Dermatological Research

Table 1. MTX concentrations in culture

Concentration [mol/l]	Group 1	Group 2	Group 3	Group 4
1.0×10^{-2}	+	+	–	–
1.0×10^{-3}	+	+	+	+
3.8×10^{-4}	+	+	+	+
3.8×10^{-5}	+	+	+	+
Without MTX [control]	+	+	+	+

Group 1, IUP treated day 0; Group 2, EUP treated day 0; Group 3, IUP treated day 6; Group 4, EUP treated day 6

divided into equal fractions according to the number of culture flasks. The tissue was subsequently cut into pieces of about 1 mm^3 and rinsed in Ca^{2+}- and MG^{2+} -free Dulbecco Solution. Following trypsinization for 20–30 min (0.02%) the fractions were placed in 25 cm^2 Falcon tissue-culture flasks. The trophoblast tissue cells were incubated in 2.5 ml Chang medium containing streptomycin/penicillin in a 5% CO_2 atmosphere at 37° C. For STC the dissected tissue was incubated in RPMI 1640 medium supplemented with 5% fetal calf serum.

Intrauterine trophoblast tissue was obtained from terminated pregnancies. Gestational age ranged from 6 to 11 weeks after menstruation. Preparation and setup were as for EUPs.

In LTC, harvesting for karyotyping was performed in untreated cultures 2–4 weeks later [7]. STC were harvested 24 h later [5]. Following GTG staining, cytogenetic evaluation was carried out.

Medium was changed every other day. HCG levels were measured in the aspirated medium. MTX was added either 12 h after culture onset (day 0) or on day 6; 48 h later the medium was removed. Concentrations of MTX ranged from 1.0×10^{-2} to 3.8×10^{-5} mol/l for treatment on day 0, and from 1.0×10^{-3} to 3.8×10^{-5} mol/l for treatment on day 6 (Table 1). For EUP or IUP, one culture flask without MTX treatment served as a control.

Results

Karyotyping was performed successfully on 49 of 59 specimens (Table 2). In 16 EUPs the amount of tissue allowed both STC and LTC. In no case was a difference between STC and LTC detected. In only one of the 49 cultured EUPs did a chromosomal aberration occur; it was demonstrated in two of four subcultures. In four cases we observed increased chromosomal fragility. The ratio of female to male karyotype was 31:18 (Table 2).

All cultures showed HCG secretion into the culture medium within the first 16–20 days. Most untreated cultures showed a slight elevation of HCG

Table 2. Cytogenetic results

	n
Patients	59
Sufficient culture growth	54
Cytogenetic results	49
STC and LTC established	16
Only LTC established	43
Differences STC versus LTC	0
46, XX	31
46, XY	18
Mosaics	0
Numerical aberrations	0
Structural aberrations	1
Increased fragility	4

values on day 4 and afterwards a constant decline. Declining HCG levels coincided with an increase of fibroblast-like cell type in culture. The majority of the cultures treated on day 0 with the highest concentration of MTX lacked this elevation. This is also observable on the graphs depicting mean HCG levels (Fig. 1).

Figure 1 displays a reduction of HCG values of cultures treated on day 0 on almost all days compared with the control. The higher the applied MTX concentration, the more pronounced was this effect. Significant reduction ($P < 0.05$) was observed only for higher MTX concentrations. The longer the interval between treatment and measurement, the lower was the concentration that led to a significant reduction of HCG levels. EUPs needed about a ten times higher concentration to reach the same level of significance as IUPs (Fig. 2).

Figure 3 shows that in groups 3 and 4 a lowering of HCG values occurred only about 6–10 days after treatment. The decline is significant, starting on day 12 for IUPs and on day 14 for EUPs (Fig. 4). Whereas most of the IUPs and EUPs showed a marked reduction of the HCG values (at least for the highest MTX concentration), few IUPs and EUPs are lacking such a reaction (Fig. 5a, b).

Conclusion

Our study shows that trophoblast tissue of IUP as well as EUP can be successfully cultured in vitro. Furthermore, it indicates that chromosomal aberrations do not seem to play an important role in ectopic implantation. Nevertheless, the ratio of 31:18 favoring the female karyotype is remarkable. Tissue cultures of EUP as well as IUP showed secretion of HCG into the culture medium within the first 16–20 days. Although we want to emphasize that one has to be very careful in drawing conclusions from in vitro

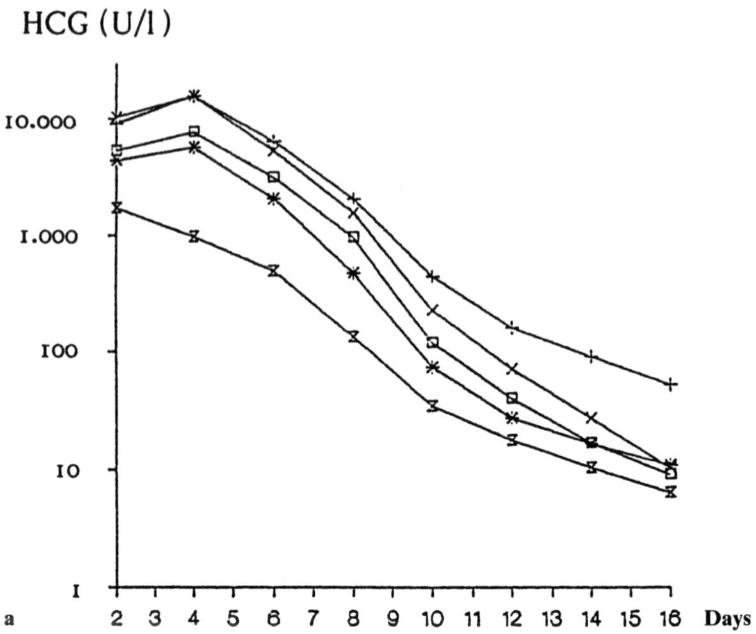

a

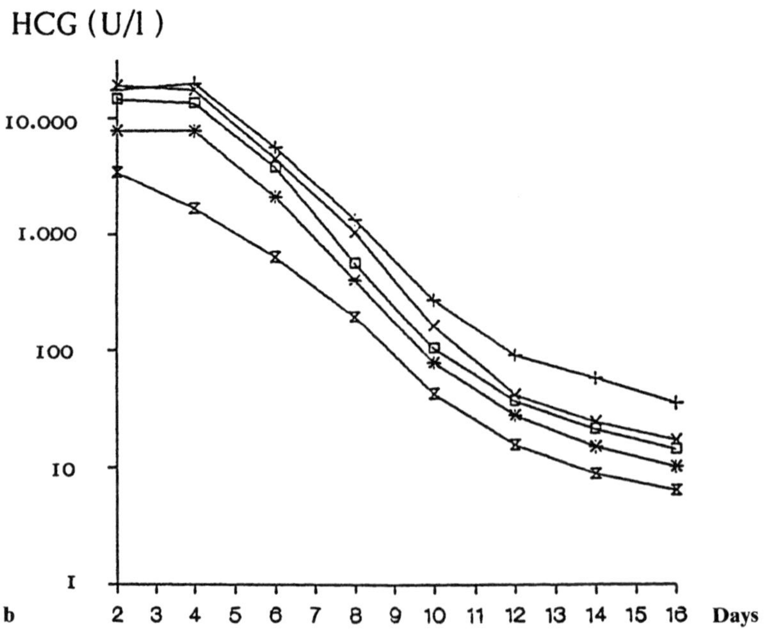

b

Fig. 1a, b. Average human chorionic gonadotrophin (*HCG*) levels *a* IUP ($n = 10$) and **b** EUP ($n = 10$). Treatment on day 0. Methotrexate (MTX) concentration: —+—, without MTX; —▼—, 1.0×10^{-2}; —*—, 1.0×10^{-3}; —□—, 3.8×10^{-4}; —×—, 3.8×10^{-5}; *n = 10.*

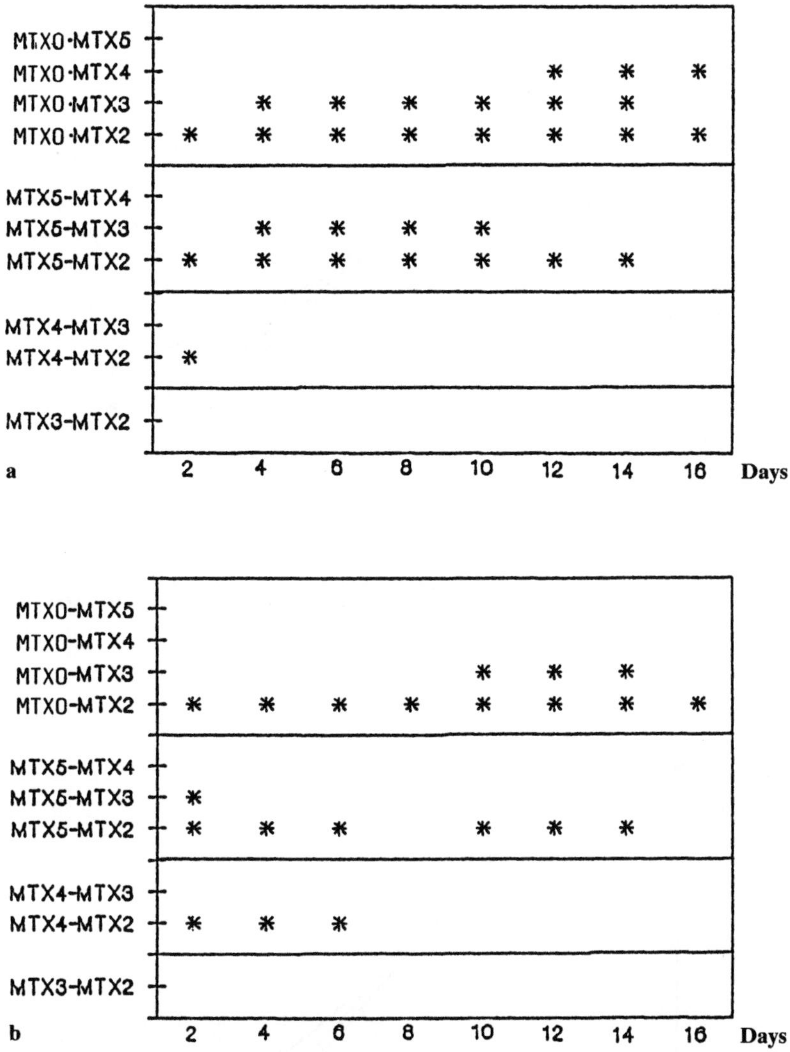

Fig. 2a, b. Multiple comparison (Wilcoxon/Wilcox) in **a** IUP (n = 10) and **b** EUP (n = 10). Treatment on day 0. *MTX0*, without MTX; *MTX2*, 1.0×10^{-2} *mol/l; MTX3*, 1.0×10^{-3} mol/l; *MTX4*, 3.8×10^{-4} mol/l; *MTX5* 3.8×10^{-5} mol/l; *, significant at $p < 0.05$

results with respect to the clinical situation, we would like to point out some aspects.

The established culture system can be used to test substances proposed for conservative treatment of EUP. Measurement of HCG levels in the aspirated culture medium is suitable to compare the effect of MTX on IUP and EUP tissue cultures, as well as to reveal differences between various regimens of application.

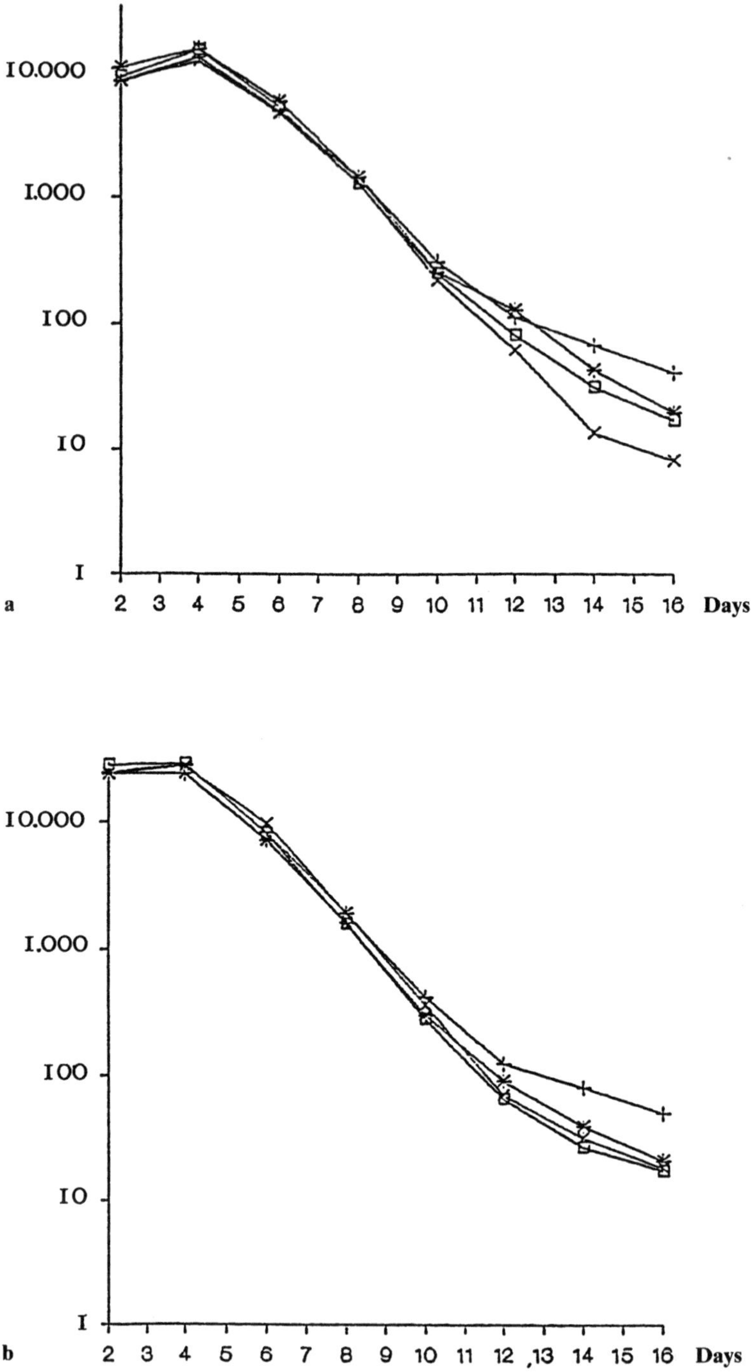

Fig. 3a, b. Average human chorionic gonadotrophin (HCG) levels, in **a** IUP (n = 15) and **b** EUP (n = 8). Methotrexate (MTX) concentration: —+—, without MTX; —×—, 1.0×10^{-3}; —■—, 3.8×10^{-4}; —*—, 3.8×10^{-5}.

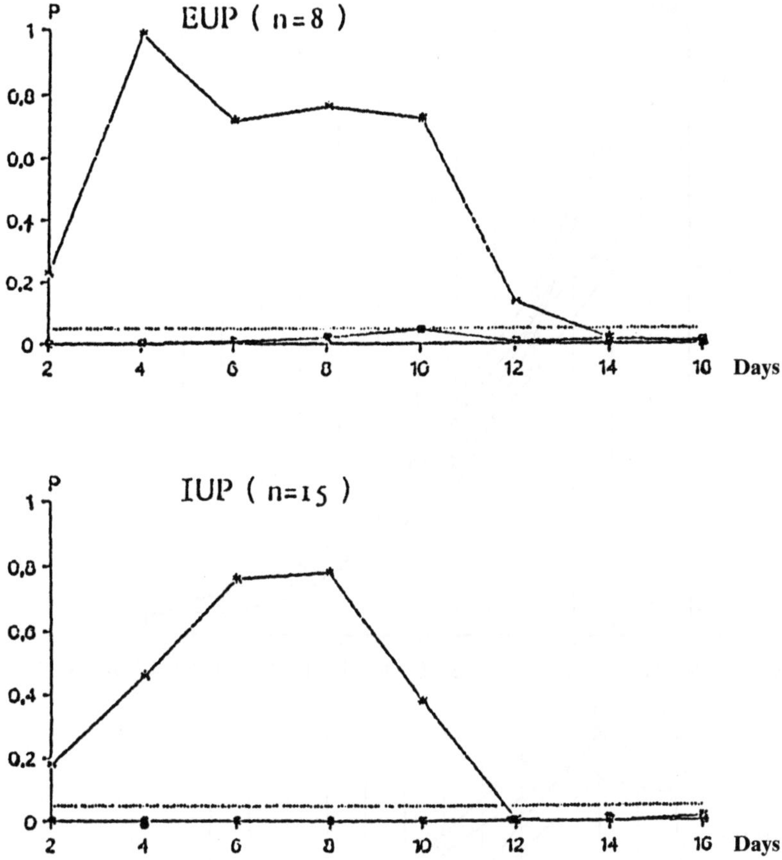

a b

Fig. 4a, b. Two way ANOVA analysis of variance (Friedman) for **a** intrauterine and **b** extrauterine data. ■, Treatment day 0, *, treatment day 6.

Trophoblast tissue cultures of EUPs need a concentration of MTX about ten times higher to lead to an equivalent reduction of HCG levels compared with IUP. This shows that one must be cautious in extrapolating from data gained with IUP to EUP.

With some regimes of MTX application it takes about 6–10 days before the effect becomes visible. In combination with our clinical data, we therefore recommend not repeating an MTX dose too early in the treatment of patients with ectopic pregnancy. Last but not least, the possibility should be borne in mind that there may be nonresponders among EUP patients.

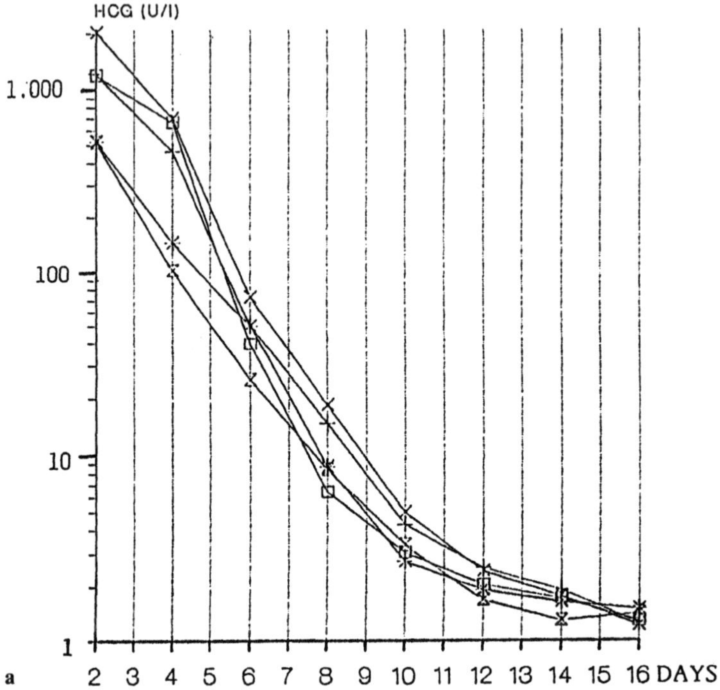

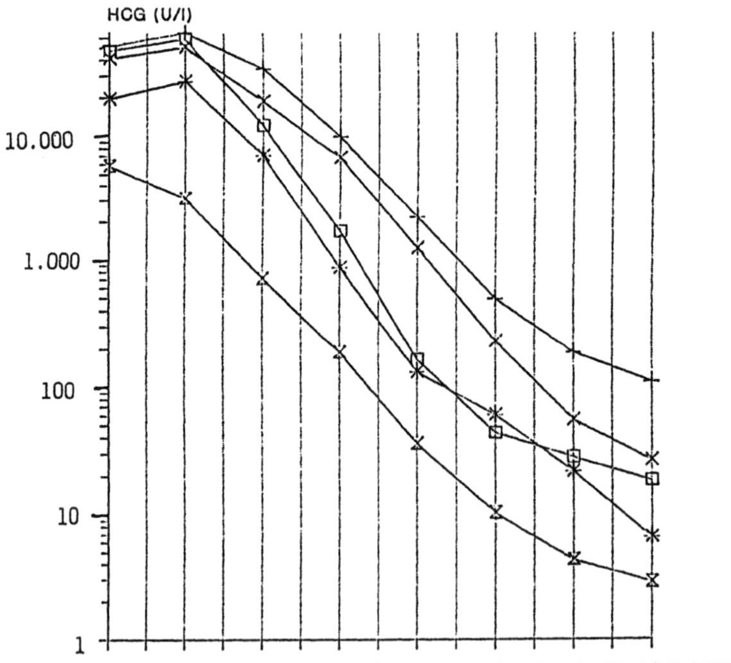

Fig. 5a, b. Responder (**a**) and nonresponder (**b**) in extrauterine pregnancies. Methotrexate (MTX) concentration: (See Fig. 1 for symbols and methotrexate concentrations)

References

1. Egarter C, Husslein P (1989) Treatment of ectopic pregnancy by means of prostaglandins. Prostaglandins Leukot Essent Fatty Acids 35:91
2. Kooi S, Kock HCLV (1990) Treatment of tubal pregnancy by local injection of methotrexate after adrenaline injection into the mesosalpinx: a report of 25 patients. Fertil Steril 54:580
3. Sand PK, Stubblefield PA, Ory SJ (1986) Methotrexate inhibition of normal trophoblast in vitro. Am J Obstet Gynecol 55:324
4. Schäfer D, Pfuhl JP, Baum R, Rößler M, Baumann R (1989) Cytogenetische, endokrinologische sowie immunologische Studien an kultiviertem Gewebe ektoper Schwangerschaften. Zentralbl Gynakol 111:1476
5. Simoni G, Terzoli G, Romitti L (1986) Fetal Karyotyping by direct chromosome preparation. In: Brambati B, Simoni G, Fabro S (eds) Chorionic villus sampling. Dekker, New York, pp 99–117
6. Stovall TG, Ling FW, Buster JE (1990) Reproductive performance after methotrexate treatment of ectopic pregnancy. Am J Obstet Gynecol 162:1620
7. Therkelsen AJ, Jensen PKA, Trier Hansen J, Smidt-Jensen S, Hahnemann N (1985) Choice of medium for cultivation and 24-hour incubation of chorionic villi: selective effects in utero. In: Fraccaro M, Simoni G, Brambati B (eds) First trimester fetal diagnosis. Springer, Berlin Heidelberg New York, pp 157–160

2. *Models in Differentiation of Keratinocytes*

Mechanical Stimulation Enhances Human Keratinocyte Differentiation in Culture: Induction of Cytokeratin 9 Synthesis

F.E. Görmar, A. Bernd, J. Bereiter-Hahn, and *H. Holzmann*

Abstract

Epidermal cells are committed to terminal differentiation, thereby undergoing a series of morphological and biochemical changes. A specific biochemical differentiation marker of keratinocytes is the expression of the cytokeratin pair K1 and K10/11 or K1 and K9 in footpad and palm keratinocytes. Considering that in vivo, keratinocytes are subjected continuously to mechanical stress, we investigated the effect of cyclic mechanical pressure on differentiaton: HaCaT cells (spontaneously transformed human keratinocytes) were cyclically stimulated by the pressure of Teflon weights. The stimulation process lasted 1–6 days. Consequent to the pressure treatment, multilayered culture growth occurred and the pattern of cytokeratin was modified. The relative amount of suprabasal keratins was increased. After 2 days of stimulation a newly synthesized keratin (64 kD, pI 5.4) was demonstrated, and in western blots it was identified as cytokeratin K9. To exclude the possibility that cytokeratin K9 was a mere laboratory contamination, autoradiography was performed. These results, gained in vitro, demonstrate that mechanical pressure is a notable factor regarding the induction of the keratin K9 synthesis, a specific differentiation marker of palmar and plantar keratinocytes.

Introduction

The epidermis is composed of a combination of proliferative and terminally differentiating cells. The proliferative cells are mostly restricted to the basal layer of the epidermis. Keratinocytes leave the basal layer, lose their proliferative potential, and undergo terminal differentiation [9]. During this process, the keratin pattern becomes altered [5]. As characterized by molecular weight (ranging from 40 to 70 kD) and isoelectric point, about 20 different human keratins are found [10] (Fig. 1). Specific heterodimers are known for various differentiation stages of the keratinocyte. Keratin K5 and K14 are synthesized in the basal layer, whereas the suprabasal keratins K1 and K10/11 are differentiation markers [15]. Keratin K9, co-expressed with K1, characterizes palmar and plantar epidermis [8, 11]. Compared with the

A. Bernd, J. Bereiter-Hahn,
F. Hevert, H. Holzmann (Eds.)
Cell and Tissue Culture Models
in Dermatological Research

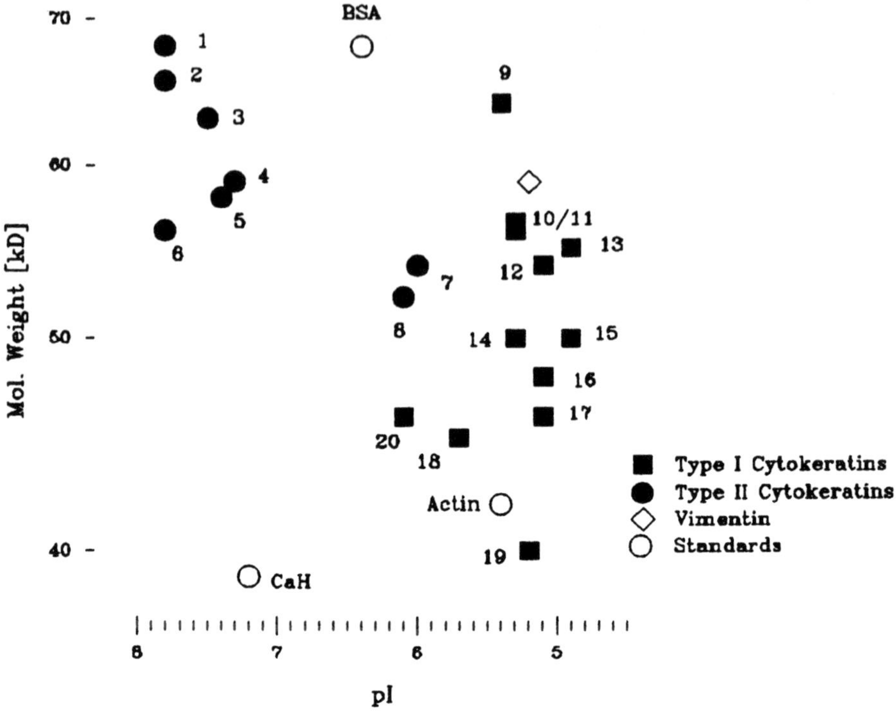

Fig. 1. Scheme of 2D-gel distribution of keratins according to molecular weights and isolelectric points of human keratin subunits. Nomenclature according to Moll et al. [10]

synthesis of cornified envelopes and filaggrin, the suprabasal keratin K1 is an early marker of keratinocyte differentiation, since it appears at the transition from the basal layer to the upper epidermal layers [12]. Many parameters influence the differentiation processes. Experiments with cell cultures revealed the significance of the Ca^{2+} concentration [7], of vitamin A, of dermal factors, of cell density, and of the cultivation time [4]. In vivo, the meaning of mechanical stress becomes evident by the observation of hyperkeratosis and tylosis of exposed epidermis. The introduction of a human epidermal cell line (HaCaT)[1] by Boukamp et al. [2] offered a unique model for studying differentiation processes in culture. We used this cell type to study the influence of cyclic mechanical stress to keratinocyte differentiation.

1 The spontaneously transformed keratinocyte line HaCaT was kindly provided by Prof. N. E. Fusening from DKFZ, Heidelberg

Methods

Cell Culture

Investigations were performed with the transformed keratinocyte line HaCaT (passage 30–50). The cultures were grown in 21 cm^2 dishes with minimal essential medium (4 × Hank's salts) supplemented with 5% fetal calf serum and 1% penicillin-streptomycin at 37°C and 5% CO_2. Media were changed twice a week.

Mechanical Stimulation of Cell Cultures

HaCaT cells were plated at a density of $2 \times 10^4/cm^2$. After 1 week of cultivation Ca^{2+} was shifted from 0.8 to 12 mmol/l and the cells were exposed to mechanical stress by cylindrical Teflon weights (m = 23.3 g ± 0.35 g) with an iron core. For mechanical stimulation, we used a modification of our method described previously [6] (Fig. 2).

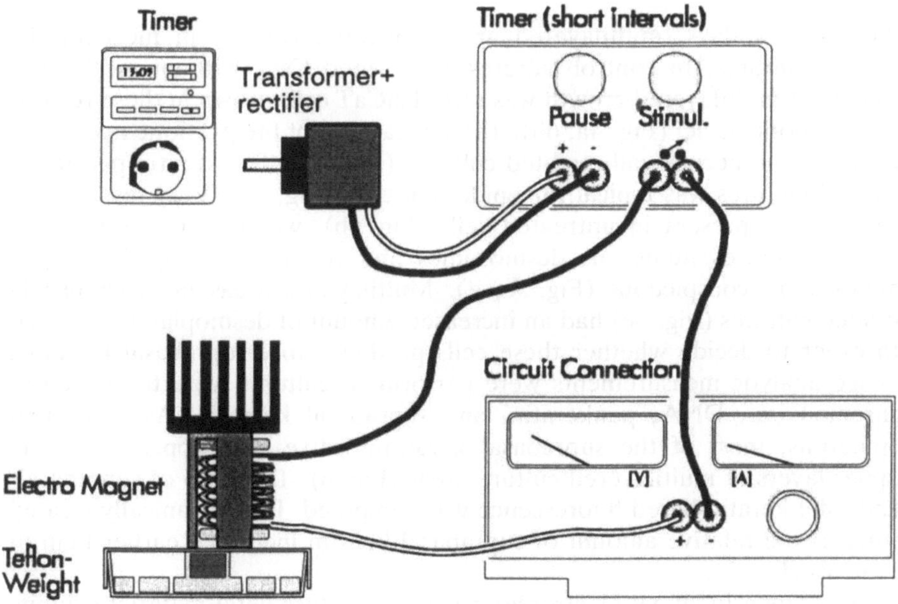

Fig. 2. Cells were exposed to mechanical stress by cylindrical Teflon weight (m = 23.3 g ± 0.35 g) with an iron core. Electromagnets on top of the Petri dishes lifted the weights from the cell layer (displacement: 1 mm) inside the culture dishes. The surface of the weights completely covered the culture area (21 cm^2). The magnets wee switched on/off every 10 s (*timer left*) for 15 min (*timer right*) of every hour. The experiments lasted 1–6 days (From [13])

Image Analysis

Cells were double stained by indirect immunofluorescence with antibodies to broad-spectrum keratin (K-4252 and 2nd AB: Anti IgG-FITC from Sigma, Deisenhofen, GERMANY) and to suprabasal keratins K1 and K10/11 (CK 8.6 and 2nd AB: Anti IgG-TRITC from Sigma, Deisenhofen, GERMANY). Homogeneous staining of the acetone/methanol-fixed cells was achieved by exposing the cultures to an equal amount of antibody solution using a $15 \times 15 \times 1\ mm^3$ staining chamber. The ratio of suprabasal keratin and total keratin fluorescence was estimated by image analysis with the Quantimet 520. For this purpose the sums of fluorescence-dependent gray levels (both for suprabasal and for total keratins) were estimated from 20 image fields per culture.

Keratin Pattern

Preparation, 2D-gel electrophoresis, autoradiography, and western blot of keratins were performed as described previously [3, 6].

Results and Discussion

After a few days, multilayered areas appeared (Fig. 3) in mechanically treated cultures. In control cultures at elevated Ca^{2+} without mechanical treatment multilayered growth was rare. HaCaT cells grown in these regions had pyknotic nuclei (Fig. 4a, 6b). The appearance of the pyknotic nuclei was enhanced in mechanically treated cultures (Fig. 4b). The keratin pattern of control cultures was typical for epidermal cells (Fig. 5a). Very few desmosomes were present in untreated cells (Fig. 5b), whereas in mechanically treated cultures numerous desmosomes and regularly arranged tonofilaments were conspicuous (Fig. 5c, d). Multilayered areas in mechanically treated cultures (Fig. 5e) had an increased amount of desmoplakin (Fig. 5f). In order to decide whether these cells produce more suprabasal keratins, image analytic measurements were performed: cultures were triple fluorochromed i.e., DNA, pankeratin, and suprabasal keratins. As in normal epidermis, most of the suprabasal keratin-positive cells appeared in the upper layers of multilayered culture areas (Fig. 6). The ratio of suprabasal- and total keratin-based fluorescence was computed. In mechanically treated cultures the relative amount of suprabasal keratin increased earlier than in controls (Fig. 7).

Two-dimensional gel electrophoresis of the cytoskeletal material exhibits the normal keratin pattern previously described for HaCaT cells by Boukamp et al. [2]. The patterns of the control cultures of the keratins K5, K6, K7, K8, K13, K14, K16, and K19 (Fig. 8a). There was no difference after 1 day of mechanical treatment (Fig. 8b), whereas after 2 days an additional protein peak appeared at the electrophoretic position of cytokeratin K9 (64

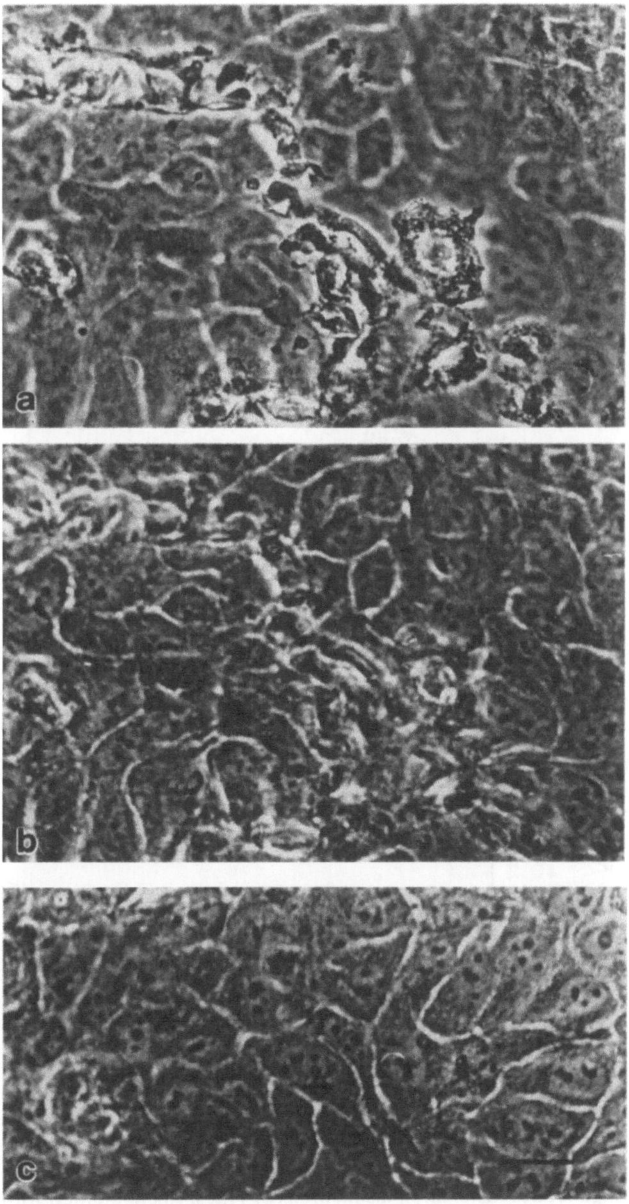

Fig. 3a–c. Mechanically stimulated HaCaT cells (2d), examined by phase-contrast microscopy: **a** focused on the upper horny cell layer; **b** same image field focused on the underlying layer; **c** monolayered HaCaT cells after incubation (2d) without stimulation (*bar* 10 m)

kD, pI 5.2; Fig. 8d) not seen in controls (Fig. 8c). In western blots this protein reacts positively with a highly specific guinea pig antiserum to cytokeratin K9 [II] (Fig. 8g, h). A preparation of human plantar epidermis (Fig. 8i), which contains a high amount of cytokeratin K9(Fig. 8h), served as a control. In order to exclude the possibility that the K9 found in stimulated cultures was a mere laboratory contamination, we performed

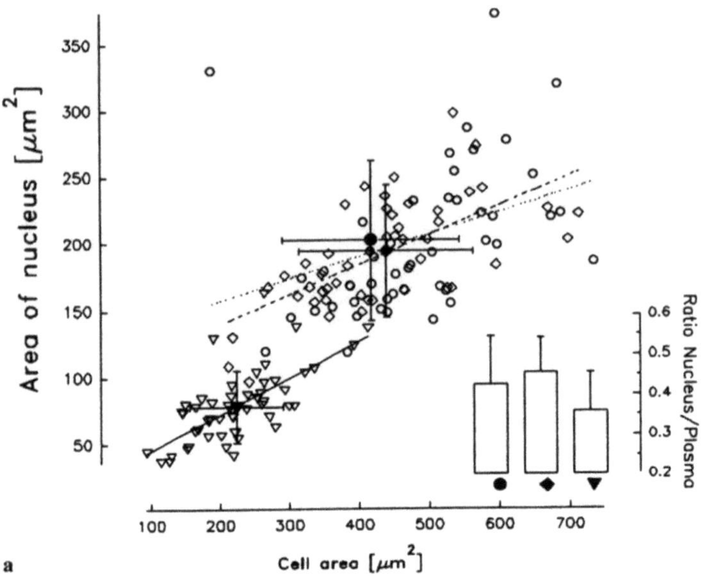

a

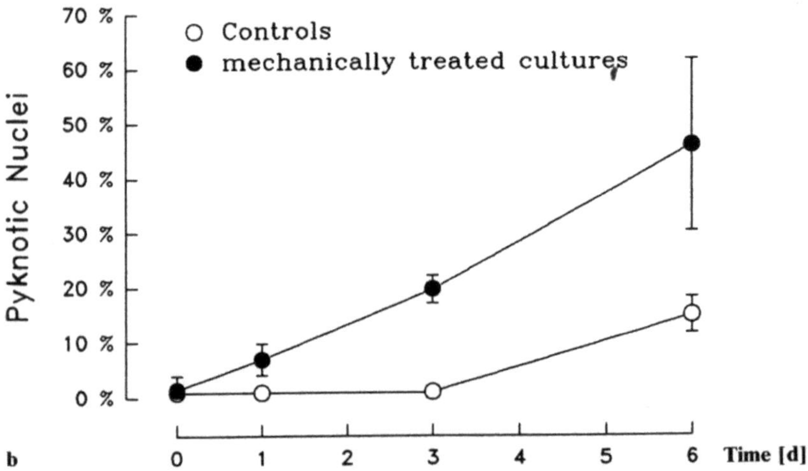

b

Fig. 4a, b. Nuclei and plasma areas of HaCaT cells stained with eosin-methylene blue were estimated with a microscope-coupled integrator (ASM, Leica). Fifty cells were randomly selected for every condition: untreated control cells were not different from mechanically treated cells (2d) in monolayered areas. Mechanically treated upper-layered cells were significantly decreased in plasma and nucleus area (U-tests). The nucleus/cytoplasma ratio was significantly (U-test) decreased in treated cells, too. Distributions are shown as box plots with 5, 10, 25, 50, 75, 90 and 95 percentiles. **b** The number of total cells and those with pyknotic nuclei were counted in cultures dyed with bisbenzimide. In mechanically treated cultures the number of pyknotic nuclei increased significantly more than in controls ($p < 0.001$, U-test)

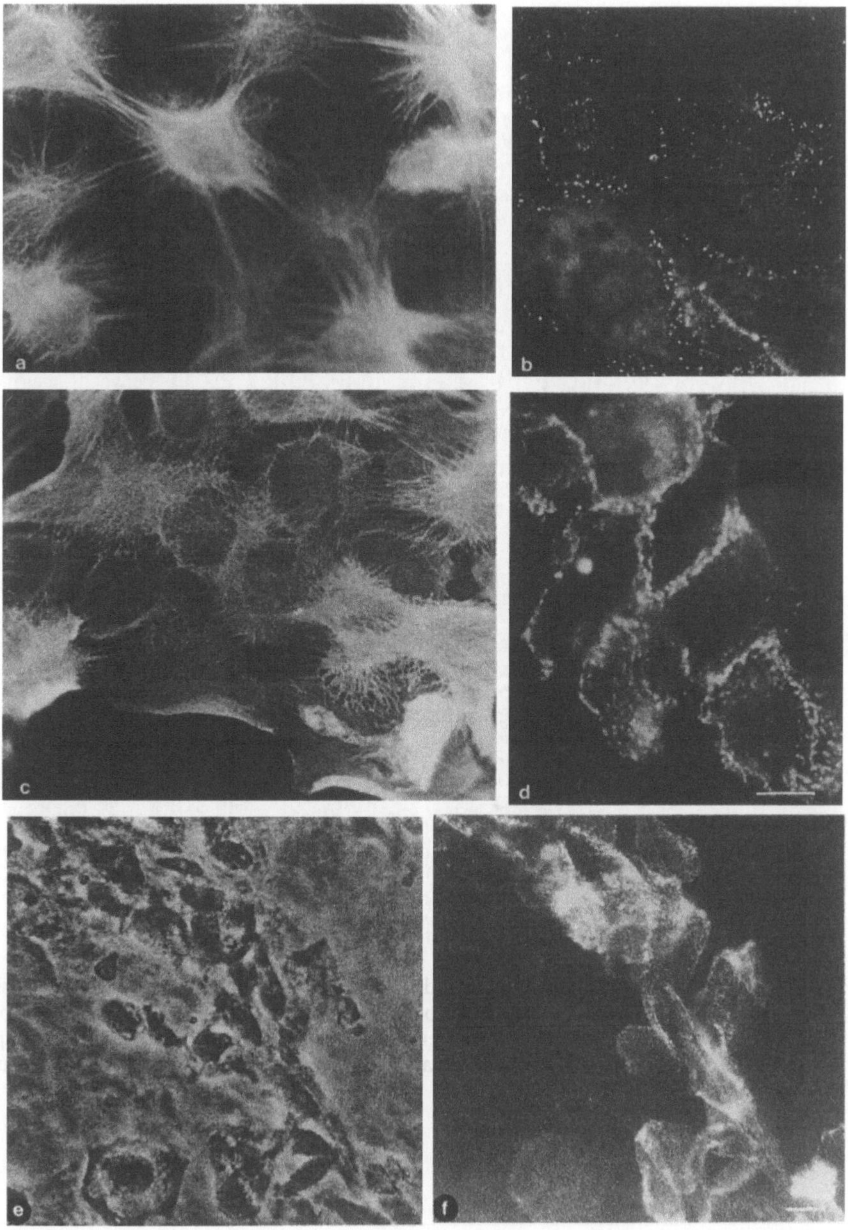

Fig. 5a–f. Demonstration of the cytokeratin pattern and the distribution of desmoplakin in HaCaT cells. **a, b** Control cultures incubated without stimulation for 60 h: **a** broad-spectrum keratin antibody LP34; **b** anti-desmoplakin I+II. Mechanically stimulated (60 h) HaCaT cells (**c–f**) observed using indirect immunofluorescence stained with (**c**) broad-spectrum keratin antibody LP34, (**d**) anti-desmoplakin II. **e, f** Phase contrast and immunofluorescence (anti-desmoplakin I+II) of mechanically treated HaCaT cells. In comparison to controls (**a, b**) the numbers of cell contacts in **c–f** are increased, and the cell-cell boundary contact is more extensive. (From [6])

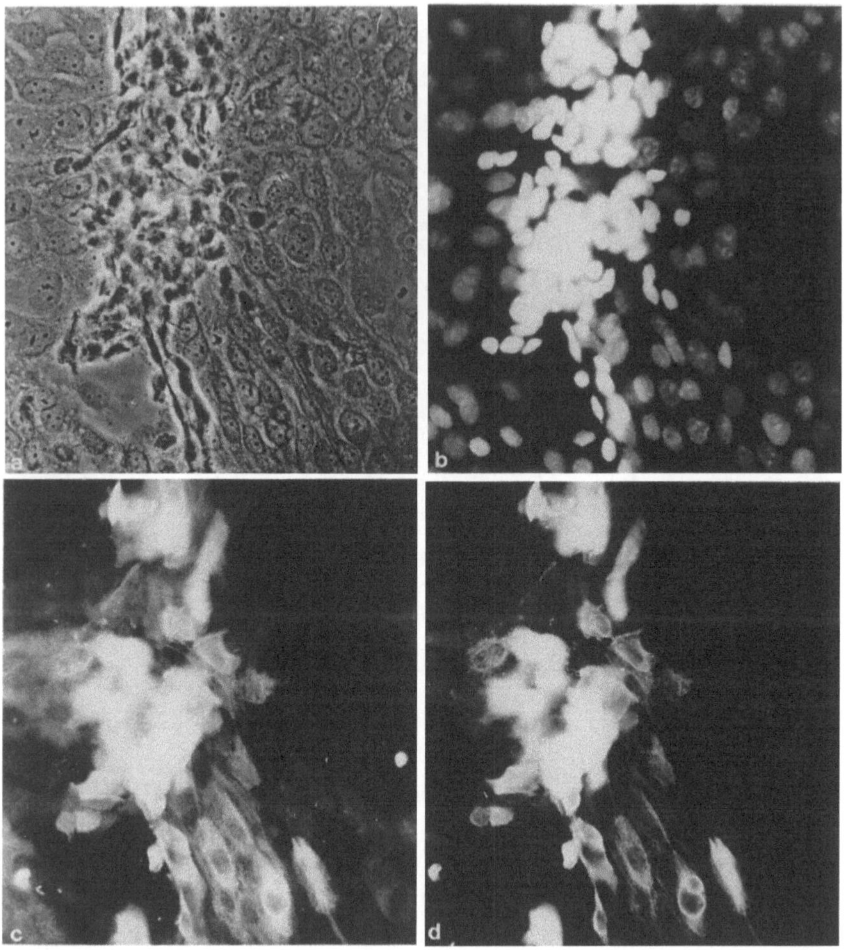

Fig. 6a–d. Cells were mechanically treated (2d) and triple marked. **a** Phase contrast shows monolayered and stratified areas. **b** Upper-layered cells had pyknotic nuclei (bisbenzimide DNA staining; **b** broad-spectrum keratin antibody marked monolayered and upper-layered cells (**c**), whereas suprabasal keratin-positive cells (**d**) appeared mainly in upper cell layers

Fig. 8a–j. Keratins, separated by two-dimensional polyacrylamide electrophoresis (NEpHGE/SPAGGE) after cytoskeletal preparation with high salt and triton buffer. **a** Controls (1d), **b** mechanically treated cultures (1d); **c** Controls (2d), **d** mechanically treated cultures (2d) with newly synthesized cytokeratin K9; autoradiographies of (**e**) controls (2d) and (**f**) mechanically treated cultures (2d). Western blots with a high specific antiserum to cytokeratin K9 of (**g**) controls (2d) and (**h**) mechanically treated cultures (2d). (**i**) Preparation of human foot sole keratins with K9 and (**j**) according to western blot with the antiserum to K9

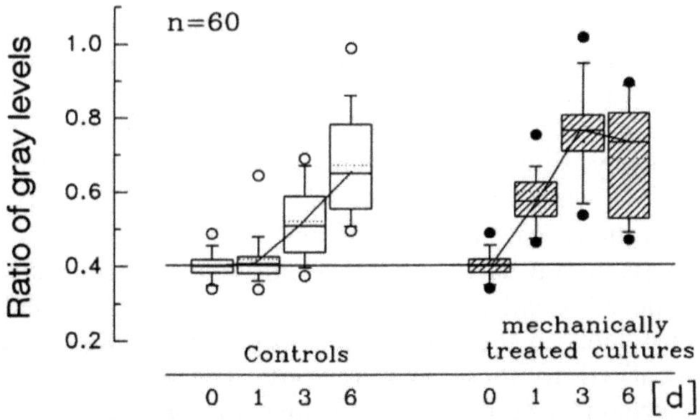

Fig. 7. Cells were double stained by indirect immunofluorescence with antibodies to broad-spectrum keratin (2nd AB: FITC) and to suprabasal keratins K1 and K10/11 (2nd AB: TRITC). The intensities of TRICT and FITC fluorescence were estimated by image analysis. The ratio of TRICT/FITC fluorescence was computed for each image field. The distributions of the ratios are shown as box plots with 5, 10, 25, 50, 75, 90 and 95 percentiles. The mean is indicated as *dotted lines*. The differences between, treated and control cultures at day 1 and day 3 are significant ($p < 0.001$, U-test)

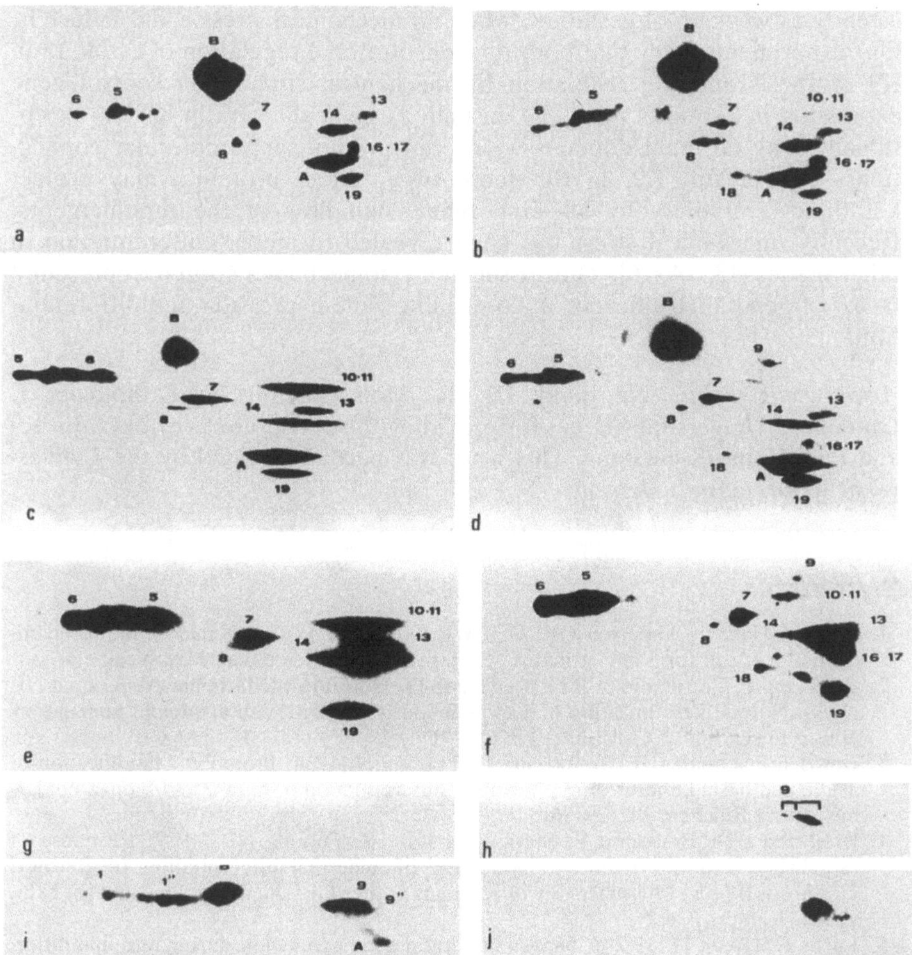

Fig. 8

autoradiographs of the 2D gels. During mechanical stimulation the cells were incubated with a ^{14}C-labeled aminoacid mixture. Autoradiography demonstrated that the protein was a genuine product of the stimulated HaCaT cells (Fig. 8f) not synthesized in controls (Fig. 8e). In situ, epidermis is always subjected to varying mechanical stress such as pressure and tension [1]. It was demonstrated that mechanical stimulation in vitro results in morphological modulations and an earlier onset of suprabasal keratin synthesis in HaCaT cells. Palmar and plantar epidermis is located at body sites exposed to extreme mechanical stress. On the molecular level this function correlates to the expression of cytokeratin K9, which is almost exclusively localized to these skin regions. The induction of cytokeratin K9 upon mechanical stimulation indicates that HaCaT cells were "switched" to a differentiation program similar to that of plantar keratinocytes. Moll et al. [11] argued against an extrinsic mechanical induction of the K9 synthesis. They demonstrated that in foot-sole skin of fetuses this protein is expressed as early as week 15 of gestation, when no mechanical stress could induce it. Our experiments show that both the constitutional regulation of cytokeratin K9 synthesis and the regulation by mechanical stimulation control gene expression in keratinocytes. The unsually large acidic keratin K9 has exceptionally long terminal domains which can promote intermolecular connections. Cytokeratin K9, in the sense of a "stress protein", may protect keratinocyte stability by allowing tighter bundling of the tonofilaments. Recently, mechanical stress has been revealed to induce differentiation in other tissues [13, 14, 16]. Thus tissue and cytomechanics seem to represent a factor of general significance in controlling gene expression and differentiation.

Acknowledgements. We thank Dr. R. Moll (Institut für Pathologie, J. Gutenberg Universität Mainz) for providing the antiserum to cytokeratin K9 and for helpful discussions. This work was partially funded by the *Landesversicherungsanstalt Hessen.*

References

1. Bereiter-Hahn J, Anderson OR, Reif WE (eds) (1987) Cytomechanics: the mechanical bases of cell form and structure. Springer, Berlin Heidelberg New York
2. Boukamp P, Petrussevska RT, Breitkreuz D, Hornung J, Markham A, Fusenig NE (1988) Normal keratinization in a spontaneously immortalized aneuploid human keratinocyte cell line. J Cell Biol 106:761–771
3. Bowden P, Quinlan R, Breitkreutz D, Fusenig N (1984) Proteolytic modification of acidic and basic keratins during terminal differentiation of mouse and human epidermis. Eur J Biochem 142:29–36
4. Breitkreutz D, Boukamp P, Stark H-J, Ryle C, Fusenig NE (1989) Response of established keratinocyte lines to modulators of epidermal differentiation. In: Reichert U. Shroot B (eds) Pharmacology of retinoids in the skin. (Pharmacology and the skin, vol 3). Karger, Basel
5. Fuchs E, Green H (1980) Changes in keratin gene expression during terminal differentiation of the keratinocyte. Cell 19:1033–1042

6. Görmar FE, Bernd A, Bereiter-Hahn J, Holzmann H (1990) A new model of epidermal differentiation: induction by mechanical stimulation. Arch Dermatol Res 282:22–32
7. Hennings H, Michael D, Cheng C, Steinert P, Holbrook K, Yuspa SH (1980) Calcium regulation of growth and differentiation of mouse epidermal cells in culture. Cell 19:245–254
8. Knapp A, Franke WW, Heid H, Hatzfeld M, Jorcano JL, Moll R (1986) Cytokeratin no. 9, an epidermal type-I keratin characteristic of a special program of keratinocyte differentiation displaying body site specificity. J Cell Biol 103:657–667
9. Maltoltsy AG (1986) Structure and function of the mammalian epidermis. In: Bereiter-Hahn J, Maltoltsy AG, Richards KS (eds) Biology of the integument, vol. 2. Springer, Berlin Heidelberg New York
10. Moll R, Franke W, Schiller D, Geiger B, Krepler R (1982) The catalog of human cytokeratins: patterns of expression in normal epithelia, tumors and cultured cells. Cell 31:11–24
11. Moll I, Heid H, Franke WW, Moll R (1987) Distribution of a special subset of keratinocytes characterized by the expression of cytokeratin 9 in adult and fetal human epidermis of various body sites. Differentiation 33:254–265
12. Ryle C, Breitkreutz D, Stark HJ, Steinert P, Roop D, Fusenig N (1989) Density-dependent modulation of synthesis of keratins 1 and 10 in the human keratinocyte line HaCaT and in RAS-transfected tumorigenic clones. Differentiation 40:42–54
13. Sumpio BE, Banes AJ, Levin LG, Johnson G (1988) Mechanical stress stimulates aortic endothelial cells to proliferate. J Vasc Surg 6:252–256
14. Thie M, Schlumberger W, Rauterberg J, Robeneck H (1989) Mechanical confinement inhibits collagen synthesis in gel-cultured fibroblasts. Eur J Cell Biol 48:294–302
15. Tseng SGC, Jarvinen MJ, Nelson WG, Huang J-W, Woodcock-Mitchel J, Sun T-T (1982) Correlation of specific keratins with different types epithelial differentiation: monoclonal antibody studies. Cell 30:361–372
16. Vandenburgh HH (1988) Computerized mechanical cell stimulator for tissue culture: effects of skeletal muscle organogenesis. In Vitro Cell Dev Biol 24:609–619

N-Alkanes in Cultured Normal Human Keratinocytes

J.C. Meyer, H.P. Grundmann, and *S. Muletta*

Abstract

Within the group of neutral lipids, n-alkanes are regularly found in scales of normal human stratum corneum. In scales from patients with congenital ichthyosiform erythroderma (CIE) n-alkanes are increased by a factor of 5–10. As these n-alkanes are ingredients of emollients and components of air pollution, it is still under discussion whether n-alkanes found in stratum corneum are of exogenous or of endogenous origin. To obtain more information on the metabolic behavior of these lipids we investigated their content in cultured normal human keratinocytes and in one experiment under the influence of an aqueous dermatophytic extract.

Primary cultures were established according to Rheinwald and Green on lethally irradiated 3T3 cells. For experiments second- or third-passage keratinocytes were cultured in the serum-free keratinocyte growth medium (KGM from Clonetics). Lipids were extracted according to Bligh and Dyer and separated and analyzed by a micromethod (TLC/FID) or by conventional TLC after incorporation of radioactive precursors.

N-alkanes were regularly found in cultured normal human keratinocytes in a range of 325 ± 6 μg/μg DNA, which corresponds to approximately 10% of total lipids. To rule out contaminations, each component of the analytical procedure from the culture medium to the chromatography equipment was tested for n-alkanes. Small amounts of n-alkanes were detected only in filters used to separate the extracted cells from the solvent extract. Trials to incorporate ^{14}C-acetate or ^{14}C-linoleic acid into n-alkanes failed. Hence, under our conditions, using just confluent cultures, synthesis of n-alkanes was not demonstrated. A dermatophytic extract from a strain of *Trichophyton mentagrophytes* induced an increase of all neutral lipid classes when incubated together with cultured normal human keratinocytes.

Introduction

All terrestrial species, being aqueous bodies, need to protect themselves from desiccation. Although plants, arthropods, reptiles, birds, and mammals all have developed different mechanisms to regulate integumental

A. Bernd, J. Bereiter-Hahn,
F. Hevert, H. Holzmann (Eds.)
Cell and Tissue Culture Models
in Dermatological Research

transepidermal water loss they have one principle in common: they require lipids for this function. High hydrophobicity is especially exhibited by the neutral lipids, among which n-alkanes, consisting mostly of very long chain aliphatic hydrocarbons without functional and oxygenated groups, exhibit the highest hydrophobic properties. Whereas in terrestrial plants and arthropods the barrier function is due largely to the imperviousness of n-alkanes, in terrestrial vertebrates a mixture of sphingolipids and nonpolar lipids seems to fulfill this requirement [4]. In this group of nonpolar lipids we find those n-alkanes which in man normally account for up to 10% of total lipids present in the stratum corneum [11]. N-alkanes are elevated up to 40% of total lipids in scales of patients with congenital ichthyosiform erythroderma (CIE) [20, 21], a finding which can be used to differentiate between lamellar ichthyosis and CIE. Elevated levels of n-alkanes were also found in neutral lipid storage disease (Chanarin-Dorfman syndrome) [6] and slightly increased contents in autosomal dominant lamellar ichthyosis [14] and in psoriasis [20]. Furthermore, increased levels of n-alkanes have been observed in the stratum corneum and in nails from atopic patients [15]. Although these findings are well documented, considerable controversy exists concerning the origin of the n-alkanes. Whereas Elias and co-workers offer good arguments in favor of an endogenous origin [7, 8], Bortz et al. recently tried to prove that these hydrocarbons are purely exogenous [2]. To find a new approach to this question, cultured normal human keratinocytes as a model for skin lipid synthesis were used. Because some dermatophytes give rise to disturbed keratinization in skin lesions, and because disordered keratinization is usually linked to changes in the metabolism of epidermal lipids [12], we wondered whether these fungus-induced disturbances are also accompanied by alterations in epidermal lipids, especially n-alkanes.

Material and Methods

Cell Cultures

Primary cultures of normal human keratinocytes were raised from foreskin obtained from the Children's Hospital or from a piece of split skin used for the coverage of leg ulcers from the Department of Dermatology (age: 85 years) according to the method of Rheinwald and Green [18] on lethally irradiated 3T3 cells in DMEM/Ham's F12 (3:1) supplemented with 10% FCS, adenine, cholera toxin, hydrocortisone, insulin, and human EGF (prepared) by Dr. C. George-Nascimento, Chiron Corp. Emeryville, USA). For experiments confluent cultures were then transferred to serum-free keratinocyte growth medium (KGM from Clonetics Corp., San Diego, USA). All lipid analyses and experiments were performed with just confluent cultures.

Extract of Trichophyton mentagrophytes

A strain of *T. mentagrophytes* isolated from a patient was grown in Sabouraud broth for 3 weeks and then centrifuged at 50000 g for 30 min at 4° C in a refrigerated centrifuge (Sorvall RC-5). The supernatant was discarded. The sediment was washed three times with 0.05 mol/l Tris-HCl pH 7.2. The washed preparation was resuspended in 2–3 ml Tris-HCl and then homogenized in a precooled (–25° C) freeze press. To prepare a cell-free extract the homogenate was centrifuged at 50000 g for 30 min at 4° C, and the clear supernatant was sterilized by filtration through a 0.2-μm filter (Millipore Corp. Bedford, USA). One milliliter of homogenate contained 416 μg protein and 133 μg DNA.

Incorporation of ^{14}C-Acetate or ^{14}C-Linoleic Acid

We added 148 kBq (4 μCi) of 2-^{14}C-acetate, sodium salt (NEC-085H from NEN, 54 mCi/mmol) in 40 μl ethanol or 37 kBq (1 μCi) of ^{14}C(U)-linoleic acid (NEC-595 from NEN, 950.8 mCi/mmol) in 50 μl to just confluent cultures in 75-cm^2 flasks in new KGM. Cells were further incubated for 6 or 24 h and then collected by trypsin/EDTA, washed twice with PBS. An aliquot was used for DNA assay [16] and the rest was analyzed for incorporation of radioactivity in neutral lipids after lipid extraction.

Analysis of Neutral Lipids

A washed suspension of trypsinized keratinocytes was centrifuged and then the sediment was extracted by a modified Bligh and Dyer extraction method according to Lampe et al. [11]. Total lipids were weighed after solvent extraction on an electronic microanalytical balance UM 3 (Mettler Instrumente AG, Greifensee, Switzerland). Neutral lipids were analyzed by the thin-layer chromatography-flame ionization detection system (TLC-FID, Iatroscan TH-10, Iatron Laboratories Inc., Tokyo) [1]; slightly modified according to Brown et al. [3]. In brief: 4–8 μg of lipid extract in 1 μl chloroform/methanol 2:1 were applied to each chromarod (type SIII) with a sample spotter Model SES 3200/IS-01 (SES GmbH, Niederolm, Germany) and chromarods were developed in two steps: (a) n-hexane/ diethylether/ formic acid (85:15:1) for 15 min and (b) n-hexane alone again for 15 min. After total evaporation of the solvent, chromarods were scanned in the Iatroscan and detected peaks were quantitated by an integrator calculator HP 3392A (Hewlett Packard) after calibration with external standards. A standard curve was used for each reference.

Analysis of Radioactivity Incorporated into Neutral Lipids

Lipid extracts were obtained as described above. Neutral lipids were then separated by TLC on Si 60 TLC plates (Merck) according to Melnik [13]. For each sample 40–80 μg of lipid extract were applied in 10 μl of chloroform/methanol 2:1 and developed in n-hexane/diethylether/acetic acid (80:20:10) as the first and petroleum ether (40–60° C boiling point) as the second solvent system. Lipids were stained by spraying the plate with 10% copper acetate in 8% *o*-phosphoric acid and heating 20 min at 180° C. A standard mixture consisting of ceramides type III and IV, cholesterol, oleic acid, triolein, cholesterol oleate, and eicosan was simultaneously applied with each run. After staining, the visualized lipids were scraped from the plate directly into counting vials and the radioactivity was counted in a Betamatic V liquid scintillation counter (Kontron Instruments AG, Zürich, Switzerland).

Results and Discussion

Thin-layer Chromatography with Flame Ionization Detection

The microchromatographic method combining thin-layer chromatography with flame ionization detection (TLC-FID) proved to be useful and sensitive for the separation and quantitation of small amounts of neutral lipids from cultured keratinocytes. It had been described for lipid analysis in stratum corneum by Brown et al. [3] Fig. 1a and b). The use of a sample spotting equipment was essential because standardized and accurate manual application of 1 μl of sample in solvent proved to be impossible.

Quantitation of n-Alkanes in Cultured Keratinocytes

Among the neutral lipds, hydrocarbons (HC) representing the n-alkanes were detected in cultured normal human keratinocytes in just confluent cultures in serum-free KGM (Fig. 2). These results are in agreement with those reported by Williams and co-workers, who investigated the density-dependent variation of lipids in cultured human keratinocytes [22]. To elucidate the possible influence of exogenous HC each step and each component of the entire procedure of lipid analysis, from the cell culture to the scanning of chromarods, was carefully checked for contamination. Among all the components tested – keratinocyte growth medium (KGM), PBS, trypsin, solvents, nitrogen (used to evaporate solvents), glass tubes, or chromarods themselves – only the filter paper used to separate extracted cells from the solvent extract yielded between 15 and 25 μg n-alkanes during each extraction procedure and had to be substracted as "background" from the final amount of n-alkanes. Nevertheless, 325 ± 6 μg/ μg DNA of n-

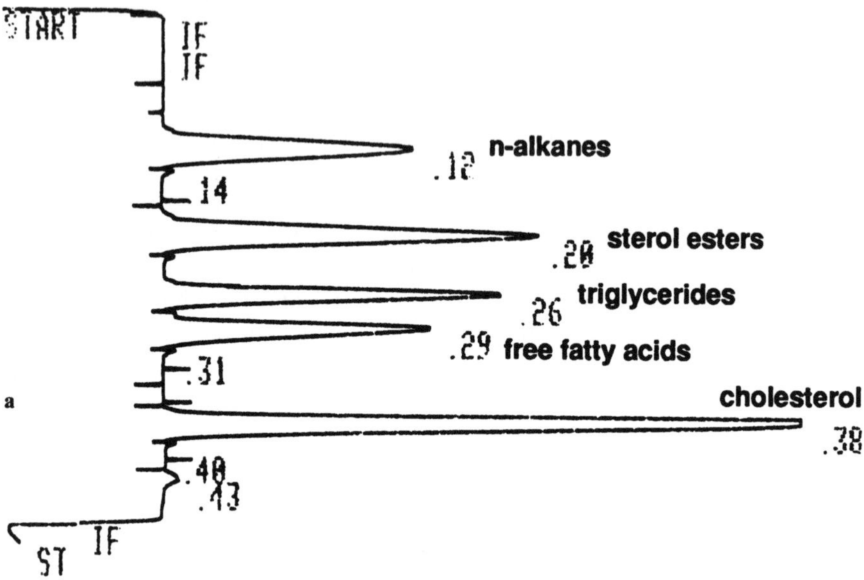

a

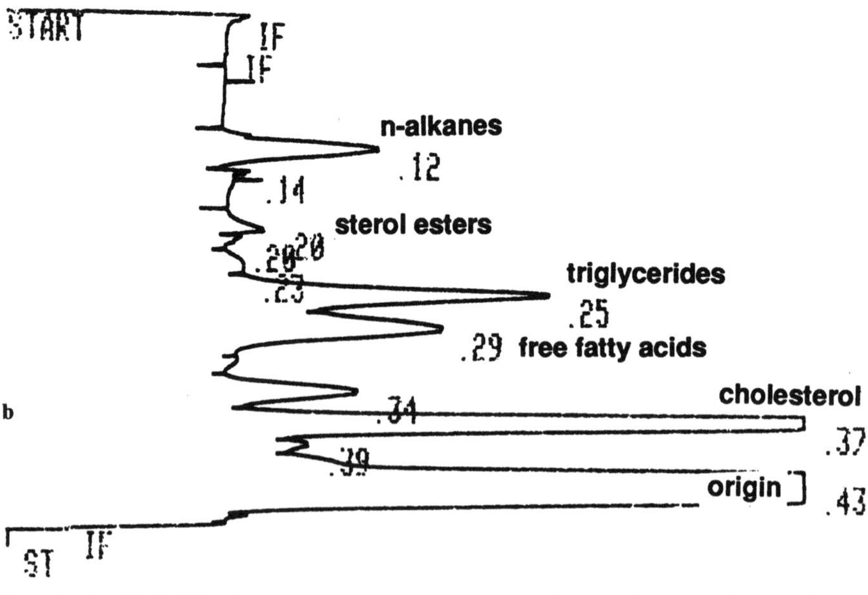

b

Fig. 1a, b. TLC-FID chromatography patterns from (**a**) a reference mixture of neutral lipids and (**b**) an extract from cultured normal human keratinocytes

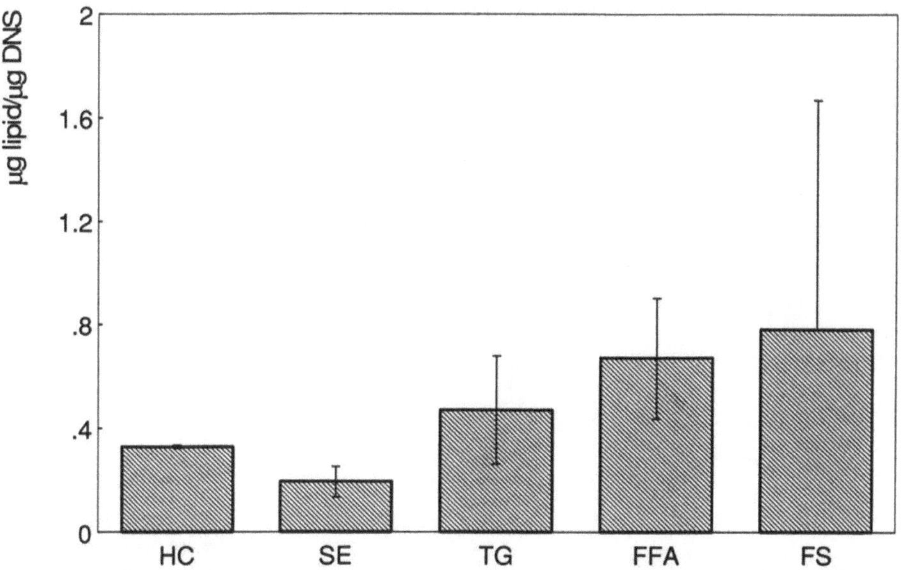

Fig. 2. Pattern of neutral lipid classes in an extract from just confluent cultured normal human keratinocytes. *HC,* Hydrocarbons; *SE,* sterol esters; *TG,* triglycerides; *FFA,* free fatty acids; *FS,* free sterols

alkanes (hydrocarbons) were detected in just confluent cultures of normal human keratinocytes. This amount corresponds to approximately 10% of total lipids in cultured keratinocytes. To see whether cultured keratinocytes could synthesize n-alkanes in vitro, we studied the incorporation of radioactive precursors into neutral lipids in this model.

Incorporation of ^{14}C-Acetate and ^{14}C-Linoleic Acid into Neutral Lipids

As seen from two experiments (Fig. 3), ^{14}C-acetate was incorporated into free sterols (cholesterol), triglycerides; a small amount was also incorporated into ceramide III and a large amount into the origin, which corresponds to polar lipids (phospholipids), but no activity was found after 6 or 24 h (not shown) in the n-alkane fraction. This corresponds to the findings of Williams et al. [22], who investigated acetate incorporation into keratinocytes cultured according to Rheinwald and Green [18].

Similarly, linoleic acid, an essential fatty acid [5, 10], which is, among other factors, required for the formation and maintenance of the epidermal barrier function [5, 9, 16] was not incorporated into the n-alkane fraction. As shown in Figure 4, in three experiments the radioactivity was detected mainly in triglycerides, ceramide type IV, and in the origin (not shown in

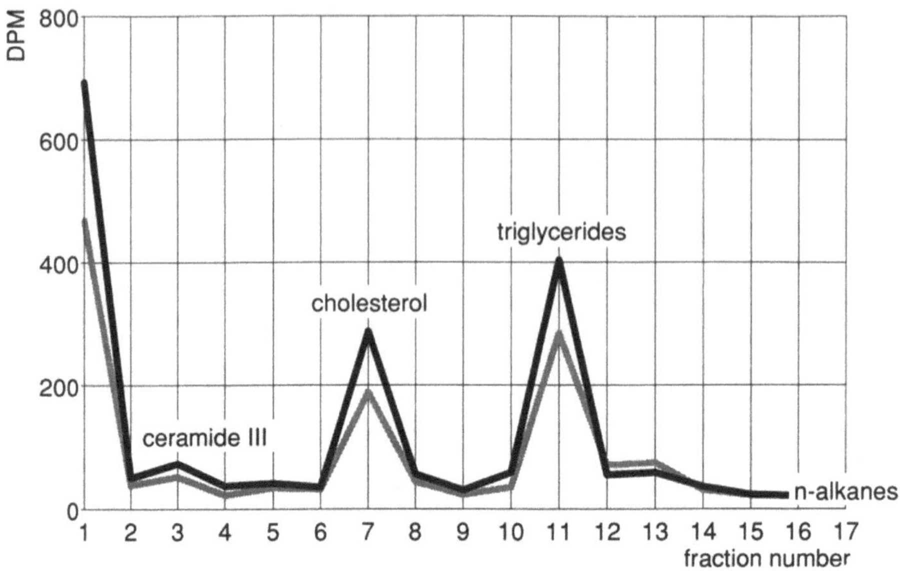

Fig. 3. Distribution of radioactivity in a chromatogram of neutral lipid classes of an extract from cultured normal human keratinocytes labeled with ^{14}C-acetate for 6 h. Just confluent cultures in 75-cm^2 plastic tissue culture flasks were incubated with 1 μCi of the isotope. Each line represents a single experiment

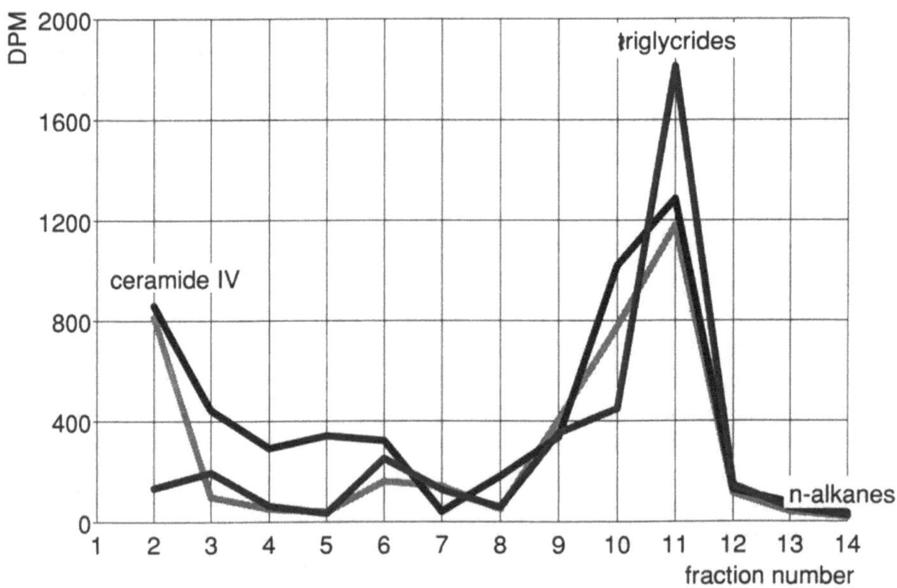

Fig. 4. Distribution of radioactivity in a chromatogram of neutral lipid classes of an extract from cultured normal human keratinocytes labeled with ^{14}C-linoleic acid for 24 h. Just confluent cultures in 75-cm^2 tissue culture flasks were incubated with 4 μCi of the isotope. Each line represents a single experiment

the figure), the latter corresponding to the polar phospholipids. These results are in accordance with a report by Wertz and Downing [19], who investigated the metabolism of linoleic acid in porcine epidermis in vivo after intradermal injection of ^{14}C-linoleic acid. Between days 1 and 3 after injection, they found a prominent incorporation of radioactivity, into phospholipids, tryclycerides, acylglucosyl- and acylceramides.

Hence, under our conditions, using just confluent cultures, no incorporation, either of ^{14}C-acetate or of ^{14}C-linoleic acid into the n-alkane fraction, was observed; this leaves open the question of whether n-alkanes are synthesized by epidermal keratinocytes or whether they are of pure exogenous origin, as proposed by Bortz et al. [2]. Despite these negative findings, the endogenous origin of n-alkanes in cultured keratinocytes must still be considered because the only exogenous source, as described in the proceding section, were the filters used in the extraction procedure, which were subtracted as "background". As n-alkanes (hydrocarbons) seem to be reduced in confluent and postconfluent cultures [22], it might be necessary to investigate the synthesis of hydrocarbons in preconfluent cultures under more proliferating conditions, a state which may also be induced by a calcium shift.

Influence of a Fungal Extract on the Lipid Content in Keratinocyte Cultures

To see whether dermatophytes influence neutral lipids of the epidermis qualitatively or quantitatively, various amounts of an extract of a strain of *T. mentagrophytes* from a patient was added to just confluent cultures of keratinocytes for 3 days. As shown in Figure 5, not only n-alkanes, but the amount of all classes of neutral lipids increased when the volume of the fungal extract was raised over 0.5 ml (corresponding to 213 µg of protein or 66.5 µg DNA). Between 0.5 and 1 ml of fungal extract a prominent increase of all neutral lipids was observed by factors between 11.2 for free fatty acids and 20 for sterol esters. Under these conditions we observed a detachment of the cells from their plastic support and simultaneously a prominent decrease of the cell number, probably due to proteinases in the fungal extract. Hence, this steep increase in neutral lipids may be a reaction to an extreme stress situation. This finding obviously demonstrates that hydrocarbons, as well as the other neutral lipids in cultured keratinocytes, were not of exogenous origin. In the case of an exogenous source, e.g, from the fungal extract itself, a steady increase of the lipids would be expected. In future experiments, again using radioactive precursors, it should be investigated whether in the presence of fungal extracts hydrocarbons and the other neutral lipids are newly synthesized.

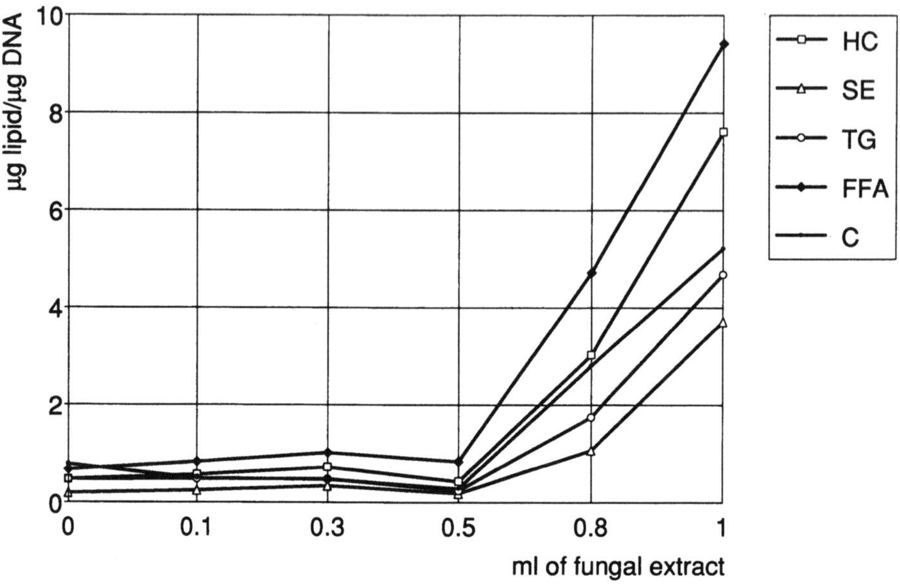

Fig. 5. Amounts of neutral lipid classes in cultured normal human keratinocytes, incubated with different volumes of an aqueous extract from the dermatophyte *T. mentagrophytes* for 3 days. *HC,* Hydrocarbons; *SE,* sterol esters; *TG,* triglycerides; *FFA,* free fatty acids; *C,* cholesterol

References

1. Ackman RG(1981) Flame ionization detection applied to thin layer chromatography on coated quartz rods. Methods Enzymol 72:205–252
2. Bortz JT, Wertz PW, Downing DT (1989) The origin of alkanes found in human skin surface lipids. J Invest Dermatol 93:723–727
3. Brown BE, Williams ML, Elias PM (1984) Stratum corneum lipid abnormalities in ichthyosis. Detection by a new lipid microanalytical method. Arch Dermatol 120:204–209
4. Grubauer M, Feingold KR, Elias PM (1989) Lipid content and lipid type as determinants of the epidermal permeability barrier. J Lipid Res 30:89–96
5. Elias PM (1981) Lipids and the epidermal permeability barrier. Arch Dermatol Res 270:95–117
6. Elias PM (1985) Neutral lipid storage disease with ichthyosis. Arch Dermatol 121:1000–1008
7. Elias PM (1990) n-Alkanes in the skin. Function or fancy. Arch Dermatol 126:868–870
8. Elias PM, Williams ML (1990) Alkanes: endogenous or exogenous? J Invest Dermatol 94:730–731
9. Elias PM, Brown PE, Ziboh VA (1980) The permeability barrier in essential fatty acid deficiency: evidence for a direct role for linoleic acid in epidermal barrier function. J Invest Dermatol 74:230–233
10. Holman RT (1968) Essential fatty acid deficiency. Prog Chem Fats Other Lipids 9:275–358

11. Lampe MA, Burlingame AL, Whitney J, Williams ML, Brown BE, Roitman E, Elias PM (1983) Human stratum corneum lipids: characterization and regional variations. J Lipid Res 24:120–130
12. Melnik B (1989) Epidermal lipids and the biochemistry of keratinization. In: Traupe H (ed) The ichthyoses. A guide to clinical diagnosis, genetic counseling, and therapy. Springer, Berlin Heidelberg New York, pp 15–42
13. Melnik BC, Hollmann J, Erler E, Verhoeven B, Plewig G (1989) Microanalytical screening of all major stratum corneum lipids by sequential high-performance thin-layer chromatography. J Invest Dermatol 92:231–234
14. Melnik B, Küster W, Hollmann J, Plewig G, Traupe H (1989) Autosomal dominant lamellar ichthyosis exhibits an abnormal scale lipid pattern. Clin Genet 35:152–156
15. Melnik BC, Hollmann J, Hofmann U, Yuh MS, Plewig G (1990) Lipid composition of outer stratum corneum and nails in atopic and control subjects. Arch Dermatol Res 282:549–551
16. Melton JL, Wertz PW, Downing DT (1987) Effects of essential fatty acid deficiency on epidermal O-acylsphingolipids and transepidermal water loss in young pigs. Biochem Biophys Acta 921:191–197
17. Meyer JCh, Grundmann HP (1984) Fluorometric determination of DNA in epidermis and cultured fibroblast, using 4'-6-diamidino-2-phenolindole (DAPI). Arch Dermatol Res 276:52–56
18. Rheinwald JG, Green H (1975) Serial cultivation of strains of human epidermal keratinocytes: the formation of keratinizing colonies from single cells. Cell 6:331–344
19. Wertz PW, Downing DT (1990) Metabolism of linoleic acid in porcine epidermis. J Lipid Res 31:1839–1844
20. Williams ML, Elias PM (1981) n-Alkanes in normal and pathological human scale. Biochem Biophy Res Comm 107:322–328
21. Williams ML, Elias PM (1984) Elevated n-alkanes in congenital ichthyosiform erythroderma: phenotypic differentiation of two types of autosomal recessive ichthyosis. J Clin Invest 74:296–300
22. Williams ML, Rutherford SL, Ponec M, Hincenbergs M, Placzek DR, Elias PM (1988) Density-dependent variations in the lipid content and metabolism in cultured human keratinocytes. J Invest Dermatol 91:86–89

Influence of Retinoids on the Keratin Profile in Normal and Spontaneously Transformed Human Keratinocytes

R. Stadler, B. Korge, and *D. Mischke*

Abstract

Normal epidermal keratinocytes cultured on 3T3-fibroblasts differentiated under the influence of nonaromatic (all-*trans* retinoic acid, 13-*cis* retinoid acid) and polyaromatic retinoids (arotinoid, arotinoid acid, arotinoid sulphone) in a concentration of $10^{-8}M$. HaCaT cells were cultured for 8 days in the presence of nonaromatic and aromatic (etretinate, etretin) retinoids in concentrations of $10^{-7}M$ and $10^{-6}M$, respectively. Control cultures differentiated only in the presence of 0.2% DMSO.

Keratin-enriched protein fractions of normal and transformed retinoid-treated keratinocytes have been analyzed by means of one- and two-dimensional SDS-gel electrophoresis, immunoblot, and later on laser densitometry. The laser densitometric evaluation of 1-D gels howed an important increase in the expression of the keratins 13, 15, and 19 under retinoid influence for normal as well as for transformed keratinocytes. In this connection all-*trans* retinoic acid, etretin, and arotinoid acid provoked the strongest effects. The enhanced induction of the keratins 13, 15, and 19 points to a more embryonic phenotype of the keratinocyte differentiation among retinoids, as the common expression of these keratins was observed only during the embryonic development of the epidermis. The basic compound arotinoid showed no effect.

Surprisingly enough, we determined repeatedly a significant reduction in the expression of keratin 16, whereas the expression of keratin 6 was unchanged or slightly increased. Here, arotinoid acid showed the strongest effect. The reduction of keratin 16 and the slight increase of keratin 6, both so-called hyperproliferation-associated keratins, point to the fact that keratin K16 is obviously a suitable parameter for determining the antiproliferative effect of retinoids in vitro.

In contrast to normal keratinocytes in culture, HaCaT cells additionally express the epidermal differentiation-specific keratins 1 and 10, as well as the keratins 7, 8, and 18. Retinoids cause an increase in the expression of keratins 7, 8 and 18 and a reduction in the expression of keratin 1 and 10. Etretin and all-*trans* retinoic acid were the most effective ones. For this reason, HaCaT cells are probably a more suitable and more handy in vitro model than normal keratinocytes to elucidate the retinoid-dependent regulation of the keratin expression.

A. Bernd, J. Bereiter-Hahn,
F. Hevert, H. Holzmann (Eds.)
Cell and Tissue Culture Models
in Dermatological Research

Introduction

Keratins belong to the large family of intermediate filaments. Table 1 shows the present state of all intermediate filaments known and their division into six types.

Keratins are subdivided into the acid type-I and the neutral-basic type-II keratins. They are found in all epithelia. So far, 15 different keratins have been described in each type. Type-III intermediate filaments include vimentin, which appears in mesenchymal tissue, desmin, which was found in muscle tissue, the glia filaments, which occur in astrocytes, and the recently described peripherin, which is found in neurons. Intermediate filaments of type IV include the three neurofilaments H, M, L and alpha-internexin, expressed in prenatal neurons. Lamins, which occur in the nuclear lamina of all eukaryotes, have been described as type-V intermediate filaments. Nestin, finally, a protein expressed in embryonic neuroepithelial basic cells, is a type-VI intermediate filament [1–4].

Keratins represent a very heterogeneous family of acid and neutral-basic proteins with a molecular weight of 40–67 kD [5–8]. Normally, they are expressed as specific pairs consisting of a type-I and a type-II protein unit each, which obviously seems to be essential for the assembling of the filaments [1,9–13]. The epithelial-specific expression of keratins depends on the respective state of development and differentiation of each epithelial cell. Thus, epidermal keratinocytes are subdivided into the differentiation-specific keratins K1/K10 [5, 14–16], the basal-cell keratins K5/K14 [17, 18], and the hyperproliferation keratins K6/K16 [19–24].

Vitamin A and its derivatives are vital to normal epidermal development and differentiation [25, 26]. A number of articles show that retinoids may

Table 1. Classification of the different and their tissue distribution

Type	Name	Tissue type	Chain number	Molecular weight (kD)
I	Acidic keratins	All epithelia	15	40–60
II	Neutral-basic keratins	All epithelia	15	50–70
III	Vimentin	Mesenchymal cell	1	53
	Desmin GFAP[a]	Myogenic cells	1	52
	GFAP[a]	Glia cell astrocytes	1	51
	Peripherin	Neuron cells	1	52
IV	Neurofilaments	Neurons	3	60–150
	Alpha-internexin	Prenatal neurons	1	57
V	Lamin	All eukaryotes	4	60– 70
VI	Nestin	Embryonal, neuro-epithelial stem cell	1	200

[a] Glial fibrillary acidic proteins.

influence the expression of keratins on different molecular levels. In cultured normal epidermal keratinocytes, nonaromatic retinoids in high concentrations suppress the expression of keratins 13 and 19 [27–29].

Although the successful therapy of psoriasis and other dermatological diseases attained by the use of retinoids is unquestionable [30], strong efforts still have to be made to elucidate the exact molecular mechanism which triggers the effectiveness of retinoids.

The present study describes the influence of different retinoids on the keratin expression of cultured normal and spontaneously transformed human keratinocytes. Keratin-enriched protein fractions have been analyzed by means of one- and two-dimensional SDS-gel electrophoresis, immunoblot, and laser densitometry (as described in [24]). We observed on enhanced expression of the keratins 13, 15, and 19 under retinoid influence in both keratinocyte cell lines. The keratins 7, 8, and 18 were induced in the HaCaT cells, and the expression of keratin 1 and 10 was reduced after retinoids were applied. Furthermore, retinoids led to a reduction of keratin 16, whereas keratin 6 remained unchanged or slightly increased. Therefore, we conclude that keratin 16 is probably a suitable parameter for determining the antiproliferative potency of retinoids in vitro, and that HaCaT cells obviously represent a more suitable and more handy in vitro model than normal keratinocytes to explain the retinoid-dependent regulation of keratin expression.

Results

Keratin-enriched fractions of cultured normal human keratinocytes were separated by means of highly dissolutive two-dimensional gel electrophoresis. On the Coomassie blue-stained gel (Fig. 1a) and the immunoblots (Fig. 1b–c) keratins 5, 6, 13, 14, 15, 16, 17, and, very slightly, also keratin 19 are demonstrable. Under the influence of arotinoid acid increased expression of the keratins 13, 15, and 19 is observed (Fig. 2). Table 2 summarizes the results of the laser densitometric evaluation of one-dimensional SDS gels. In a concentration of 10^{-8} retinoids induce a considerable increase in

Table 2. Effect of nonaromatic and polyaromatic retinoids (10^{-8} *M*) on keratin expression

	K6	K16	K13	K15	K19	K14	K17
all-*trans* RA	− 9	− 3	+ 10	+ 31	+ 79	− 18	+ 6
13-*cis* RA	+ 6	− 2	+ 5	+ 20	+ 4	− 29	− 11
Arotinoid	+ 7	− 3	n.d.	n.d.	n.d.	+ 1	− 20
Arotinoid sulphone	+ 11	− 2	n.d.	+ 18	+ 20	+ 21	+ 39
Arotinoid acid	+ 31	− 34	n.d.	+ 33	+ 60	+ 28	+ 12

Laser densitometric evaluation of keratin synthesized in cultures of human epidermal keratinocytes in the presence of 10^{-8} *M* retinoids. Values indicate changes (absorption units) as compared with DMSO (control) cultures. nd., Not detectable.

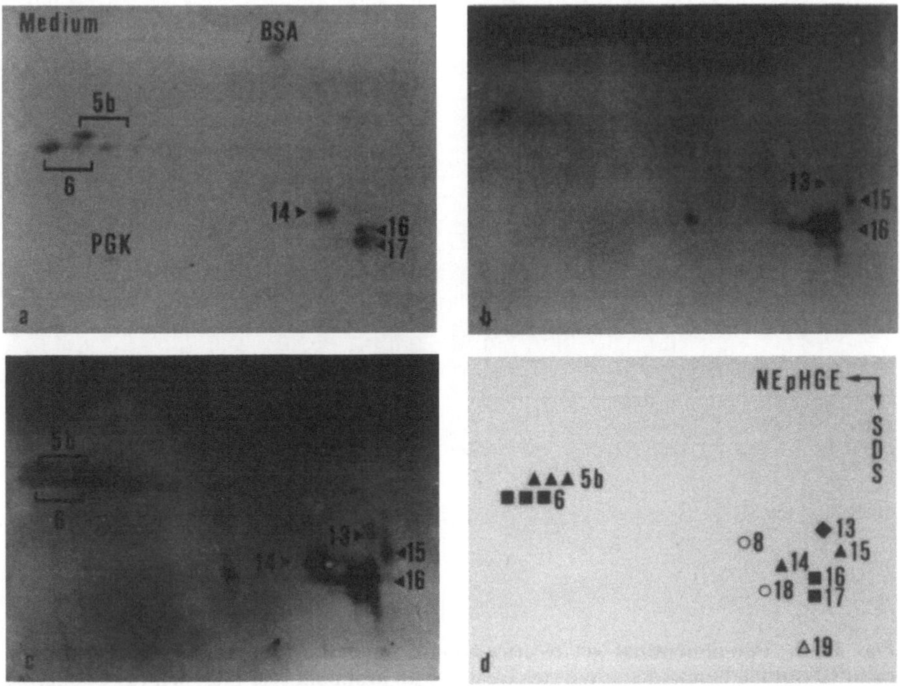

Fig. 1a–d. Two-dimensional gel electrophoretic splitting of keratins expressed from cultured untreated normal human keratinocytes (control). (**a**) Coomassie blue-stained gel; (**b**) immunoblot, reaction with monoclonal antibodies Ck 4.62 (marks keratin 19) and Ck 8.12 (marks K13 and K16); (**c**) immunoblot, reaction with 8–2/4 (mainly marks type I keratins, [14]); (**d**) schematic description of the keratins expressed by normal human keratinocytes cultured on 3T3 fibroblasts. The keratins are named according to the catalogue of keratins [7] in view of the polymorphism of some keratins [14]. Keratins that are expressed in pairs have the same symbols

the expression of keratins 13, 15 and 19. In this connection, all-*trans* retinoic acid and arotinoid acid showed the strongest effects. Unexpectedly, we repeatedly determined a significant reduction in the expression of keratin 16, whereas the expression of keratin 6 was mostly increased. Here, arotinoid acid was the most effective.

To verify this observation we cultured normal human keratinocytes in the presence of all-*trans* retinoic acid and 13-*cis* retinoid acid, two retinoids which are clinically relevant, in concentrations of $10^{-9}M$ – $10^{-7}M$. The extracted keratins were resolved one-dimensionally (Fig. 3; see also Table 3).

On the Coomassie blue-stained gel (Fig. 3a) and the respective immunoblot (Fig. 3b) the considerable increase in the expression of keratins 13 and 19 under the influence of the two retinoids (lines 3–8) in comparison with the DMSO control becomes visible (line 2). Such one-dimensional gels were scanned with a laser densitometer. On the integration curves obtained, the

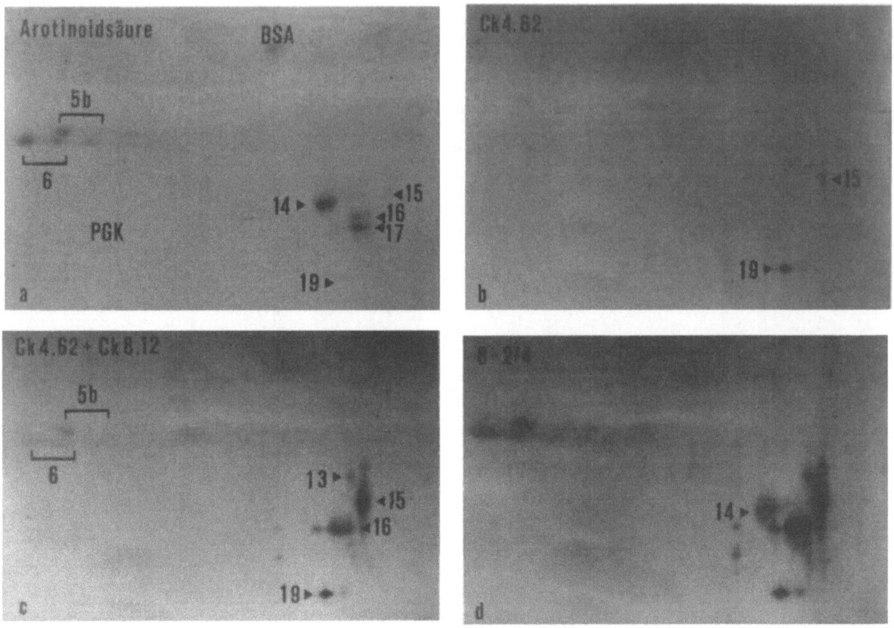

Fig. 2a–d. Two-dimensinal gel electrophoretic separation of keratins expressed from cultured normal human keratinocytes treated with arotinoid acid (10^{-8} *M*). (**a**) Coomassie blue-stained gel; (**b**) immunoblot, reaction with the monoclonal antibody Ck 4.62; (**c**) immunoblot, additional reaction with Ck 8.12; (**d**) immunoblot, additional reaction with 8-2/4

Table 3. Concentration-dependent changes in keratin expression

	K6	K16	K13	K15	K19	K14	K17
all-*trans* RA, 109 *M*	+12	+ 4	+ 2	+17	+ 8	−23	− 3
all-*trans* RA, 108 *M*	− 9	− 3	+10	+31	+ 79	−18	+ 6
+ all-*trans* RA, 10^{7} *M*	+15	− 8	+40	+67	+168	−18	+ 4
13-*cis* RA, 10^{-9} *M*	+17	+17	+ 2	n.d.	− 18	0	+21
13-*cis* RA, 10^{-8} *M*	+ 6	− 2	+ 5	+20	+ 4	−29	−11
13-*cis* RA, 10^{-7} *M*	0	−11	+15	+32	+ 83	−17	−18

Quantification of changes in keratin expression in human epidermal keratinocytes cultured in the presence of different concentrations of all-*trans* RA and 13-*cis* RA as compared with DMSO (control) cultures. n.d., Not detectable.

increase of K13 and K19 under 13-*cis* retinoic acid can clearly be demonstrated (Fig. 4a) in comparison with the control culture Fig. 4b). The laser densitometric analysis shows a concentration-dependent increase in the expression of K13, K15, and K19. After the application of all-*trans* retinoic acid the expression of K6 and K16 is reduced, whereas 13-*cis* retinoic acid in a concentration of 10^{-7}*M* again showed only a reduction of K16 and an

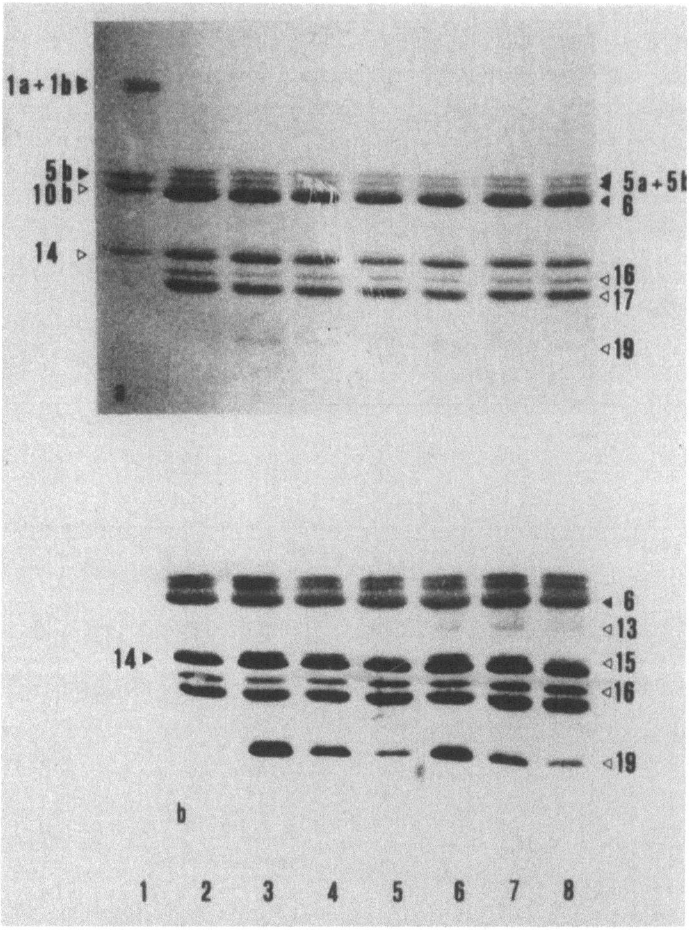

Fig. 3a–b. One-dimensional SDS gel of the keratins extracted from cultured normal human keratinocytes. (**a**) Coomassie blue-stained gel; (**b**) corresponding immunoblot, reaction with a mixture of the polyclonal rabbit-antibodies 8-2/4 and 10-2/2 mark almost all keratins, [14]). *Line 1*, human epidermis; *line 2*, DMSO (control), *line 3*, all-*trans RA*, 10^{-7} *M*; *line 4*, all-*trans* RA, 10^{-8} *M*; *line 5*, all-*trans* RA, 10^{-9} *M*; *line 62*, 13-*cis* RA, 10^{-7} *M*; *line 7*, 13-*cis* RA, 10^{-8} *M*; *line 8*, 13-*cis*, 10^{-9} *M*. *Blank arrows* represent type-I, *filled arrows* type-II keratins

unchanged or slightly increased K6 expression, as was seen with arotinoid acid. This reinforced our assumption that keratin K16 is a suitable parameter for determining the antiproliferative effect of retinoids. To verify this, we treated subconfluent, hardly proliferating HaCaT cultures with nonaromatic (all-*trans* retinoic acid, 13-*cis* retinoic acid) and monoaromatic (etretinate, etretin) retinoids in a concentration of 10^{-b} for 8 days. In contrast to normal cultured keratinocytes, untreated HaCaT cells additionally expressed the epidermal differentiation-specific keratins 1 and 10, as well as keratins 7, 8 and, very slightly, also keratin 18, which occur in simple epithelia (Fig. 5a, c). Retinoids increase not only the expression of keratins 13, 15, and 19, but also that of keratins 7, 8, and 18 (Fig. 5b, d). The expression of keratins 1 and 10 was reduced. Etretin and all-*trans* retinoic acid were the most effective substances. The laser densitometric data in Table 4 show the considerable reduction of keratin 1 and, suprisingly, a

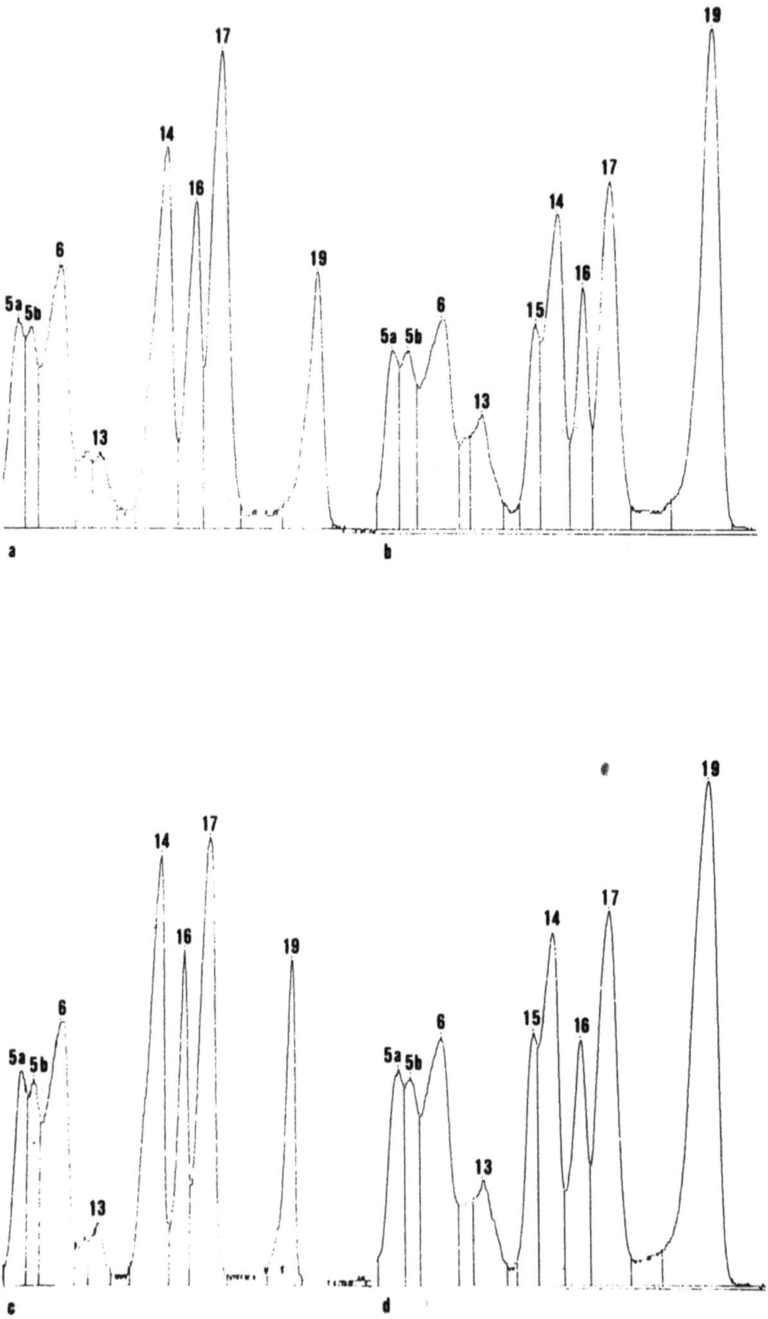

Fig. 4a–d. Laser densitometric integration graphs of one-dimensional immunoblots which reacted with a mixture of the polyclonal rabbit antibodies 8-2/4 and 10-2/2. (**a**) DMSO-control; (**b**) 13-*cis* RA, 10^{-7} M; (**c**) all-*trans* RA, 10^{-9} *M*; (**d**) all-*trans* RA, 10^{-7} *M*. The keratins are named as described in Fig. 1

Table 4. Effect of nonaromatic and monoaromatic retinoids (10^{-6} *M*) on the keratin expression of HaCaT cells

	K1	K14	K16	K13	K19	K8	K18
all-*trans* RA	−38	−43	− 1	+ 8	+12	+ 5	+ 6
13-*cis* RA	−22	−32	− 2	0	+ 6	− 1	+ 4
Etrinate	−27	−21	+ 1	− 2	+ 8	0	+ 8
Acitretin	−37	−42	+ 3	+ 6	+22	+12	+11

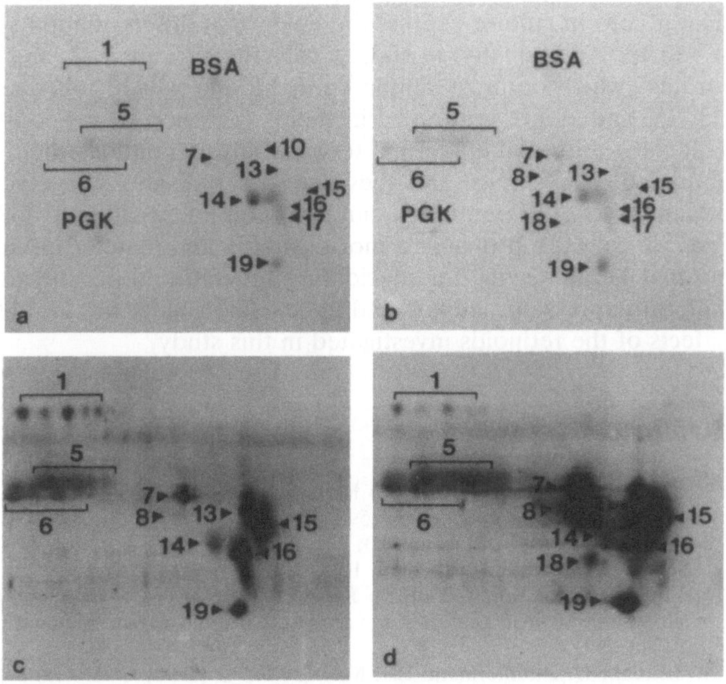

Fig. 5a–d. Two-dimensional gel electrophoretic splitting of the keratins expressed from untreated (**a** and **c**) HaCaT cells and those treated with all-*trans* RA (10^{-6} M) (**b** and **d**). **a** and **b** Coomassie blue-stained gel; **c** and **d** corresponding immunoblot reaction with a mixture of the antibodies 8-2/4 and 10-2/2

reduction of keratin 14 as well. Keratin 16 showed only a slight decrease or even an increase. This result may be related to the low proliferative activity of the cultures used in this series of experiments. Retinoids reinforce the expression of keratin 8, 13, 18 and 19.

Conclusion

Retinoids – especially the polyaromatic ones – are effective modulators of epidermal development and differentiation. The enhanced induction of keratins 13, 15, and 19 under retinoid influence points to a more embryonic phenotype of keratinocyte differentiation, as the common expression of these keratins has been observed only during the embryonic development of the epidermis. The reduction of keratin 16 and the slight increase of keratin 6, both so-called hyperproliferation-associated keratins, indicate that keratin 16 is probably a suitable parameter for determing the antiproliferative effect of retinoids in vitro. This is supported by corresponding results attained by the autoradiographic measurement of ^{3}H-thymidine incorporation into the DNS under retinoid treatment [31]. In contrast to normal keratinocytes, Hacat cells in culture express the epidermal differentiation-specific keratins 1 and 10. Additionally, in HaCaT cells the keratins 7, 8, and 18 are demonstrable, which occur in simple epithelia and which, together with keratins 13, 15, and 19 are reinforced in their expression by retinoids. HaCaT cells still show a normal epidermal terminal differentiation when they are transmitted to naked mice and they show as well an embryonic phenotype of keratinocyte differentiation under retinoid treatment [32]. Therefore, HaCaT cells are probably a more suitable and handier in vitro model than normal keratinocytes for elucidating the retinoid-dependent regulation of keratin expression and for finding starting points for the different clinical effects of the retinoids investigated in this study.

References

1. Steinert PM, Roop DR (1988) Molecular and cellular biology of intermediate filaments. Annu Rev Biochem 57:593–625
2. Klymkowsky MW, Bachant JB, Domingo A (1989) Function of intermediate filaments. Cell Motil Cytoskeleton 14:1–23
3. Fliegner KH, Ching GY, Liem RKH (1990) The predicted amino acid sequence of alpha-internexin is that of a novel neuronal intermediate filament protein EMBO J 9:749–759
4. Lendahl U, Zimmermann LB, MacKay RDG (1990) CNS stem cells express a new class of intermediate filament protein. Cell 60:585–595
5. Fuchs E, Green H (1980) Changes in keratin gene expression during terminal differentiation of the keratinocyte. Cell 19:1033–1042
6. Franke WW, Schiller DL, Moll R, Winter S, Schmid E, Engelbrecht I, Denk H, Krepler R, Platzer B (1981) Diversity of cytokeratins: differentiation-specific expression of cytokeratin polypeptides in epithelial cells and tissues. J Mol Biol 153:933–959
7. Moll R, Franke WW, Schiller DL, Geiger B, Krepler R (1982) The catalog of human cytokeratins: patterns of expression in normal epithelia, tumors and cultured cells. Cell 31:1–24 (1984)
8. Eichner R, Bonitz P, Sun T-T (1984) Classification of epidermal keratins according to their immunoreactivity, isoelectric point and mode of expression. J Cell Biol 98:1388–1396
9. Fuchs E, Coppock S, Green H, Cleveland D (1981) Two distinct classes of epidermal keratin genes and their evolutionary significance. Cell 27:75–84
10. Fuchs E, Marchuk D (1983) Type K and type II keratins have evolved from lower eukaryotes form the intermediate filaments in mammalian skin. Proc Natl Acad Sci USA 80:5857–5861
11. Kim KH, Theinwald JG, Fuchs EV (1983) Tissue-specificy of epithelial keratins: differential expression of mRNAs from two multigene families. Mol Cell Biol 3:494–502

12. Eichner R, Sun T-T, Aebi U (1986) The role of keratin subfamilies and keratin pairs in the formation of human epidermal intermediate filaments. J Cell Biol 102:1767–1777
13. Hatzfeld M, Maier G, Franke WW (1987) Cytokeratin domains involved in heterotypic complex formation determined by in-vitro binding assays. J Mol Biol 197:237–255
14. Mischke D, Wild G (1987) Polymorphic keratins in human epidermis. J Invest Dermatol 88:191–197
15. Zhou X-M, Idler WW, Steven AC, Roop DR, Steinert PM (1988) The complete sequence of the human intermediate filament chain keratin 10. J Biol Chem 263:15584–15589
16. Rieger M, Franke WW (1988) Identification of an orthologous mammalian cytokeratin gene. High degree of intron sequence conservation during evolution of human cytokeratin 10. J Mol Biol 204:841–856
17. Neson WG, Sun T-T (1983) The 50- and 58-kD keratin classes as molecular markers for stratified squamous epithelial cell culture studies. J Cell Biol 97:244–251
18. Sun T-T, Eichner R, Schermer A, Cooper D, Nelson WG, Weiss RA (1984) Classification, expression and possible mechanisms of evolution of mammalian epithelial keratins: a untying model. In: The transformed phenotype. Levine A , Topp W, Vande Woude G, Watson JD (eds) (Cancer cells, vol I) Could Spring Harbor Laboratory, Cold Spring Harbor, pp 169–176
19. Baden HP, McGilvray N, Cheng CK, Lee LD, Kubilus J (1978) The keratin polypeptides in normal and psoriatic epidermis. J Invest Dermatol 70:294–297
20. Weiss RA, Eichner R, Sun T-T (1984) Monoclonal antibody analysis of keratin expression in epidermal diseases: a 48- and 56-kdalton keratin as molecular markers for hyperproliferative keratinocytes J Cell Biol 98:1397–1406
21. McGuire J, Osber M, Lightfoot L (1984) Two keratins of MW 50,000 and 56,000 are synthesized by psoriatic epidermis. Br J Dermatol Suppl 27:27–37
22. Stoler A, Kopan R, Duvic M, Fuchs E (1988) Use of monospecific antisera and cRNS probes to localize the major changes in keratin expression during normal and abnormal epidermal differentiation. J Cell Biol 107:427–446
23. Korge B, Stadler R, Mischke D (1990) Effect of retinoids on hyperproliferation-associated keratins K6 and K16 in cultured human keratinocytes: a quantitative analysis. J Invest Dermatol 95:450–455
24. Lotan R (1980) Effects of vitamin A and its analogs (retinoids) on normal and neoplastic cells. Biochim Biophys Acta 605:33–91
25. Elias PM, Williams ML (1981) Retinoids, cancer and the skin. Arch Dermatol 117:160–180
26. Fuchs E, Green H (1981) Regulation of terminal differentiation of cultured human keratinocytes by vitamin A. Cell 25:617–625
27. Kopan R, Traska G, Fuchs E (1987) Retinoids as important regulators of terminal differentiation: examining keratin expression in individual epidermal cells at various stages of keratinization. J Cell Biol 105:427–440
28. Eckert RL, Green H (1984) Cloning of cDNAs specifying vitamin A-responsive human keratins. Proct Natl Acad Sci USA 81:4321–4325
29. Orfanos CE, Ehlert R, Gollnick H (1987) The retinoids: a review of their clinical pharmacology and therapeutic use. Drugs 34:459–503
30. Stadler R, Korge B, Müller K, Mischke D, Detmar M, Orfanos CE (1989) New potent arotinoid derivatives modify growth and differentiation of cultured murine and human keratinocytes. In: Reichert U, Shroot B (eds), Pharmacology of retinoids in the skin. Karger, Basel, pp 15–19 (Pharmacology of the strin, vol 3)
31. Boukamp P, Petrussevka RT, Breitkreutz D, Hornung J, Markham A, Fusenig NE (1988) Normal keratinization in a spontaneously immortalized anouploid human keratinocyte cell line. J Cell Biol 106:–761–771
32. Breitkreuz D, Boukamp P, Stark H-J, Ryle C, Fusenig NE (1989) Response of established keratinocyte lines to modulators of epidermal differentiation. In: Reichert U, Shroot B (eds) Pharmacology of retinoids in the skin. Karger, Basel, pp 8–14 (Pharmacology of the skin, vol 3)

Cytological Characterization of Three-dimensional, Graftable Human Keratinocyte Cultures

E. Lang, and *K. Maier*

Abstract

Although the successful application of graftable cultured human keratinocytes depends on the cytological state, they still lack exact qualitative and quantitative characterization. Our investigations revealed that graftable keratinocyte cultures can be quantitatively characterized using a new cell-size analyser (PPSA) and the radioactive EGF-ligation technique. With both techniques, typical changes in culture development can be proven.

Introduction

For several years, we have successfully applied three-dimensional human keratinocyte cultures as epidermal substitutes on burn patients or as wound-healing stimuli on chronic wounds. However, the optimal cytological and biochemical properties of these cultures, essential for successful grafting, have not yet been precisely defined. Usually, barely confluent cultures are used for grafting, since they are considered to be the most vital and proliferative ones.

During the process of differentiation, keratinocytes undergo major biochemical and morphological modifications. The analysis of these altered cell parameters permit the characterization of cell cultures with regard to their differentiation status and their proliferative capacity.

In order to gage the quality of cultured grafts we have investigated morphological and cytological properties, i.e., the distribution of cell sizes and the expression of high-affinity EGF/TGF alpha-receptors in the course of culture development. This experimental approach was based on the results of Barrandon and Green [1–3], which showed the causal coherence of cell size and proliferative capacity, and the results of Ponec and Boonstra [4, 5], demonstrating an inverse relationship between differentiation stage and EGF/TGF alpha-receptor expression on keratinocytes.

Taking into consideration the aspects of cell size distribution, we analyzed cultured normal human keratinocytes during six passages with a pulse plane surface analyzer PPSA (CASY) and compared the results with those obtained with a laser cell scanner (FACS) in different stages of confluency.

A. Bernd, J. Bereiter-Hahn,
F. Hevert, H. Holzmann (Eds.)
Cell and Tissue Culture Models
in Dermatological Research

In addition, the expression of the EGF/TGF alpha receptor, which is regarded as a marker of undifferentiated keratinocytes with the capability of cell division, was quantified by a radioactive ligation analysis.

Materials and Methods

Cell Culture

Keratinocytes were isolated from a skin biopsy of a 4-year-old girl by overnight trypsinization at 4° C. Cells were cultivated using the standardized feeder-layer technique according to Rheinwald and Green [6] in DMEM containing 10% fetal calf serum (FCS) and supplements [7].

For analysis of the cell size distribution and quantification of the EGF/TGF alpha receptor expression, keratinocyte cultures were harvested in subconfluent, confluent, and postconfluent stages by trypsinization with a 0.05% Trypsin/0.04% EDTA mixture in PBS without calcium an magnesium.

EGF/TGF alpha Receptor Quantification by Ligation with [^{125}I]-EGF

According to cell preparation techniques for flow cytometry, suspended keratinocytes were incubated with iodinated EGF in 96-well microtiter plates. This method allows uninhibited access of the ligand even to the receptors expressed in the lowermost cell layers. Unspecific binding was measured by adding excessive amounts of unlabeled EGF.

Analysis of Cell Size Distribution with Laser Scanner and Pulse Plane Surface Analyzer

A subfunction of the laser cell scanner (FACS, Becton Dickinson) permits the measurement of cell size and cell granulation. Aliquots of keratinocytes harvested by trypsinization were fixed in 0.7% paraformaldehyde solution and successively scanned.

For analysis with the PPSA, aliquots of trypsinized keratinocyte cultures were suspended in a 0.9% sodium chloride solution. A single measurement represents the cell size distribution of 4000–10000 single cells. Mean values of comparable culture states were calculated using the graphical statistics program CASYSTAT (Schärfe Systems).

Results

The laser cell scanner analysis reflects the cell size distribution in keratinocyte cultures measuring differentiated inactive as well as vital cells. The distribution elucidates the dependence of cell size on the degree of confluency and the number of passages (Fig. 1).

During the culture development from 80% to a full confluency lasting several days, the population of small, poorly granulated cells undergoes a

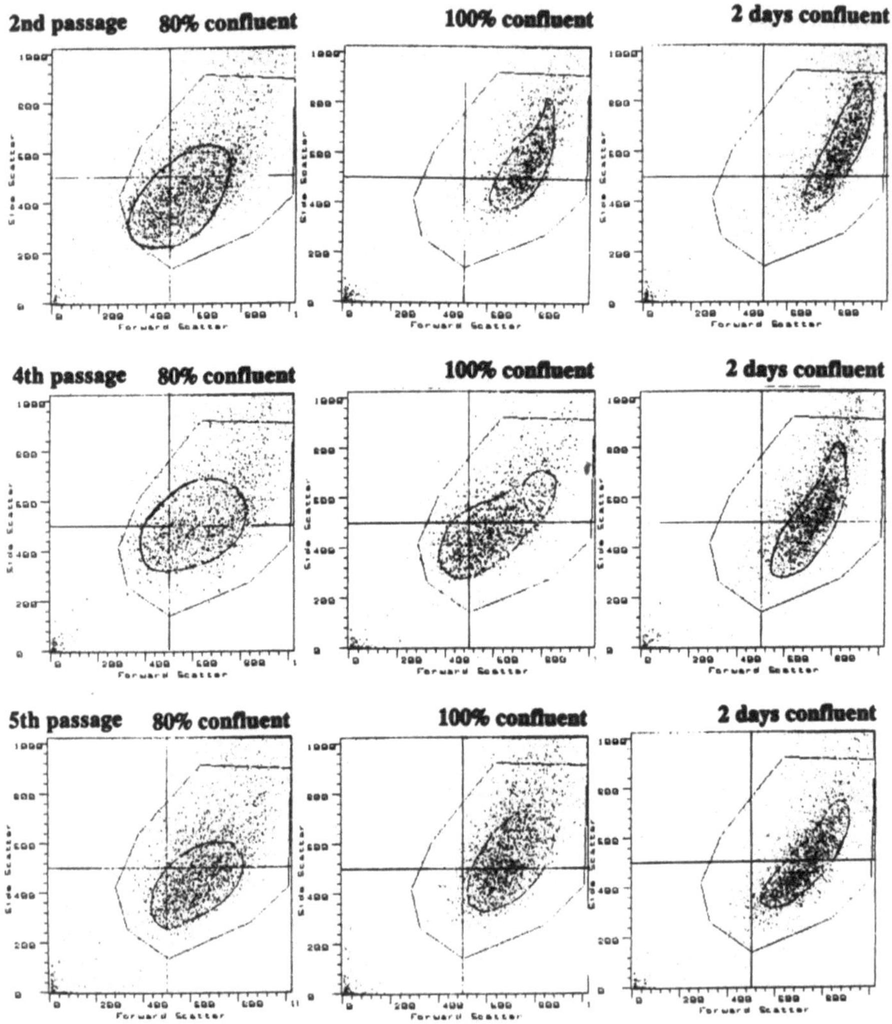

Fig. 1. Laser cell scanner analysis histograms of human keratinocytes cultures in different passages and stages of confluence. *Forward scatter,* cell size; *side scatter,* cell granulation; bitmaps show the dynamic range (0–50 μm) of a trypsinized culture including debris and clumps; cell populations and highest population densities are encircled

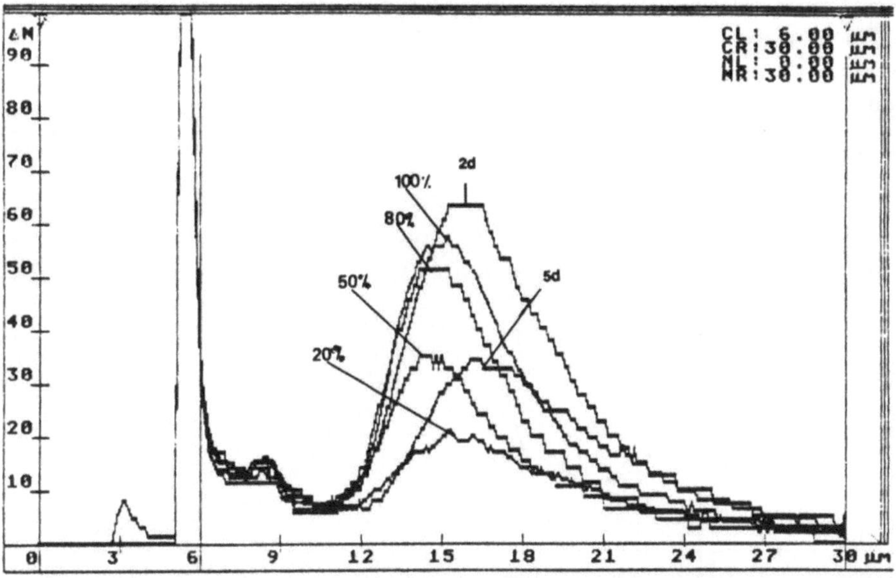

Fig. 2. PPSA 2-D overlay of averaged cell size distributions in keratinocyte cultures between a 20% and a 5-day confluence. 2-D overlay is standardized on equal peak maxima (range: 6–30 µm). Maximum peaks show the relative amounts of middle-sized cells in relation to the number of 9 µm cells. **d**, days of confluence

noticeable reduction, and in the confluent stage they are virtually not detectable, whereas the share of large cells and their state of granulation increases with confluency. The reduction of small, poorly granulated cells also takes place in the course of subsequent passages and becomes obvious in subconfluent cultures as well (Fig. 1).

For analysis of the general development of cell size using the PPSA (CASY), the raw data of many comparable single runs were averaged and compared in 2-D overlays. The PPSA detects only cells with intact membranes, but no terminally differentiated cells or cells that are injured during preparation.

The development of cell size distribution shown in Fig. 2 clearly reflects the changes in cell size of the main population of intact cells during the course of culture development from 20% confluency until confluency lasting 5 days. The increase of cell size in a keratinocyte population – in relation to the increasing age of culture – was proven (Fig. 2).

Likewise it is shown that the reduction of smaller cells observed with the laser cell scanner does not represent a complete loss but rather and underrepresentation resulting from the systemic principle of laser cell scanner measurements as described above.

The 2-D overlays shown in Fig. 3 demonstrate the development of cell size within the distinct passages. Generally, increasing numbers of larger cells are detectable with increasing duration of confluency. However, some

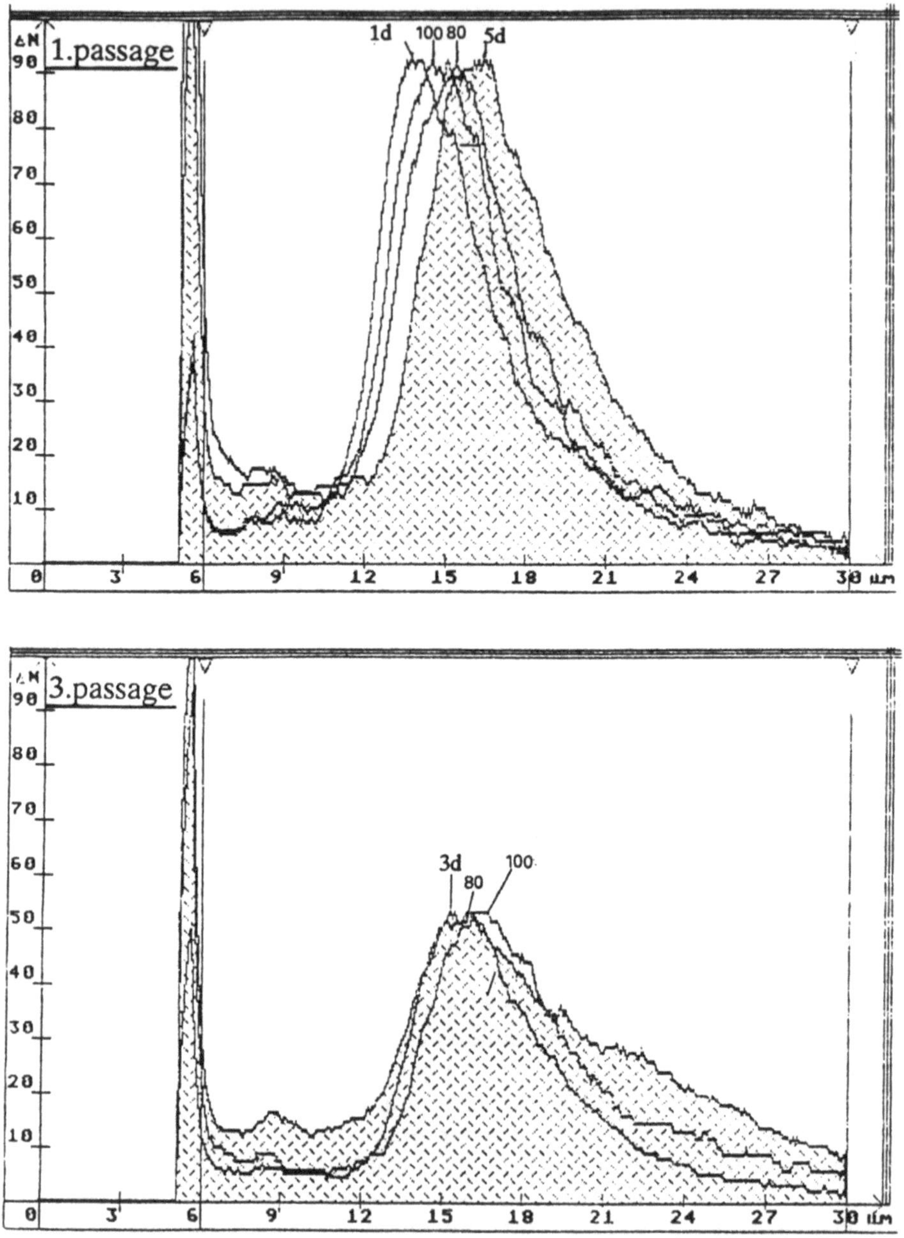

Fig. 3. PPSA 2-D overlays of cell size distribution in passages *1, 3, 4, and 5.* In each passage, cell size in the main cell population undergoes synchronous alterations, visible in shifts of the maximum peaks. A shift of about 2 μm reflects a reduction of cellular volume by half. d, days of confluence; *numbers,* state of confluence

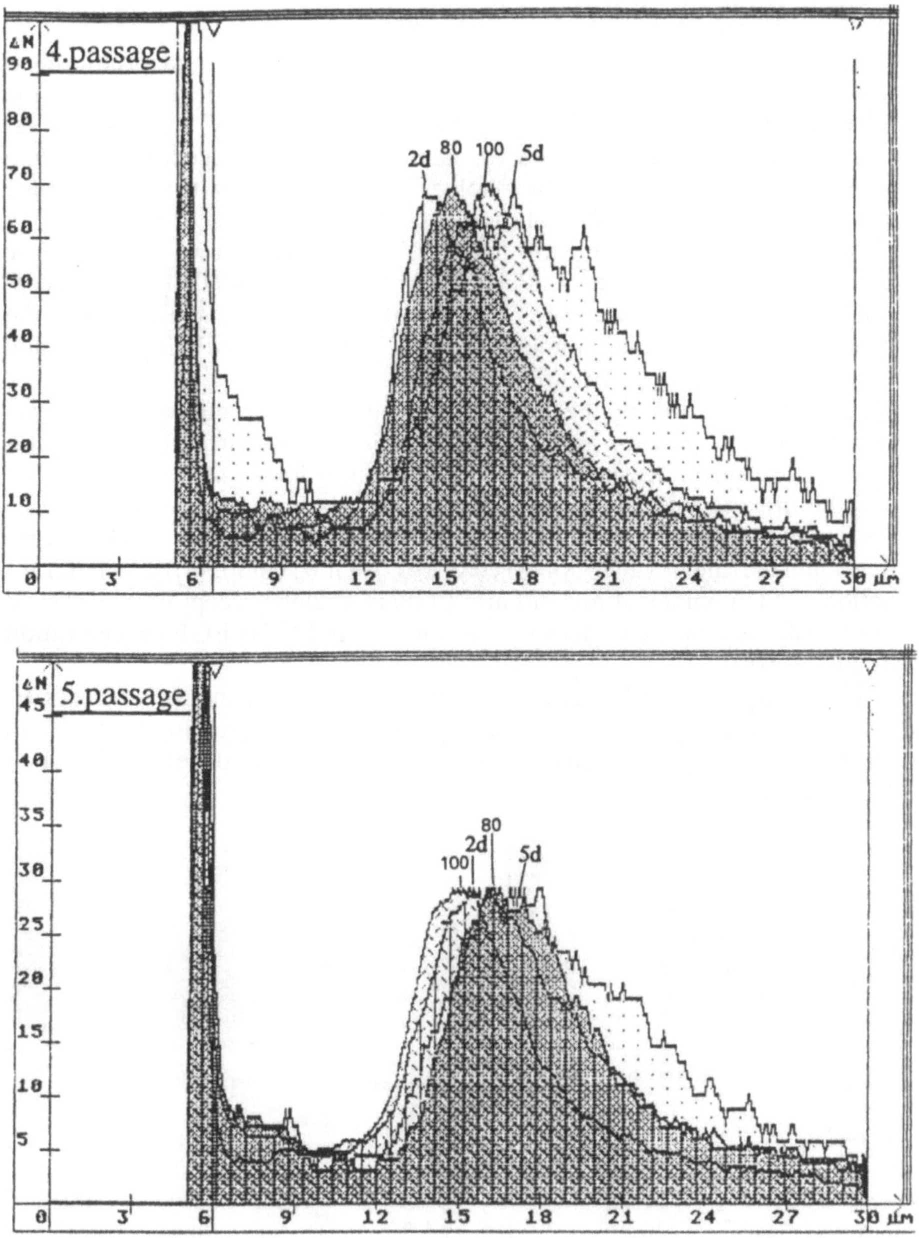

Fig. 3

postconfluent cultures that still contain proliferating cells exhibit reductions of cell size. These results suggest a synchronized cell cycle in growing cultures, since the observed shifts of the maximum peaks of 2 µm correspond with a bisection of cell volumes (Fig. 3).

EGF/TGF alpha Receptor Expression

EGF and TGF alpha are regarded as the main regulators of proliferation and differentiation of epidermal cells. Their effect is dependent on the presence and the functional stage of the specific EGF/TGF alpha receptors.

Immunofluorescence photographs demonstrated that receptor expression is confined to cells of intact morphology and is reduced with the increasing stage of differentiation (unpublished data).

Application of the receptor ligation method has shown characteristic quantitative changes in the expression of functional EGF/TGF alpha receptors in dependence on the state of confluency, allowing conclusions as to the condition of graftable keratinocyte cultures with regard to their proliferation capacity and their stage of differentiation.

For determination of the EGF/TGF alpha receptor affinity, concentration-dependent receptor assays were established with several selected cultures. The values measured were entered into a linearizing Scatchard plot and the dissociation constant was determined via linear regression analysis. As the dissociation constants showed no significant variations during culture development (mean value: $K_d = 0.028 \pm 0.014$ nM; comp. [4]), the determination of the number of high-affinity EGF/TGF alpha receptors was carried out for all the analyzed stages with a standardized [^{125}I]-EGF concentration of 1.67×10^{-10} M.

The quantitative course of the receptor expression during the six analyzed passages is shown in Fig. 4. In early passages, corresponding increases in receptor expression are visible as confluency of cultures develops (Fig. 4).

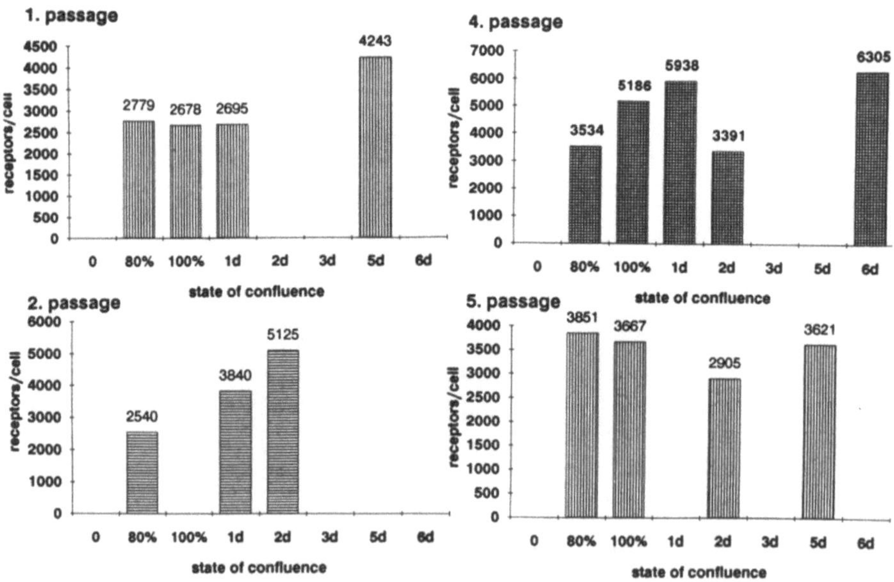

Fig. 4. EGF/TGF alpha receptor expression on cultured keratinocytes in passages *1, 2, 4, 5* and different states of confluence

Discussion

The parameters analyzed are related to the qualitative development of graftable keratinocyte cultures. The shifts of cell size observed in laser cell scanner analysis are a result of increasing numbers of differentiated cells compared with a relatively constant number of vital cells. The results show a slight decrease in the number of terminally differentiated cells, indicating a reduction of the cell proliferation rate in higher passages (in early passages proliferation is faster, resulting in more cells able to undergo differentiation). This is associated with a reduction of the small, poorly granulated cell population. PPSA analysis showed, however, that the observed reductions do not imply a complete loss of the 9- to 15-μm population, but rather a relatively constant vital cell population even in later stages of confluency.

Additionally, the analysis of the single passages showed that growing cultures are characterized by a synchronous volume bisectioning, reflecting the proliferation activity of the cells. Both methods of analysis confirm a characteristic increase in cell size with longer duration of confluency, concerning only a subpopulation and not the whole keratinocyte population.

The receptor quantification analysis verifies these findings as well. According to the literature ("... the competence to differentiate is inversely related to EGF receptor expression" [4]), a reduction in the number of EGF/TGF alpha receptors was expected in postconfluent stages. However, our results showed that especially postconfluent stages expressed higher amounts of receptors, which should be able to react to physiological doses of stimulating factors likely present in wounds. The increase of functional high-affinity receptors on the surface of intact cells indicates that those grafts might be particularly reactive on wounds. The resulting increase in cell activity and mitotic rate should have positive influences on speed and quality of the graft-take behavior. In future, these assumed advantages of postconfluent keratinocyte sheets in clinical application will have to be further evaluated in appropriate in vivo model systems.

References

1. Barrandon Y, Green H (1985) Cell size as a determinant of the clone-forming ability of human keratinocytes. Proc Natl Acad Sci USA 82:5390–5394
2. Barrandon Y, Green H (1987) Three clonal types of keratinocytes with different capacities for multiplication. Proc Natl Acad Sci USA 84:2302–2306
3. Barrandon Y, Green H (1987) Cell migration is essential for sustained growth of kertinocytes colonies: the role of TGFalpha and EGF. Cell 50:1131–1137
4. Boonstra J,de Laat S, Ponec M (1986) Epidermal growth factor receptor expression related to differentiation capacity in normal and transformed keratinocytes. Exp Cell Res 161:421–433
5. Boonstra J, Ponec M (1989) Proliferating, transformed keratinocytes cultured under low Ca^{2+} conditions exhibit high-affinity EGF-receptors. Cell Biol Int Rep 13:609–617
6. Rheinwald JG, Green H (1975) Serial cultivation of strains of human epidermal keratinocytes: the formation of keratinizing colonies from single cells. Cell 6:331–344
7. Rheinwald JG, Green H (1977) Epidermal growth factor and the multiplication of cultured human epidermal keratinocytes. Nature 256:421–424

Fluorometric Measurement of Suprabasal Cytokeratins and DNA of Human Keratinocytes Cultivated in Microtiter Plates

F. E. Görmar, C. Theilig, K. Dold, A. Bernd, and *H. Holzmann*

Abstract

Keratinocytes undergo many differentiation stages while moving up from the basal to the cornified layer of epidermis. The synthesis of suprabasal keratins (K1 and K10/11) is an early marker of keratinocyte differentiation, found in the keratinocyte line HaCaT, too. We demonstrated that with fluoroscan it was possible to measure the time-dependent relative content of suprabasal keratins of adhesive-grown microtiter cultures. The advantages of the method are the small amount of cell material needed, the high number of simultaneous tests, and the avoidance of other time-consuming methods such as electrophoresis and image analysis.

Introduction

The epidermis is in a steady state of a continuous renewal and desquamation of keratinocytes. In the basal cell layer of the epidermis the keratinocytes are highly proliferative. While moving up to the higher strata they undergo many differentiation steps, including the modification of their keratin pattern and calcium content [4]. At least 20 human keratins have been described, occurring in different combinations reflecting the location and differentiation state of keratinocytes [5]. Cultivated epidermal cells are not able to follow entirely all steps of this differentiation process. The human keratinocyte cell line HaCaT[1] [1] synthesizes and unusually broad spectrum of cytokeratins. Under normal culture conditions they produce the keratins K5, K6, K7, K8, K13, K14, K16/17, K18, and K19. Which of these keratins are synthesized depends on the culture conditions. In case of high cell density, HaCaT cells synthesize the suprabasal keratins K1 and K10/11 [6]. These keratins are important markers of keratinocyte differentiation [2, 3]. In this paper we describe a fast method of measuring the content of suprabasal keratins relative to the total keratin amount. The content of suprabasal keratins was estimated in adhesively grown HaCaT cell cultures

[1] The spontaneously transformed keratinocyte cell line was kindly provided by Prof. N. E. Fusenig from the DKFZ, Heidelberg

A. Bernd, J. Bereiter-Hahn,
F. Hevert, H. Holzmann (Eds.)
Cell and Tissue Culture Models
in Dermatological Research

by indirect immunofluorescence measurement with a microtiter plate fluoroscan. We demonstrated the detection of very small antigen amounts in microtiter plate cultures.

Material and Methods

Cell Culture

Investigations were performed with the transformed keratinocyte line HaCaT (passage 30–50). The cultures were cultivated in 96-well microtiter plates with Minimal Essential Medium (4× Hank's salts) supplemented with 5% fetal calf serum and 1% penicillin-streptomycin at 37°C and 5% CO_2. Media were changed twice a week.

Immunofluorescence

After cultivation, cells were washed twice in PBC and fixed for 2.5 min in acetone/methanol (1:1) at room temperature. Then unspecific binding sites were blocked by 1% BSA/PBS. Keratins were double-marked by 1:100 K8.6 (monoclonal antibody against suprabasal keratins raised in the mouse) and 1:100 antiserum to keratin (broad-spectrum antiserum raised in the guinea pig) in 1% BSA/PBS for 12 h at 4°C. After being washed twice with PBS, second antibodies (1:100 antimouse-IgG-TRITC and 1:100 anti-guinea pig-IgG-FITC in 1% BSA/PBS) were incubated for 12 h at 4°C. Subsequently, the cultures were incubated with 5 μg/ml (0.5 μg/ml) bisbenzimide (H33342) for 10 min and washed. Then each well was filled with 100 μl PBS. Fluorescence was measured with the Fluorscan II (Fig. 1; Flow Co.). The following

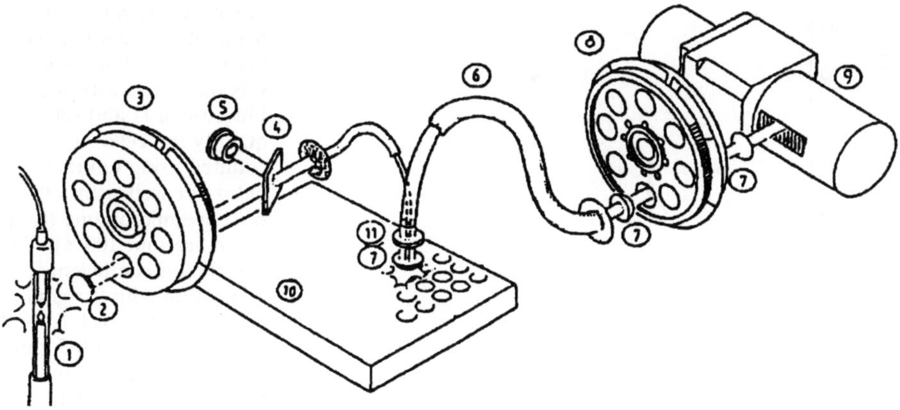

Fig. 1. Optical system of Fluorscan II (Biorad): *1*, Xenon lamp (300-to-1000-nm wavelength scale; 40 W); *2*, plancovenx lens; *3*, exitation filter wheel; *4*, mirror; *5*, reference detector; *6*, quartz filament optic; *7*, planconvex lens; *8*,emission filter wheel; *9*, photomultiplier; *10*, microtiter plate; *11*, KV-glass e

filter combinations were used: exitation weavelength at 355 nm (T0%, HBW35+4) and emission 460 nm (T60%, HBW 25+3) for bisbenzimide; exitation at 485 nm (T50%, HBW 14+2) and emission at 538 nm (T65%, HBW 25+3) for FITC, exitation at 544 nm (T60% HBW 15+2) and emission at 590 nm (T30%, HWB 14+2) for TRITC measurement.

Estimation of Cell Number

Cells were seeded in microtiter plates at different densities (5–4000 cells/mm^2; each density in eight parallel microtiter wells) and cultivated for 24 h. The cultures were washed with PBS and fixed 10 min at 4°C with acetone/methanol (1:1) and subsequently incubated for 10 min with 5 μg/ml and 0.5 μg/ml bisbenzimide (H33342). Then the cell number of the cultures was counted for ten image fields of the fluorescence microscope (Axioscop, Zeiss) per microtiter well at a magnification of × 100. The results were compared with the amount of bisbenzimide fluorescence in Fluoroscan II.

Measurement of the Relative Keratin Content

HaCaT cells were seeded at a density of 100/mm^2 and cultivated in high Ca^{2+} (1.2 mmol/l) medium. Cells were incubated for 2, 5, 7, 9, 12, 16, and 22 days. Then the cultures were fixed and marked by indirect immunofluorescence as described above.

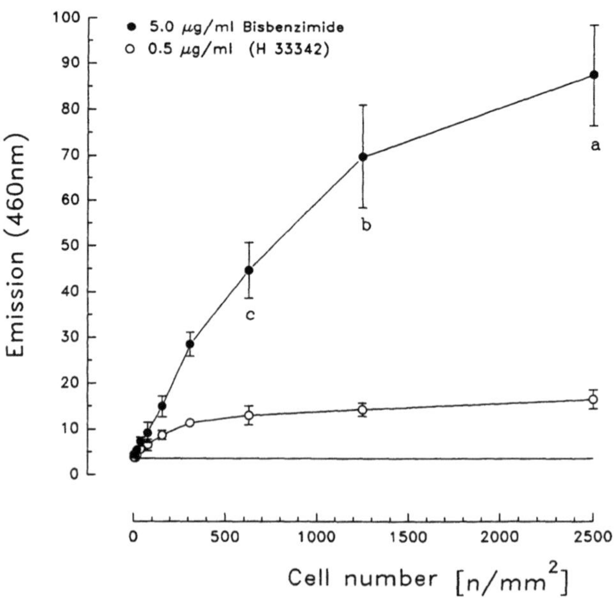

Fig. 2. Correlation of bisbenzimide fluorescence with the number of cells. HaCaT cells were seeded in various densities and fixed after spreading. The bisbenzimide fluorescence was measured with the Fluoroscan II and subsequently the cells were counted using ten image fields per well with a fluorescence microscope (total magnification: 100 ×). The correlation coefficient for incubation with 0.5 g/ml bisbenzimide was 0.85 and for incubation with 5 g/ml for regression lines with the highest cell number. *a*, 2500 cells/mm^2, $r = 0.970$; *b*, 1250 cell/mm^2, $r = 0.990$, *c*, 625 cells/mm^2, $r = 0.996$

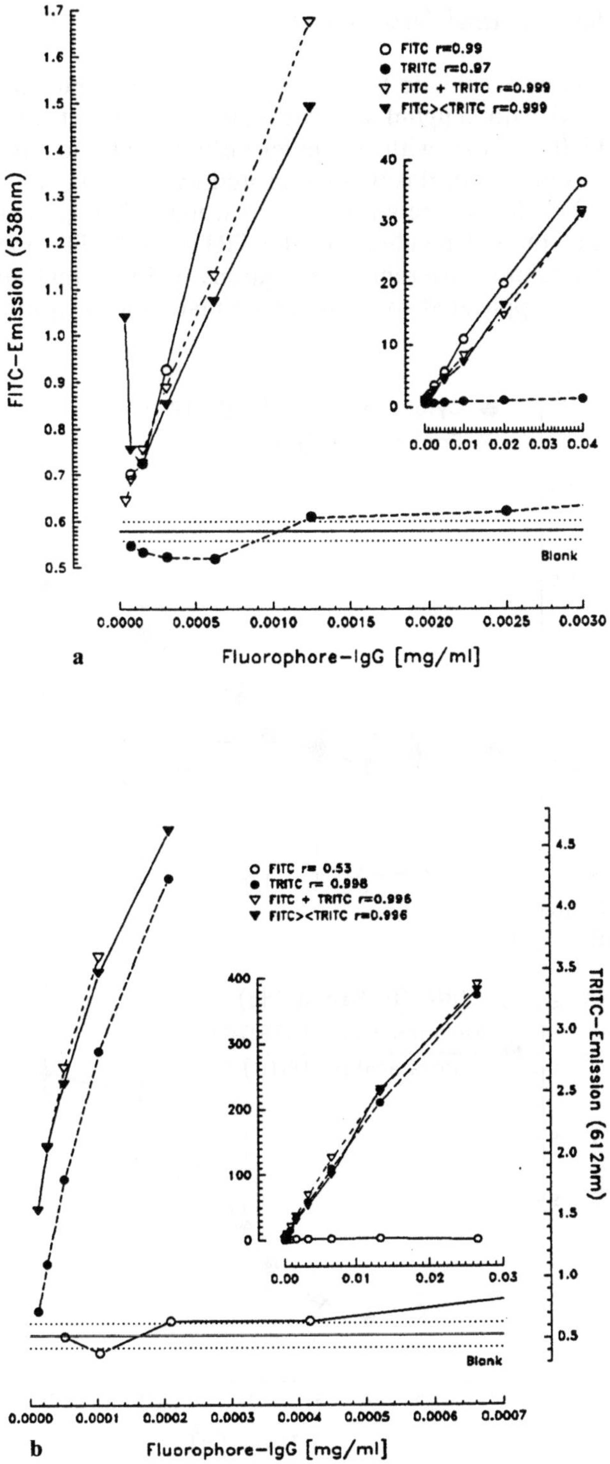

Fig. 3a, b. The linear correlation (r = correlation coefficient of linear regression) between the emissions of FITC (538 nm) and TRITC (612 nm) to the fluorophore concentration was demonstrated. In set figures give the whole measurement scale. FITC-IgG was diluted from 0.04 mg/ml and TRITC from 0.03 mg/ml to 0.1 µg/ml. The value of PBS without fluorophore-containing wells are given as *blank line* with standard deviation (*dotted lines*). *Open circles:* Dilution of FITC-conjugated anti-guinea pig-IgG; *filled circles:* dilution of TRITC-conjugated anti-mouse-IgG, *open triangles:* diluted mixture of both fluorophores; *filled triangles:* mixture of contrary fluorophore dilutions: the lowest concentration of the measured fluorophore was mixed with the highest concentration of the second fluorophore (in case of FITC emission there is a sudden increase because of the high amount of TRITC in the lowest concentrations of FITC-IgG)

Results and Discussion

The bisbenzimide fluorescence and the cell number correlated if DNA was dyed with 5 µg/ml bisbenzimide (Fig. 2). At very high cell densities (> 1500/mm^2) or with 0.5 µg/ml bisbenzimide no saturated marking of nuclei was obtained; this was monitored by fluorescence microscopy.

For the double immunofluorescence labeling of suprabasal and total keratins, the independency of the TRITC and FITC measurements was proven. Therefore, different combinations of FITC-IgG and TRITC-IgG solutions were applicated to a microtiter plate and measured in fluoroscan (Fig. 3).

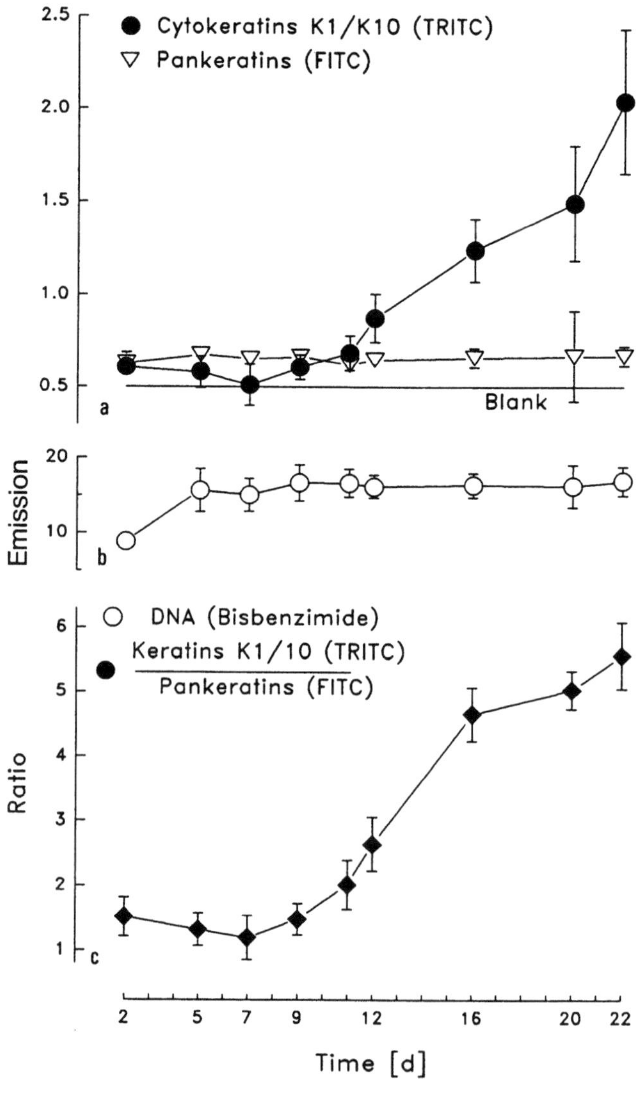

Fig. 4a–c. HaCaT cells were seeded (100 cells/mm^2) in microtiter plates and fixed after different cultivation times. Then the cells with indirect immunofluorescence were stained to "total keratin" (2. AB Anti IgG-FITC), to suprabasal keratins (2. AB Anti IgG-TRITC) and dyed with bisbenzimide demonstrating DNA. **a** On the 12th day the amount of the anti-mouse-IgG-TRITC increases, whereas the FITC fluorescence corresponding to total keratin and **(b)** the bisbenzimide fluorescence remained constant. **c** The computation of the suprabasal-to-total keratin ratio for each microtiter plate well reduces the noise level

The correlation coefficients for the FITC (Fig. 3a) and TRITC (Fig. 3b) measurements were greater than 0.99. There was a good correlation for low fluorescence units down to 0.6. In FITC measurement, a slight decrease of fluorescence appeared in the case of fluorophore mixtures (Fig. 3a: TRITC + FITC, TRITC < > FITC). This fluorescence decrease is probably due to reabsorbtion of the FITC emission by TRITC.

To estimate the ratio of suprabasal keratin to total keratin, HaCaT cells were cultivated for 3 weeks in medium containing high Ca^{2+} (1.2 mmol/l). On the 12th day the amount of anti-mouse-IgG-TRITC significantly increased (Fig. 4a), whereas the FITC fluorescence corresponding to total keratin and the bisbenzimide fluorescence (Fig. 4b) remained constant. The signal-to-noise ratio increased for the well/well ratio of TRITC and FITC, which is an index for the relative content of suprabasal keratins (Fig 4c).

Conclusion

The time-dependent measurement of suprabasal keratins – keratinocyte differentiation markers – in HaCaT microtiter plate cultures was demonstrated. The advantages of this method are the small amount of cell material needed, the high number of simultaneous tests performed and, last but not least, the possibility of measuring the relative amount of suprabasal keratins without time-consuming methods such as 2-D gel electrophoresis or image analysis.

References

1. Boukamp P, Petrussevska RT, Breitkreutz D, Hornung J,Markham A, Fusenig NE (1988) Normal keratinization in a spontaneously immortalized aneuploid human keratinocyte cell line. J Cell Biol 106:761–771
2. Franke WW, Schiller DL, Moll R, Winter S, Schmid E, Engelbrecht I (1981) Diversity of cytokeratins. Differentiation-specific expression of cytokeratin polypeptides in epithelial cells and tissues. J Mol Biol 153:933–959
3. Hatzfeld M, Franke WW (1985) Pair formation and promiscuity of cytokeratins: formation in vitro and intermediate-sized filaments by homologous and heterologous recombinations of purifed polypeptides. J Cell Biol 101:1826–1841
4. Hennings H, Michael D, Cheng C, Steinert P, Holbrook K, Yuspa SH (1980) Calcium regulation of growth and differentiation of mouse epidermal cells in culture. Cell 19:245–254
5. Moll R, Franke W, Schiller D, Geiger B, Krepler R (1982) The catalog of human cytokeratins: patterns of expression in normal epithelia, tumors and cultured cells. Cell 31:11–24
6. Ryle C, Breitkreutz D, Stark HJ, Steinert P, Ropp D, Fusenig N (1989) Density-dependent modulation of synthesis of keratins 1 and 10 in the human keratinocyte line HaCaT and in RAS-transfected tumorigenic clones. Differentiation 40:42–54

3. Models in Pigmentation

Biochemistry of Mammalian Pigmentation: Enzymatic Regulation of Melanogenesis

F. Solano Muñoz

Abstract

This article covers a biochemical approach to melanogenesis. It is divided into four parts: The first two parts are a general description of melanogenesis and the structure of the intermediates of the pathway. The two types of melanin, eumelanin and phaeomelanin, are described. The third part is devoted to the properties of tyrosinase, the main enzyme of the pathway. The last part is devoted to own our contribution to the knowledge of melanogenesis, the characterization of dopachrome tautomerase, its properties, and the role of 5,6-dihydroxyindole (DHI) versus 5,6-dihydroxyindole-2-carboxylic acid (DHICA) in the regulation of eumelanogenesis. It is shown that the melanin biosynthesis is a complex pathway whose regulation is not as simple as was thought a few years ago. The first part of the pathway is the well-known oxidation of tyrosine to dopaquinone. Tyrosinase is the key enzyme at these stages, but the level of dopa is crucial, since the tyrosine hydroxylation greatly depends on this co-factor. Dopaquinone is the switch point to form eumelanin (absence of glutathione) or phaeomelanin (presense of this thiol compound). In the eumelanogenesis pathway, dopachrome is also an important intermediate. Although it was thought for many years that dopachrome yielded melanin through DHI, it is now well established that dopachrome tautomerase catalyzes the tautomerization of dopachrome into DHICA. This indole is much more stable thatn DHI, but it can also be incorporated to the melanin polymer. The regulation of the distal phase of melanogenesis should also be pointed out as a new aspect in melanogenesis control.

Introduction

Melanin pigmentation is a physiological process which has drawn the attention of man for a long time. Colors decorate all animals, and human races differ in the color of their skin. Biologically, the pigment system is an important component of the skin, mucous membranes, eyes, ears, and meninges in mammals, and particularly in human beings. The current interest in melanin is due to its importance as a protective agent against damage

A. Bernd, J. Bereiter-Hahn,
F. Hevert, H. Holzmann (Eds.)
Cell and Tissue Culture Models
in Dermatological Research

by UV light, as well as to the social, cosmetic, and sometimes pathologic significance of pigmentation. Malignant melanoma is one of the two types of cancer (the other is breast cancer in smoking women) with the highest rate of increase among Caucasians.

Melanin formation occurs in the melanocytes. These are melanin-synthesizing-and -secretory cells derived from the neural crest. During embryonic development, the precursor cells, named melanoblasts, migrate to the skin, where they differentiate into mature melanocytes in the hair bulbs. A typical mammalian melanocyte is a cell with a small body, with two or more slender dendrites, and it can respond to hormonal, chemical, or physical stimuli by increasing its synthesis of melanin pigment, the dendritic branching, and the transfer of pigment to its neighboring keratinocytes of the skin [6].

Melanocytes are a good model, not only for dermatological research purposes, but also for the study of cellular differentiation and cell growth. In cellular differentiation this is because both melanin or tyrosinase are easily measured, and they can be used as very good markers of differentiation, in the case of cell growth because there are a number of malignant melanocyte lines which are standardized by cell culture methods. As a result, melanosynthesis has attracted the attention of scientists from many different fields.

The Melanogenic Pathway

Basis understanding of the melanosynthetic pathway developed during the first half of this century. In the 1920s and 1930s Raper [21], and in the 1940s Mason and Lerner [14, 15], postulated the pathway in a fairly exact way (Fig. 1). Most of their ideas about the structure of the intermediates of this pathway are still valid and they have been incorporated in Fig. 1. The main precursor is the naturally occurring L-tyrosine. This amino acid is oxidized to L-dopaquinone by the action of atmospheric oxygen in two different steps, both catalyzed by tyrosinase.

These activities of the enzyme are referred to tyrosine hydroxylase and dopa oxidase, respectively. The characteristics of both reactions are quite different. Tyrosine hydroxylation is a slow and difficult reaction, and tyrosinase requires L-dopa as a co-factor. Note that L-dopa is the product of the reaction, so that this step seems to be autocatalytic. Dopa oxidase is quite straightforward, and L-dopa is rapidly oxidized into L-dopaquinone by the action of tyrosinase without any co-factor requirements. The reason for such differences between the two reactions is found in the mechanism and active site of tyrosinase. This point will be discussed later on.

L-dopaquinone is a very reactive and important intermediate of melanogenesis. According to the Raper-Mason scheme, this chemical spontaneously undergoes a series of chemical reactions that finally produce a black polymeric pigment called eumelanin (Fig. 1, upper part). These reactions involve first a cyclization to build a complete indolic unit, the leukodopachrome. Both dopaquinone and leukodopachrome are very reactive, and

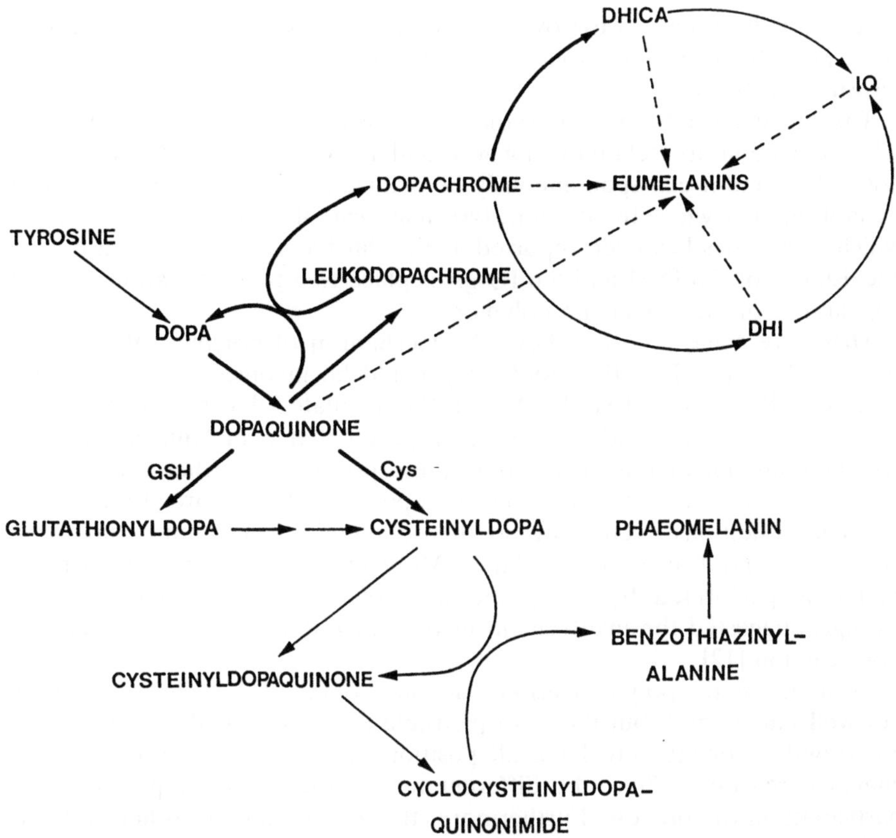

Fig. 1. Melanogenesis pathway. Tyrosinase catalyzes the first two reactions, from L-tyrosine to L-dopaquinone. This intermediate reacts with thiol compounds (phaeomelanogenesis) or undergoes an internal rearrangement which leads to L-dopachrome (eumelanogenesis). L-Dopachrome is tautomerized to DHICA by dopachrome tautomerase, or it is transformed to DHI by spontaneous decarboxylation. The proportion between carboxylated and decarboxylated indole units can regulate the formation rate and structure of the final pigment

their coexistence gives place, very rapidly, to half-regeneration of L-dopa and half-production of L-dopachrome [7]. The L-dopa reformed is again oxidized by tyrosinase, so that finally all material is oxidized to L-dopachrome. This is the first colored product of the melanin biosynthetic pathway. It is orange-red, with a maximal absorbance peak at 475 nm, and it is the product which is measured spectrophotometrically in the dopa oxidase assay of tyrosinase. L-dopachrome is fairly stable, depending on the conditions of the solution, but in a few hours it is converted to eumelanin. It is important to note that the first spontaneous reaction of L-dopachrome is a decarboxylation yielding DHI. Since all these processes occur spontaneously, for many years it has been postulated that the only enzyme regulating melanosynthesis is tyrosinase. Moreover, the determination of the whole melanogenic potential of melanocytes has also been used as a measurement of the tyrosinase activity.

The mechanism for the subsequent oxidative polymerization from 5,6-dihydroxyindole to melanin is not well understood. This dihydroxyindole is not stable, and it can spontaneously be oxidized to 5,6-indolequinone (IQ) by molecular oxygen. In addition, tyrosinase can also catalyze at this stage, which sometimes has been reported as the third activity of tyrosinase. The coexistence of 5,6-DHI and 5,6-IQ gives place to a polymerization and the appearance of the eumelanin polymer.

There are some theories about the mechanism of polymerization. Positions 2, 3, 4 and 7 of the indole ring could be involved in the reaction. However, Bu'Lock and Harley-Mason [5] postulated that positions 3 and 7 were more reactive, and they even proposed that eumelanin could be a regular homopolymer involving only these positions. Further experiments carried out by Prota [20] have shown that positions 2 and 4 are also reactive, in some conditions even more so than 3 and 7. Thus, dimers involving positions 2 and 4 have been isolated. More recently, experiments of pulse radiolysis performed by the group of Prof. Land at Manchester have strongly suggested the involvement of trihydroxyindoles in the polymerization reaction [12].

In summary, the polymerization reactions from 5,6-DHI to eumelanin are not well understood, but the most plausible mechanism is the formation of an irregular polymer involving all positions and more than one different chemical reaction. Thus, the different mechanisms so far reported could participate in the process. Furthermore, the incorporated of other previous melanogenic intermediates (dopachrome, dopa, etc.) into the growing polymer cannot be ruled out.

Phaeomelanogenesis

Nevertheless, mammalian coloration is determined mainly by two different types of melanin: the black *eumelanin,* formed by the classical Raper-Mason pathway, and *phaeomelanin* [10]. This type of pigment has lighter colors, which range from numerous shades of brown to orange, reddish, and yellow. The formation of phaeomelanin was investigated principally by Prota in 1980 [19]. They proved that phaeomelanosynthesis involves L-tyrosine and also an other amino acid, L-cysteine, so that the final pigment has sulfur in its structure (Fig. 1, lower part). The first two steps in the biosynthesis of phaeomelanin are common to eumelanin formation, and they consist in the tyrosinase-catalyzed oxidation of L-tyrosine to L-dopaquinone. As mentioned before, this quinone is an important intermediate of melanosynthesis and is considered to be the major switch point in the global melanin pathway. In the presence of cysteine, or any other thiol-containing compound such as the tripeptide glutathione, L-dopaquinone does not form leukodopachrome; rather, it combines rapidly with thiol groups to produce cysteinyldopa or glutathionyldopa adducts. The thiol groups react with several positions of the aromatic ring of dopaquinone, the most favored position being 5. Thus, the yield of the 5-cysteinyldopa or the corresponding analogue with glutathione, 5-glutathionyldopa is higher than 70% of the total amount of isomers formed [19].

5-Glutathionyldopa is the main adduct formed in vivo, since the level of glutathione inside melanocytes is much higher than the cysteine level. However, the subsequent action of γ-glutamyltranspeptidase, an enzyme which is very active in melanocytes [16], and a dipeptidase leads to 5-cysteinyldopa. This intermediate can be oxidized by tyrosinase in the presence of catalytic amounts of L-dopa, yielding 5-cysteinyldopaquinone. In the light of present knowledge, it has been thought that this compound is spontaneously transformed in to 1,4-benzothiazinylalanine, the structural unit of phaeomelanin and a family of related compounds called trichochromes that exist in avian feathers and in human red hair. Alternatively, some intermediates of the phaeomelanin pathway can be oxidized in a way similar to the eumelanin pathway to form mixed melanin. These mixed melanins, containing both eumelanogenic and phaeomelanogenic precursors, are found in all sources, as proven by the presence of sulfur in melanin from all organisms analyzed.

Thus, the formation of eu- or phaeomelanin seems to be dependent on the availability of thiol compounds to combine with L-dopaquinone in the early stages of melanogenesis. Although this process is genetically determined, the exact mechanism is unclear. As an indication of the complexity at the genetic level, there are over 100 mutations affecting mouse pigmentation, and these occur at more than 40 different genetic loci [22]. In some mammals, such as the agouti mouse, alternating cycles of eu- and phaeomelanosynthesis take place, so that in the hair of these animals, bands of black and yellow colors are formed.

Tyrosinase

The enzyme tyrosinase is a multifunctional, copper-containing glycoprotein. In mammals, active tyrosinase can be found in a variety of subcellular organelles of the melanocyte, but in vivo, melanogenic activity occurs only in specialized melanocytic organelles, termed melanosomes or melanin granules. Microsomal and cytosolic tyrosinase isozymes are the result of the normal metabolic production of protein. Tyrosinase is synthesized at the rough endoplasmic reticulum, glycosylated at the GERL system (a complex structure of Golgi, endoplasmic reticulum and lysosomes) and transferred to premelanosomes in vesicles [8, 10]. Formation and accumulation of melanin are restricted to melanosomes. The obliteration of the premelanosome to the melanin granule is a four-step process. Finally, the melanin granule, totally black, is transported to the dendrites and then transferred to the keratinocytes.

The principal reason for thinking that only melanosomal tyrosinase is active in vivo is that some melanogenic intermediates, mainly quinones, are very reactive species, as stated above. Therefore, the formation of such substances in the cytosol of melanocytes would be cytotoxic. One approach to killing specifically malignant melanocytes is the addition to the cells of possible substrates of tyrosinase (*p*-hydroxyanisole, dihydroxybenzylamine) in order to provoke cytosolic melanogenesis. Under these conditions, cytosolic-formed quinones react with a great variety of biomolecules, such as thiol-containing enzymes, leading to cell death.

Tyrosinase is bound to the melanosomal membrane. The smaller active form has a molecular weight of approximately 70 kD and pI of 3.3. In the deglycosylated state it is only 55 kD and the pI is 4.2 [8]. The tridimensional structure has not yet been ascertained. Most of the information on the active site of tyrosinase and the molecular mechanism of the catalysis comes from the work of Lerch [13], on the *Neurospora* and nushroom tyrosinases. The data can be extrapolated to the mammalian enzyme.

The most important feature of tyrosinase is the involvement of two copper atoms. These coppers are bound to the polypeptidic chain by at least three histidines. Depending on the redox state of the copper, tyrosinase can be found in three different forms, called met-, oxy-, and desoxy-tyrosinase.

Met-tyrosinase is the resting state of the enzyme. Both coppers are as Cu(II), and not substrates are bound to the enzyme catalytic site at this state. Oxy-tyrosinase has one molecule of oxygen bound to the active site, and the copper atoms are Cu(II). Finally, the desoxy form is an intermediate of the enzyme during the catalytic cycle, where the copper atoms are as Cu(I) in a reduced state.

In vivo, all tyrosinase can be found as met-tyrosinase, although a small amount could be present as oxy-tyrosinase. The distribution between these two forms is very important to account for the different behavior of tyrosinase in regulating the tyrosine hydroxylase and dopa oxidase activites. As mentioned above, tyrosine hydroxylation reguires the presence of a certain amount (often referred to as the catalytic amount) of dopa as a-co-factor. Otherwise, this activity requires an induction period (lag period) before the steady state is reached [14]. The lower the L-dopa concentration, the higher is the lag period. If exogenous L-dopa is not added, the lag period lasts several hours, depending on the amount of enzyme and the concentration of L-tyrosine used in the assay. However, the rate of reaction once the steady state has been reached is the same in all cases, and it does not depend on the initial amount of L-dopa added to the reaction media.

Figure 2 shows the catalytic cycles for both activities, tyrosine hydroxylase (on the right) and dopa oxidase (on the left). During tyrosine hydroxylation, in the absence of L-dopa, met-tyrosinase binds the substrate, L-tyrosine, to form an inactive complex (bottom). Only the small amount of oxy-tyrosinase which occurs in the reaction medium is active. In the catalytic cycle, oxy-tyrosinase oxidizes L-tyrosine to dopaquinone. Two important features of this cycle should be noted:

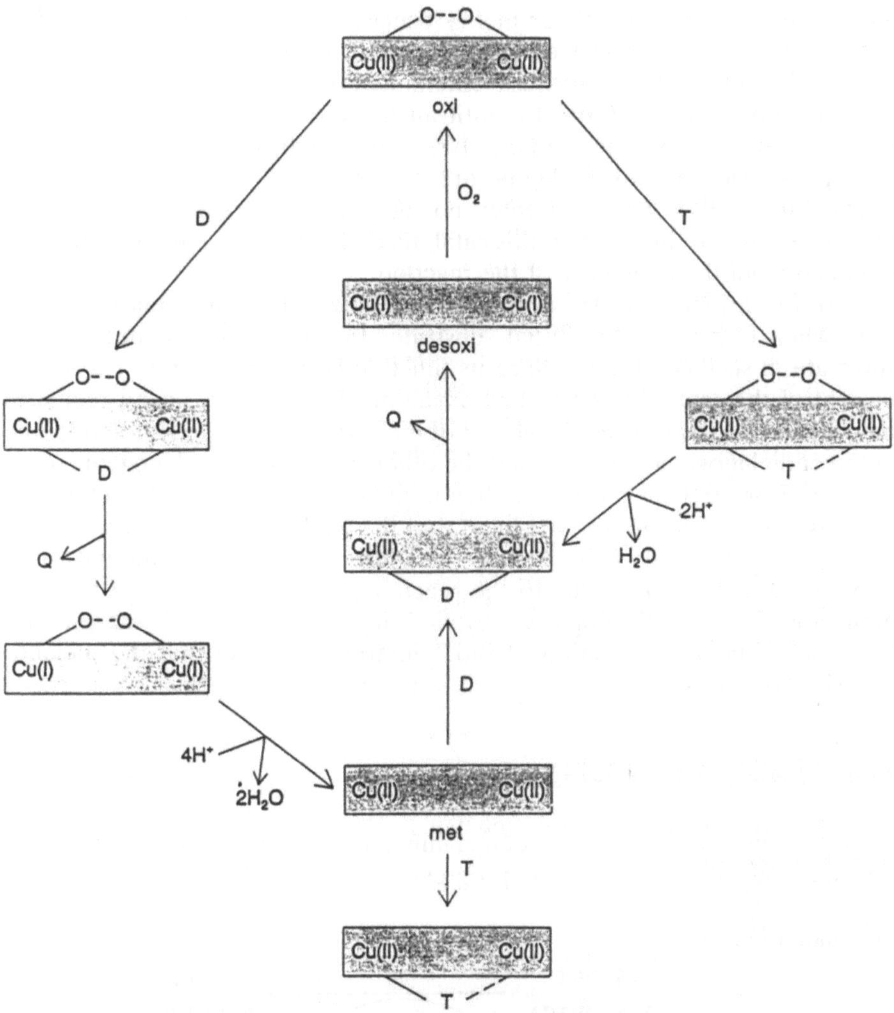

Fig. 2. Mechanism of catalysis by tyrosinase. Inside melanocytes, this enzyme occurs mainly as met-tyrosinase, but a small amount occurs as oxy-tyrosinase. Tyrosinase can act on L-tyrosine (*T, right cycle*, tyrosine hydroxylase activity) or on L-dopa (*D, left cycle*, dopa oxidase activity). It is worth noting that met-tyrosinase can bind L-tyrosine, and this leads to an inactive form of the enzyme. This explains the lag period observed in the tyrosine hydroxylase activity of the enzyme in the absence of the co-factor, L-dopa

1. met-Tyrosinase is not an intermediate of the catalytic cycle.
2. L-dopa is not actually formed in tyrosine hydroxylation.

The enzyme picks up L-tyrosine and releases directly dopaquinone instead of dopa. Thus, what is the origin of L-dopa to act as co-factor?

L-dopa is produced as result of the chemical reaction between dopaquinone and leukodopachrome (Fig. 1). This dopa appears during the reaction and competes with tyrosine for met-tyrosinase (Fig. 2), so that this form of the enzyme is slowly recovered as an active form and the rate of catalysis is increased. As more met-tyrosinase enters the cycle, more dopa is formed, so that the process is autocatalytic until all the tyrosinase reaches the active form, oxy-tyrosinase. Then, no met-tyrosinase is left, the steady state of the reaction is reached, and the lag period is over.

The dopa oxidase cycle is simpler; no lag period exists since both oxy- and met-tyrosinase are involved in this catalytic cycle. Thus, maximum activity is obtained from the beginning of the reaction.

Therefore, dopa is a very important compound in melanogenesis regulation, when tyrosine is the initial substrate. In fact, L-dopa is not really a substrate in such conditions, since its function is rather to act as a co-factor to control the rate of tyrosine hydroxylation. Furthermore, dopa regulates the rate of two steps at the distal part of melanogenesis (Fig. 3): the first one in the eumelanogenic pathway and the other in the phaeomelanogenic pathway. DHI or DHICA on the one hand, and cysteinyldopa on the other, all are very poor substrates of tyrosinase, but the oxidation to their corresponding quinones is greatly enhanced by the addition of catalytic amounts of L-dopa. Essentially, this is due to the reaction of dopaquinone with those *o*-diphenols. In this way, dopa is reformed and the net effect is the pseudo-tyrosinase-catalyzed oxidation of those intermediates, so that the melanogenesis progresses more quickly than in the absence of dopa.

Dopachrome Tautomerase

As mentioned above, in the eumelanin pathway after the formation of dopaquinone all reactions can progress spontaneously to yield melanin.

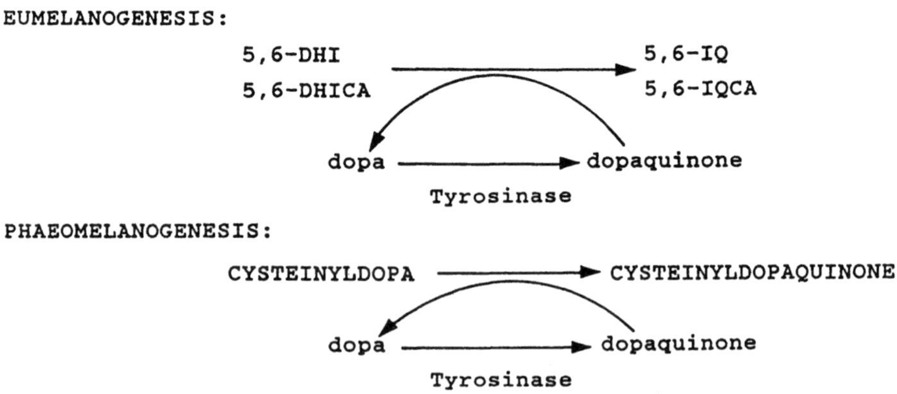

Fig. 3. Pseudo-activities of tyrosinase. L-Dopaquinone, formed by the action of tyrosinase on L-dopa, is able to react with DHI, DHICA, or cysteinyldopa to yield indolequinone or cysteinyldopaquinone, respectively. The final effect is the oxidation of melanogenic intermediates which are poor substrates of tyrosinase. Their direct oxidation, in the absence of L-dopa, would be difficult

Therefore, for many years tyrosinase was considered the only enzyme controlling the melanin biosynthesis. However, in 1980 the existence of three other factors which could be involved in the regulation of melanogenesis was postulated [11, 18]. These three postulated factors were: (a) the dopachrome conversion factor, which accelerated the dopachrome decomposition; (b) the dihydroxyindole accelerating factor, which accelerated melanin formation from DHI; and (c) the dihydroxyindole blocking factor, which prevented the polymerization on DHI, and therefore melanin formation. Two years later, the dihydroxyindole accelerating activity was attributed to tyrosinase. In fact, this is one of the pseudo-activites of this enzyme, as discussed above. Then, only two possible new factors were postulated. Furthermore, Barber et al. [4] characterized the dopachrome conversion factor as a protein, and they proposed the name of dopachrome oxidoreductase for the new enzyme, since its action of dopachrome gave way to:

1. An acceleration in the rate of dopachrome decomposition
2. An inhibition of melanin formation in comparision to a control assay made in the absence of the enzyme

Thus, it seemed that dopachrome oxidoreductase joined the activity of two factors: dopachrome conversion and indole-blocking activities. At this stage, three major points were noted:

1. It was thought that the rearrangement reaction of dopachrome, catalyzed by dopachrome oxidoreductase, was identical to the spontaneous decarboxylation of dopachrome to yield DHI.
2. The fact that other agents such as metal ions accelerated the dopachrome decomposition and the instability of this intermediate gave rise to doubts about the existence of protein factors regulating the rate of dopachrome decomposition. The melanosome is rich in Cu(II), Fe(II), and Zn(II), and some researchers [17] attributed the accelerating effect on dopachrome decomposition to these cations.
3. However, the isolation of the indole-blocking factor was not achieved, and metal ions could not account for the inhibition of melanin polymerization from indoles. On the contrary, they also accelerated this step.

During the years 1988–1990, our group isolated a protein from mouse melanoma with dopachrome conversion activity [1, 2, 9]. When the reaction catalyzed by this protein was analyzed by spectrophotometric and radiometric measurements, we observed the following properties: The evolution of the UV-Vis spectra of dopachrome in the presence of dopachrome-converting factor was different from those obtained in the absence of this protein. In both cases, a decoloration of dopachrome was observed, decreasing the absorbance in the visible region. The decoloration rate was obviously higher in the presence of dopachrome-converting factor. However, the situation was different in the UV region. The absorbance increased in the presence of dopachrome-converting factor and decreased in its absence. This indicated that the products formed were different, depending on the action of the protein on dopachrome. Spectroscopic interpretation of these data strongly suggested that the increase in the absorbance by around 300 nm should be due to the appearance of carboxylated indole units from dopachrome [2].

Similar experiments carried out with dopachrome labeled in the carboxyl group showed that the action of the enzyme inhibited CO_2 evolution. In the absence of dopachrome-converting factor, more than 80% of the carboxyl group from dopachrome was released as CO_2 during melanin formation, and this percentage was lower, the higher the amount of enzyme in the reaction medium.

All these results supported the hypothesis that the reaction catalyzed by the dopachrome-converting factor is a tautomerization (a special type of isomerization involving a keto/enol transformation) of dopachrome to yield DHICA, instead of the spontaneous decarboxylative rearrangement of dopachrome to yield DHI. This is why we renamed the enzyme 'dopachrome tautomerase' (EC 5.3.2.3), since this is more precise and informative than 'dopachrome-converting factor', 'dopachrome oxidoreductase', or 'dopachrome isomerase', used before the characterization of the enzyme-catalyzed reaction was achieved.

Once the nature of the reaction product was identified, the apparent indole-blocking activity associated with the enzyme could also be explained without postulating a new protein factor. DHICA is much more stable than DHI. This decarboxylated indole is easily oxidized to indolequinone, and its polymerization occurs quite rapidly, since positions 2, 3, 4, and 7 of the indole ring might be involved in the reactions. The situation is quite different for DHICA. Position 2 is blocked by the carboxyl group, and position 3 is greatly inactivated by the mesomeric effect of the neighboring carboxyl group. Thus, this carboxylated indole is not easily oxidized to its corresponding quinone, and the practical consequence is that DHICA is fairly stable in neutral solutions. Taking into account that dopachrome tautomerase yields DHICA instead of DHI, the apparent indole-blocking activity postulated as associated with dopachrome tautomerase can be easily explained without the existence of an extra protein factor. The prevention of melanin polymerization in the presence of dopachrome tautomerase could be due only to the greater stability of the product, DHICA, in comparison to DHI.

The enzyme shows a great specificity [3]. First, it is stereospecific. It tautomerized-L-dopachrome, but no D-dopachrome. Second, the carboxyl group on the substrate is essential for enzymatic recognition. Among some dopachrome analogues, only dopachrome methyl ester is an alternative substrate, and not dopaminochrome (Table 1). Furthermore, all inhibitors found for the enzyme (Table 2) contain a carboxyl group at position 2 or in a certain position which can somehow mimic the sterical location of position 2. The product of the reaction, DHICA, and some indoles such as L-Trp and indole-3-propionic, are so far the strongest inhibitors found for the enzyme. Thus, it can be inferred that the active site of the enzyme recognizes the indole ring and the carboxyl group in a very specific way.

Table 1. Structure of possible substrates of dopachrome tautomerase

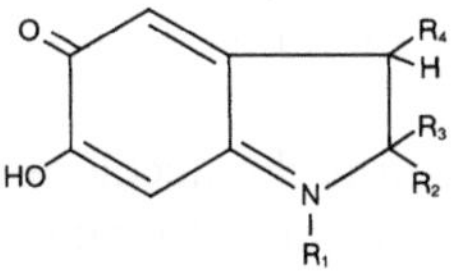

	R_1	R_2	R_3	R_4	Substrate
D-Dopachrome	H	COOH	H	H	No
L-Dopachrome	H	COOH	H	H	Yes
L-Dopachrome methyl ester	H	$COOCH_3$	H	H	Yes
L-α-Methyl dopachrome	H	COOH	CH_3	H	No
Dopaminochrome	H	H	H	H	No
Deoxyadrenochrome	CH_3	H	H	H	No
Adrenochrome methyl ether	CH_3	H	H	OCH_3	No

Table 2. Inhibitors of dopachrome tautomerase

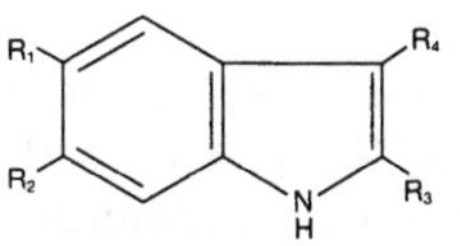

Inhibitors	(concentration)	R_1	R_2	R_2	R_1	% Inh.
I-2-carboxylic	(1 m*M*)	H	H	COOH	H	11.5
5-HI-2-carboxylic	(1 m*M*)	OH	H	COOH	H	18.5
5,6-DHI-2-carboxylic	(1 m*M*)	OH	OH	COOH	H	24
L-Tryptophan	(2 m*M*)	H	H	H	$CH_2.CH(NH_2).COOH$	49
I-3-propionic	(2 m*M*)	H	H	H	$(CH_2)_2.COOH$	47
5-Hydroxy-L-Trp	(2 m*M*)	OH	H	H	$CH_2.CH(NH_2).COOH$	25

If melanogenesis from dopachrome to DHI occurs without any enzymatic catalysis, what are the physiological roles of dopachrome tautomerase in the melanocyte? These might be:

1. An increase the content of carboxyl groups in the melanin polymer, since the major intermediate of melanogenesis in the presence of the enzyme is DHICA instead of DHI
2. The formation of a more ordered polymer. As discussed above, the positions for the cross-linking among dihydroxyindole units are four in the case of DHI and only two in the case of DHICA.
3. The formation of a melanin polymer with a higher chelating capacity. It sould be taken into account that one of the main functions of melanin in vivo is to act as a scavenger of metal ions and radicals which could be toxic for the cell. The chelating capacity of melanin is sited at positions 5

and 6 (hydroxyl groups, and in the case of incorporation of DHICA units, also at positions 1 (indolic nitrogen) and 2 (the carboxyl group). This new complexing center would permit the binding of a greater amount of metal ions to the polymer.

However, the major function of dopachrome tautomerase is the acceleration of melanin synthesis from dopachrome. This might be perplexing, since DHICA is fairly stable, but the study of the evolution of melanin polymerization from dopachrome in different conditions was very informative. When dopachrome was treated only with dopachrome tautomerase, tyrosinase, or both enzymes, the polymerization was rather slow. But when dopachrome was treated with dopachrome tautomerase, tyrosinase, and a catalytic amount of L-dopa (10 μM) all together – conditions very similar to those found in vivo – the situation was different. The polymerization took place quite rapidly, and melanin formation was greatly accelerated in comparison to all other conditions. This is due to the accelerating effect of dopachrome tautomerase on dopachrome conversion to DHICA and the accelerating effect of tyrosinase plus L-dopa on DHICA conversion (this is one of the above-mentioned pseudo-activities of tyrosinase).

The same confirmatory results were obtained when radioactive melanin formation was measured. Incubation of uniformly labeled ^{14}C-tyrosine with tyrosinase in the presence of several amounts of dopachrome tautomerase showed that melanin formation was higher when the amount of tautomerase in the reaction medium was increased [9]. Thus, enzymatic DHICA formation should be considered a way of regulating melanin formation, in contrast to spontaneous DHI formation which leads to melanin polymerization without any regulation in this phase of the pathway.

Acknowledgements. I thank all may colleagues at the Department of Biochemistry and Molecular Biology, School of Medicine, University of Murcia; particularly, I want to mention P. Aroca, J. C. Garcia-Borron, J. R. Jara, J. H. Martinez-Liarte, and J. A. Lozano, head of the research group. This article would not have been possible without their invaluable help.

References

1. Aroca P, Garcia-Borron JC, Solano F, Lozano JA (1990a) Regulation of mammalian melanogenesis. I. Partial purification and characterization of a dopachrome-converting factor: dopachrome tautomerase. Biochim Biophys Acta 1035:266–275
2. Aroca P, Solano F, Garcia-Borron JC, Lozano JA (1990b) A new spectrophotometric assay for dopachrome tautomerase. J Biochem Biophys Methods 21:35–46
3. Aroca P, Solano F, Garcia-Borron JC, Lozano J (1991) Specificity of dopachrome tautomerase and inhibition by carboxylated indoles. Biochem J 277:393–397
4. Barber JI, Townsend D, Olds D, King RA (1984) Dopachrome oxidoreductase: a new enzyme in the pigment pathway. J Invest Dermatol 83:145–149
5. Bu'Lock JD, Harley-Mason J (1951) Melanin and its precursors. II. Model experiments on the reactions between quinones and indoles, and consideration of a possible structure for the melanin polymer. J Chem Soc 703–712

6. Fitzpatrick TB, Lener AB (1954) Biochemical basis of human melanin pigmentation. Arch Dermatol Syph 69:133–149
7. Garcia-Carmona F, Garcia-Canovas F, Iborra JL, Lozano JA (1982) Kinetic study of the pathway of melanization between L-dopa and dopachrome. Biochim Biophys Acta 717:124–131
8. Hearing VJ, Jimenez M (1987) Mammalian tyrosinase. The critical regulatory control point in melanocyte pigmentation. Int J Biochem 19:1141–1147
9. Jara JR, Solano F, Garcia-Borron JC, Aroca P, Lozano JA (1990) Regulation of mammalian melanogenesis. II. The role of metal cations. Biochim Biophys Acta 1035:276–285
10. Jimbow K, Fitzpatrick TB, Quevedo WC (1986) Formation, chemical composition and function of melanin pigment. In: Bereiter-Hahn J, Matolby AG, Richards KG (eds) Biology of the integument, vol. 2. Springer, Berlin Heidelberg New York, pp 278–291
11. Körner A, Pawelek JM (1980) Dopachrome conversion: a possible control point in melanin biosynthesis. J Invest Dermatol 75:192–195
12. Lambert C, Chacon JN, Chedekel MR, Land EJ, Riley PA, Thomson A, Truscott TG (1989) A pulse radiolysis investigation of the oxidation of indolic melanin precursors: evidence for indolequinones and subsequent intermediates. Biochim Biophys Acta 993:12–20
13. Lerch K (1981) Cooper monooxygenases: tyrosinase and dopamine β-monooxygenase. In: Sigel H (ed) Copper proteins. Dekker, Basel, pp 143–186 (Metal ions in biological systems, vol 13)
14. Lerner AB, Fitzpatrick TB (1950) Biochemistry of melanin formation. Physiol Rev 30:91–126
15. Mason HS (1948) The chemistry of melanin. III. Mechanism of the oxidation of dihydrophenylalanine by tyrosinase. J Biol Chem 172:83–90
16. Mojamdar M, Ichihashi M, Mishima Y (1982) Effect of dopa loading on glutathione-dependent 5-S-cysteinyldopa genesis in melanoma cells in vitro. J Dermatol 9:73–77
17. Palumbo A, Ischia M, Misuraca G, Prota J, Schultz TM (1988) Structural modifications in biosynthetic melanins induced by metal ions. Biochim Biophys Acta 964:193–199
18. Pawelek J, Körner A, Bergstrom A, Bologna J (1980) New regulators of melanin biosynthesis and the autodestruction of melanoma cells. Nature 286:617–619
19. Prota G (1980) Recent advances in the chemistry of melanogenesis in mammals. J Invest Dermatol 75:122–127
20. Prota G (1988) Progress in the chemistry of melanins and related metabolites. Med Res Rev 8:525–556
21. Raper HS (1928) Aerobic oxidases Physiol Rev 8:245–266
22. Silvers WK (1979) The coat colors of mice. A model for mammalian gene action and interaction. Springer, Berlin Heidelberg New York

Note added to the proofs: This article is part of a lecture delivered in the Frankfurter Gespräch "Zellkulturmodelle in der Dermatologischen Forschung", hold in Bad Nauheim, Germany, on March 8, 1991. Since that date, some new data about regulation of mammalian melanogenesis, which are not included the bibliography, have been published. Readers are addressed to Aroca et al. (1992) "Regulation of the distal phase of mammalian melanogenesis", Eur. J. Biochem., 208, 155–163; Hearing & Tsukamoto (1991) "Enzymatic control of pigmentation in mammals", 5, 2902–2909, FASEB J., and Pawelek (1991) "After dopachrome?", Pigment Cell Res. 4, 53–62 for further recent details on that topic.

Cultured Epidermal Melanocytes from Patients with Neurofibromatosis 1 (Recklinghausen's Disease): a Cell Culture Model for Neurofibromatosis 1

D. Kaufmann, I. Eisenbarth, S. Wiandt, and *J. Veser*

Abstract

As derivatives of the neural crest, epidermal melanocytes belong to the cell populations which are directly affected by gene defects causing neurofibromatosis 1 (NF1): increased café-au-lait macules (CALM), shifted melanin unit, and increased melanin-macroglobules in vivo. In vitro, melanocytes from the normally pigmented skin of NF 1 patients show an increased melanin content, and an additional increase of melanin content is found in melanocytes derived from CALM of these patients. In a similar two-step manner, NF 1 melanocytes contain a higher number of large melanosome complexes and of melanosomes, and tyrosine hydroxylase activity is increased. NF 1 melanocytes incorporate higher amounts of dopa, their growth rate is lower, and most of them do not achieve a normal in vitro life span. Omission from the culture medium of the tumor promotor PMA, a strong inducer of protein kinase C (PKC), brings about a comparable increase in melanin content in all three kinds of melanocyte cultures. Under this condition NF 1 melanocytes alter their morphology faster than control melanocytes. We condlude that melanogenesis and growth in NF 1 melanocytes are effected by defective alleles of the NF 1 gene, and that differences in melanogenesis between NF 1 melanocytes and controls persist also in the absence of PMA-mediated PKC stimulation.

According to the recommandations made by the National Institues of Health (NIH), at the consensus Development Conference in July, 1987, there are seven distinct forms of neurofibromatosis (NF): NF 1–NF 7 [26]. NF 1 has also been referred to as von Recklinghausen's neurofibromatosis, or peripheral NF. NF 1 is a heritable disorder with an autosomal dominant mode of transmission (frequency 1:4000), high penetrance, and variable expressivity. The NF 1 gene has been localized at chromosome 17q11.2 [22], and its size, structure, and expression have recently become amenable to detailed molecular analysis [4, 5, 27–30]. NF 2 (frequency 1:40000) has been(frequency 1:40000) has been called central NF, or called central NF, or bilateral acoustic NF. The gene for NF 2 is located on the long arm of chromosome 22 [24]. The other forms of NF are very rare (NF 3, a mixed form with symptoms from NF 1 and NF 2; NF 5, also called segmental neurofibromatosis).

A. Bernd, J. Bereiter-Hahn,
F. Hevert, H. Holzmann (Eds.)
Cell and Tissue Culture Models
in Dermatological Research

The diagnosis of NF 1 is based on clinical criteria [26]:
- Six ore more café-au-lait macules
- Two or more neurofibromas of any type or one plexiform neurofibroma
- Freckling in the axillary or inguinal regions
- Two or more Lisch nodules (iris hamartomas)
- Optic glioma
- A distinctive osseous lesion such as sphenoid dysplasia or thinning of long bone cortex, with or without pseudoarthrosis
- A first-degree relative with NF 1

Derivatives of the neural crest, like Schwann cells (neurofibromas) and melanocytes (hyperpigmentations and iris hamartomas), are usualy considered the primary targets of the gene defects of NF 1 [23].

The first clinical symptom of NF 1 derived from altered epidermal melanocytes in most cases is the appearance of café-au-lait macules (CALM) [13, 23]. These discrete macules are well circumscribed, 2–20 cm, uniformly pale brown and appear at or soon after birth. CALM are not pathognomonic for NF 1. It is not uncommon to see up to six CALM in people without NF 1 and in people with other diseases, but the presence of six or more such macules makes the diagnosis of NF 1 mandatory until proven otherwise. Café-au-lait macules are variable in size, configuration, and number. There is no relationship between their location and the occurrence of other NF 1-related lesions or between their location and the severity of NF 1. They tend to disappear with age and are not known to undergo malignant change.

Freckles – another clinical symptom of NF 1 deriving from epidermal melanocytes – are small pigmented macules 2–3 mm in diameter. With NF 1 they are found in areas of the skin not exposed to sun: axillae and other intertriginous areas. They occur in children not before the age of 7 years and, in contrast to café-au-lait macules, their formation is inducible by pressure and moisture.

Lisch nodules or iris hamartomas also derive from altered melanocytes. They are almost never present at birth. After the age of 6 years, about 95% of individuals with NF 1 develop these small tan bumps on the surface of the iris. The number of these nodules also increases with age.

A common feature of melanocytes from NF 1 café-au-lait macules is the occurrence of melanin macroglobuli [15]. These appear under the light microscope as dark-brown granules about ten times the size of normal melanosomes. These pleomorphic intracellular particles are autolysosomes formed from the fusion of autophagosomes that contain variable numbers of melanosomes with primary and/or secondary lysosomes [21]. They can occur in a variety of disease conditions, but their number is increased in melanocytes from NF 1 patients: more of these structures were found in CALM of NF 1 patients compared with CALM melanocytes in controls and NF 2 patients, but in normally pigmented skin of NF 1 patients their number is increased as well [20].

In contrast to patients without NF 1, the café-au-lait macules of NF 1 patients contain more dopa-positive melanocytes per area than the surrounding skin. But in the normally pigmented skin of NF 1 patients (NF S) is the number of melanocytes is also increased compared with controls: the composition of the epidermal melanin unit (melanocyte: keratinocytes) is shifted in favor of the melanocytes already in the normally pigmented skin of NF 1 patients, and even more so in the dark-brown NF 1 CALM [7, 14].

Melanocytes are also directly involved in NF 1 gene defects:

- Increased number of café-au-lait macules, iris hamartomas, and axillary freckling
- Increased number of melanin macroglobules in epidermal melanocytes: control < NF S < NF CALM
- Shifted melanin unit: control < NF S < NF CALM

We are interested in the following questions:

- Why are neural crest-derived cells the primary targets of the underlying defects of NF 1?
- What is the reason for the variable expressivity of NF 1, and how can theis expressivity be influenced?

To make these questions amenable to experimental investigations we cultured melanocytes derived from the normally pigmented epidermis and from dark-brown café-au-lait macules of patients with NF 1 and compared these cells with melanocytes derived from controls. Cultures of pure melanocytes were prepared from epidermal cells as described by Eisinger and Marko [6] and Halaban et al. [12]. Briefly, the culture medium (TIC) contained Ham F10, 8% fetal bovine serum, 8% NU serum, 85 n*M* PMA, 0.1 n*M* IMBX, and 2.5 n*M* cholera toxin. Contaminating fibroblasts were eliminated by geneticin treatment [11].

Cultivation of melanocytes succeeded in all samples from NF 1 patients, and some of these melanocytes survived in culture for about 60 weeks, but their growth rate is lower than the growth rate of controls, and most of them did not achieve the life span of cells from healthy donors of the same age [18].

In all tested melanocyte populations sharply contasted brown particles 3–8 m in diameter were found. These were easy to count by bright-field microscopy [16], and they were lying mostly at the periphery of the cells. When examanined by electron microscopy these particles turned out to be large melanosome complexes (LMC), aggregates of melanosomes at various stages of melanization, putative precursors of the melanin macroglobules (Fig. 1). Electron-microscopic studies revealed the presence of tyrosinase and acid phosphatase activity in these structures [17]. We scored the cultured melanocytes for LMC: all tested melanocytes from the normally pigmented skin of NF 1 patients ($n = 9$) contained a higher number of LMC compared with controls, and an additional increase in LMC was found in five of six examined melanocytes derived from café-au-lait macules of NF 1

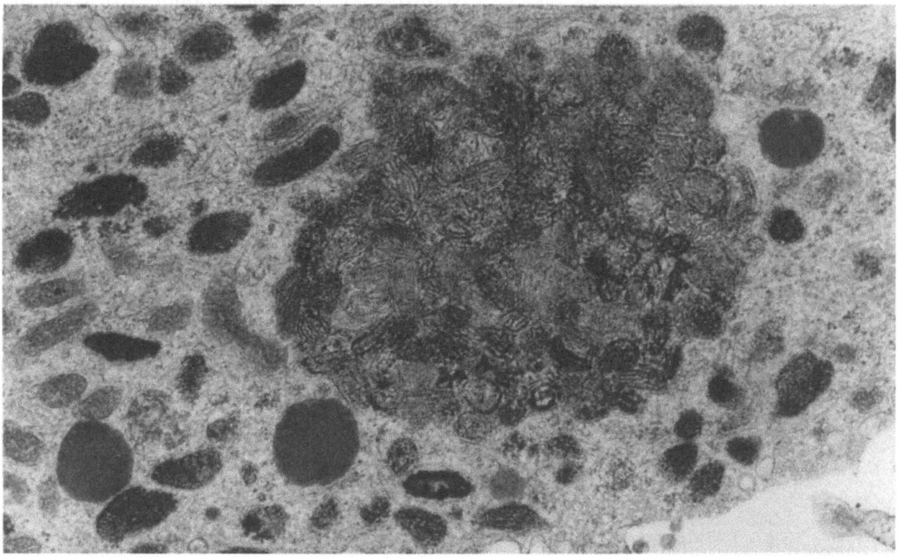

Fig. 1. Electron-microscopic picture of a large melanosome complex in a melanocyte cultured from café-au-lait macule of an NF 1 patient. The melanosomes within this complex are at various stages of melanization

patients. The differences in LMC counts between the three melanocyte cultures were maintained with increasing number of culture passages. Since the development of melanosomes and melanogenesis are coordinately regulated, it was also of interest to investigate biochemical parameters of melanogenesis.

First, we measured the melanin content of the three kinds of melanocyte cultures to be compared. For these determinations a modification of the method described by Friedmann was used [8, 18]. In all experiments the lowest melanin content was found in the cells from healthy donors. In seven of nine experiments a higher melanin content was found in NF 1 melanocytes derived from café-au-lait macules than in melanocytes derived from the normally pigmented skin of NF 1 patients (Table 1). The highest melanin contents were found in the cells from the most severely affected patients (NF 17 and NF 29).

Second, the tyrosine hydroxylase activity of tyrosinase determined in extracts prepared from the three kinds of melanocytes as described [1] was increased in a similar two-step manner (control < NF S < NF CALM).

Third, we determined the rate of melanin synthesis by measuring the incorporation of ^{14}-C-dopa into the TCA-insoluble fraction of the melanocytes [8]. Under our experimental conditions the rate of dopa incorporation was always higher in the NF 1 melanocytes than in those of the healthy donors, but there was no statistically significant difference between the two kinds of NF 1 melanocytes (NF S and NF CALM) in this respect.

Table 1. Melanin content of melanocytes from the skin of healthy donors (S), from the normally pigmented skin of NF 1 patients (NF S), and from café-au-lait macules of these patients (NF CALM)

Healthy donors				NF 1 patients			
				Melanin content (ps/cell)			
Experiment no.	In vitro passages	Donor no.	Melanin content (pg/cell)	Normally pigmented skin		Café-au-lait macules	
1	2	S 7	5.5	NF 17 S	145		
2	2	S 3	3.8	NF 31 S	40.4		
3	7	S 3	4.5	NF 31 S	13.1		
4	11	S 3	2.7	NF 31 S	28.7		
5	2	S 3	3.8	NF 71 S	44.5		
6	5	S 4	7.3	NF 94 S	24.3		
7	8	S 4	3.8	NF 94 S	36		
8	6	S 1	8.2	NF 25 S	47	NF 25 CALM	122.5
9	8	S 6	9.0	NF 25 S	47	NF 25 CALM	65
10	6	S 6	11	NF 29 S	80	NF 29 CALM	135
11	4	S 3	12	NF 31 S	43	NF 31 CALM	52
12	11	S 3	3.7	NF 31 S	20.5	NF 31 CALM	30
13	6	S 5	8.2	NF 97 S	20.9	NF 97 CALM	14.5
14	8	S 5	2.7	NF 97 S	16	NF 97 CALM	10.8
15	3	S 7	6.4	NF 92 S	32	NF 92 CALM	52
				NF 106 S	21.5	NF 106 CALM	39.2

In vivo, the expression of the pigmentary anomalies typical of NF 1 may depend on the interaction between melanocytes and keratinocytes. The fact that the increased melanogenesis exists in pure melanocyte cultures clearly excludes a direct involvement of keratinocytes; the differences that occur must be autonomously expressed properties of these cells.

Protein kinase C (PKC) has recently been identified as one of the regulators of melanogenesis [9]. This enzyme is stimulated strongly by the tumor promotor PMA. Omission of PMA from the culture medium induced a number of drastic changes in the behavior of the melanocytes: they assumed a flatter, more extended shape, developed an increased number of dendrites, stopped growing, and, after 6 weeks, detached from the substratum and died. Under this condition NF 1 melanocytes altered their morphology faster than control melanocytes. Addition of PMA reversed these changes within 2 days. Upon removal of PMA, all three kinds of melanocytes showed a comparable increase of their melanin content. Since the differences between the melanin contents of NF 1 melanocytes and control melanocytes persisted in the absence of PMA-mediated PKC stimulation, it can be concluded that the differences are independent of the stimulation of melanogenesis by PKC.

The NF 1 gene, localized on chromosome 17q11.2 [22], seems to be almost ubiquitously expressed [27, 28]. However, the gene defects involve predominantly the derivatives of the neural crest. In the case of melanocytes we do not know yet how the dysregulated cell subpopulations arise, leading to café-au-lait macules, Lisch nodules, and freckles. The NF 1 gene shares similarities with the mammalian RAS-GTPase activating protein (GAP) [29]. The NF 1 gene product has been shown recently to exert the function of a GAP, which causes the inactivation of wild-type RAS by stimulating RAS-GTPase activity [2, 19, 30]. In some neural crest cells, RAS stimulates differentiation and blocks proliferation (3, 10, 25). If melanocytes belong to the types of cells in which differentiation rather than proliferation is stimulated by a RAS oncogene, its defective inactivation in NF 1 may bring about the effects observed in this study: increased melanogenesis and reduced growth in NF 1 melanocytes.

Molecular biological investigations into NF 1 gene defects will help to solve these problems in the near future. Cultures of melanocytes may provide a valuable system for this kind of study.

References

1. Auböck J, Köfler D, Sifter M, Fritsch P (1983) Application of the tyrosinase assay to normal melanocytes in culture. Br J Dermatol 109:413–419
2. Ballester R, Marchuk D, Boguski M, Saulin A, Letcher R, Wigler M, Collins F (1990) The NF 1 locus encodes a protein functionally related to mammalian GAP and yeast IRA proteins. Cell 63:851–859
3. Bar-Sagi D, Feramisco JR (1985) Microinjection of the *ras* oncogene protein into PC12 cells induces morphological differentiation. Cell 42:841–848
4. Buchberg AM, Cleveland LS, Jenkins NA, Copeland NG (1990) Sequence homology shared by neurofibromatosis type-1 gene and IRA-1 and IRA-2 negative regulators of the RAS cyclic AMP pathway. Nature 347:291–294
5. Cawthon RM, Weiss R, Xu G, Viskochil D, Culver M, Stevens J, Robertson M, Dunn D, Gesteland R, O'Connell P, White R (1990) A major segment of the neurofibromatosis type-1 gene: cDNA sequence, genomic structure, and point mutations. Cell 62:193–201
6. Eisinger M, Marko O (1982) Selective proliferation of normal human melanocytes in vitro in the presence of phorbol ester and cholera toxin. Proc Natl Acad Sci USA 79:2018–2022
7. Frenk E, Marazzi A (1984) Neurofibromatosis of von Recklinghausen: a quantitative study of the epidermal keratinocyte and melanocyte populations. J Invest Dermatol 83:23–25
8. Friedmann PS, Gilchrest BA (1987) Ultraviolet radiation induces pigment production by cultured human melanocytes. J Cell Physiol 133:88–94
9. Gordon PR, Gilchrest BA (1989) Human melanogenesis is stimulated by diacylglycerol. J Invest Dermatol 93:700–702
10. Guerro I, Wong H, Pellicer A, Burstein DE (1986) Aktivated N-ras gene induces neuronal differentiation of PC12 rat pheochromocytoma cells. J Cell Physiol 129:71–76
11. Halaban R, Alfano FD (1984) Selective elimination of fibroblasts from cultures of normal human melanocytes. In Vitro 20:447–450
12. Halaban R, Ghosh S, Duray P, Kirkwood JM, Lerner AB (1986) Human melanocytes cultured from nevi and melanomas. J Invest Dermatol 87:95–101
13. Huson SM, Harper PS, Compston DAS (1988) Von Recklinghausen neurofibromatosis. Brain 11:1355–1381

14. Ishida O (1987) A computed image-analyzing system for quantitation of melanocyte morphology in café-au-lait macules of neurofibromatosis. J Invest Dermatol 88:287–291
15. Jimbow K, Horikowski T (1982) The nature and significance of macromelanosomes in pigmented skin lesions. Am J Dermatopathol 4:413–420
16. Kaufmann D, Krone W, Hochsattel R, Martin R (1989) A cell culture study on melanocytes from patients with neurofibromatosis 1. Arch Dermatol Res 281:510–513
17. Kaufmann D, Wiandt S, Krone W (1990) Increased melanin content in cultured melanocytes from patients with neurofibromatosis 1 (M. Recklinghausen, NF 1). Eur J Cell Biol 51:77
18. Kaufmann D, Wiandt S, Veser J, Krone W (1991) Increased melanogenesis in cultured melanocytes from patients with neurofibromatosis 1 (NF 1). Hum Genet 87 (2):144–150
19. Martin GA, Viskochil D, Bollag G, McCabe PC, Crosier WJ, Haubruck H, Conroy L, Clark R, O'Connell P, Cawthon RM, Innis MA, McCormick F (1990) The GAP-related domain of the neurofibromatosis type-1 gene product interacts with ras p 21. Cell 63:843–849
20. Martuza RL, Philippe I, Fitzpatrick TB, Zwaan J, Seki Y, Lederman J (1985) Melanin macroglobules as a marker of neurofibromatosis: a quantitative study. In Invest Dermatol 85:347–350
21. Nakagawa H, Hori Y, Sato S, Fitzpatrick TB, Martuza RL (1984) The nature and origin of the melanin macroglobule. J Invest Dermatol 83:134–139
22. O'Connell P, Leach R, Cawthon RM, Culver M, Stevens J, Viskochil D, Fournier REK, Rich DC, Ledbetter DH, White R (1989) Two NF 1 translocations map within a 600-kilobase segment of 17q11.2. Sience 244:1087–1088
23. Riccardi VM, Eichner JE (1986) Neurofibromatosis: phenotype, natural history, and pathogenesis, Johns Hopkins Press, Baltimore
24. Rouleau GA, Wertelecki W, Haines JL (1987) Genetic linkage of bilateral acoustic neurofibromatosis to a DNA marker on chromosome 22. Nature 329:246–248
25. Ridley AJ, Paterson HF, Noble M, Land H (1988) Ras-mediated cell cyle arrest is altered by nuclear oncogenes to induce Schwann cell transformation. EMBO J 7:1635–1645
26. Stumpf AD, Alksne JF, Annegers JF (1988) Neurofibromatosis. Arch Neurol 45:575–578
27. Viskochil D, Buchberg AM, Xu G, Cawthon RM, Stevens J, Wolff RK, Culver M, Carey JC, Copeland NG, Jenkins NA, White R, O'Connel P (1990) Deletions and a translocation interrupt a cloned gene at the neurofibromatosis type-1 locus. Cell 62:187–192
28. Wallace MR, Marchuk DA, Andersen LB, Letcher R, Odeh HM, Saulino AM, Fountain JW, Brereton A, Nicholson J, Mitchell AL, Brownstein BH, Collins FS (1990) Type-1 neuro-fibromatosis gene: identification of a large transcript disrupted in three NF 1 patients. Science 249:181–186
29. Xu G, O'Connell P, Viskochil D, Cawthon R, Robertson M, Culver M, Dunn D, Stevens J, Gesteland R, White R, Weiss R (1990) The neurofibromatosis type-1 gene encodes a protein related to GAP. Cell 62:599–608
30. Xu G, Lin B, Tanaka K, Dunn D, Wood D, Gesteland R, White R, Weiss R, Tamanoi F (1990) The catalytic domain of the neurofibromatosis type-1 gene product stimulates ras GTPase and complements ira mutatns of *S. cervisiae.* Cell 63:835–841

Pigmentation in Cultures of Human Melanocytes

A. Ramirez-Bosca, A. Bernd, C. Theilig, K. Dold,
and *H. Holzmann*

Abstract

Cultivated human melanocytes were irradiated by 50, 100, and 200 mJ UVA/B per cm^2. Three days later we determined the tyrosinase activity, the melanin content, and the number of cells. Human melanoma cell lines were used as controls. In all cases the number of cells decreased according to the UVA/B dose. In human melanocytes we found a marked increase of tyrosinase activity after irradiation with 200 mJ/cm^2 (from 28 to 43 nmol ^{3}HOH/24 h $\times$ 10^6 cells). The tyrosinase activity resulted in pigment formation induced by UVA/B in human melanocytes. The melanin content was correlated with the tyrosinase activity (increase from 11 to 22 μg/10^6 cells).

Introduction

Epidermal and follicular melanocytes form part of the epidermal melanization system, where complex cellular interactions take place. The pigmentary cells are situated in the basal layer of the epidermis, i.e., at the dermo-epidermal junction, and their dendritic processes keep contact with epidermal cells in the basal and the first suprabasal layer. Recent progress in the culture of epidermal cells and of melanocytes has made it possible to study a number of factors (endocrines, growth factors, cytokines, inflammation mediators, and so on) regulating the function of the pigment cell population [1]. Melanocytes in the human epidermis account for only 2%–4% of the whole epidermal cell population. Furthermore, these types of cells are difficult to cultivate because of their high state of differentiation and a corresponding low mitotic activity. Some substances have to be added to the culture medium to stimulate the proliferation without affecting the fundamental cell physiology. The subsequent promoters are found to work properly: bovine pituitary extract [15], 4-0-methyl-12-0-tetradecanoyl-phorbol-13-acetate (TPA) and cholera toxin [2, 9], basic fibroblast growth factor [5, 6], and isobutylmethyl xanthine (IBMX) [2, 9]. With a combination of these supplements, growth of melanocytes is possible even in serum-free medium [11]. One of the main functions of melanocytes is the production of pigment. Tyrosinase (E.C. 1.14.18) ranks as the key enzyme of pigment metabolism;

A. Bernd, J. Bereiter-Hahn,
F. Hevert, H. Holzmann (Eds.)
Cell and Tissue Culture Models
in Dermatological Research

it is a multifunctional enzyme and catalyzes at least two different reactions in this metabolic pathway: the hydroxylation of tyrosine to 3,4-dihydroxyphenylalanine (dopa) and the oxidation of dopa to dopachinone. The second step is the oxidation of dopachrome resulting in the production of melanin [3] (see Solano, this volume). A tautomerase seems to be involved in this process [8].

The tyrosinase can be influenced via modulators like phorbolesters, stimulators of adenylate cyclase, peptide hormones, and UV radiation. Most of the findings on the metabolism of melanocytes are known from research on animal cells or malignant transformed cells. We studied the effect of UV radiation on cultures of normal human melanocytes.

Methods

Cell Culture

Primary cultures of melanocytes taken from human foreskin of adults were initiated. Skin biopsies were taken under sterile conditions and put into Hank's medium containing 1% penicillin/streptomycin, but no calcium or magnesium. The skin was washed three times with PBS without calcium and magnesium, and then incubated for 2 h in a 0.25% trypsin solution at 37° C to remove the epidermis from the dermis. By pipetting, the epidermal cells were suspended into Hank's medium with 10% fetal calf serum and 1% penicillin/streptomycin. The suspension was centrifuged (10 min, 200 g, room temperature) and the cell sediment was suspended in F10-nutrient medium supplemented with 5% fetal calf serum, 1% penicillin/streptomycin, 85 n*M* TPA, 0.1 m*M* IBMX, 2.5 n*M* cholera toxin, and 100 μg/ml geneticin [4] and incubated at 37° C in a 10% CO_2 atmosphere. The medium was renewed three times a week. From the first renewal on, geneticin-free medium was used. After 2 weeks the confluent cultures were passaged for the first time using Hank's without calcium and magnesium with 0.025% trypsin and 0.01% EDTA. All our observations were done with cells of the third passage. In order to control the cell proliferation the number of cells was determined in a Burker chamber. Only living cells were counted; the viability was checked by trypan blue exclusion.

Human melanoma cells (CRL 1424) supplied from Flow Laboratories were cultivated in RPMI 1640 medium with 10% fetal calf serum and 1% penicillin/streptomycin at 37° C in a 5% CO_2 atmosphere. The medium was changed twice a week. Amelanotic human melanoma cells of a primary tumor (which we thankfully received from Dr. Tilgen, Universitäts-Hautklinik Heidelberg, FRG) were kept under the same conditions as CRL 1424 cells [14].

Determination of Tyrosinase Activity

To determine the tyrosinase activity 37 kBq/ml ^{3}H-L-tyrosine (spec. act. 1.85 GBq/mmol) and 50 μ*M* L-dopa were added to the culture medium [12]. After incubation at 37° C, 200 μl of the medium was used to measure the amount of tritiated water as a product of the enzymatic reaction. The tritiated L-tyrosine that had not reacted was removed by adding 200 μl charcoal slurry (20% w/v in 0.1 mol/l HCl). The samples were shaken for 15 min at room temperature and centrifuged for 5 min (8000 g, room temperature), and then the radioactivity was measured in 200 μl supernatant of each sample. The spontaneous exchange of tritium was taken into consideration by subtracting the values of fibroblast cultures tested in parallel.

Determination of Melanin

The cells were pelleted (10 min, 200 g, room temperature) and boiled in 1 ml 0.85 *M* KOH at 95° C for 3–4 h. After centrifugation (5 min,11 000 g, room temperature) the extinction of the supernatant fluid was determined at a wavelength of 400 nm.The values were compared with a standard curve which was made with synthetic melanin.

UV Irradiation

The cells were irradiated with 50, 100, or 200 mJ/cm^2 using a Psorilux 3050 A lamp (Original Hanau, FRG) with a broad range of UV light (UVA: 340–390 nm; UBV. 290–310 nm). These lamps do not emit biologically significant amounts of UVC. Before irradiation the culture medium was replaced with PBS (with Ca and Mg) to avoid formation of medium-derived toxic photoproducts. Immediately after irradiation the PBS was replaced with complete medium.

Results

During the first week of primary culture with TPA the melanocytes were dendritic, as in vivo; keratinocytes were found only rarely. After passaging, the melanocytes became bipolar with a small seam of cytoplasm (Fig. 1A). Three hours after passaging, the melanocytes were already attached to the substrate (plastic flasks, Nunc, Wiesbaden, FRG). The morphology of both melanoma cell lines (CRL 1424 and MMLI) was spindle shaped (Fig. 2A, D).

We investigated the effect of UVA/B on the proliferation rate of human melanocytes. The cells were irradiated in culture flasks. After 3 days the number of cells was counted. In all cases there were fewer cells in irradiated

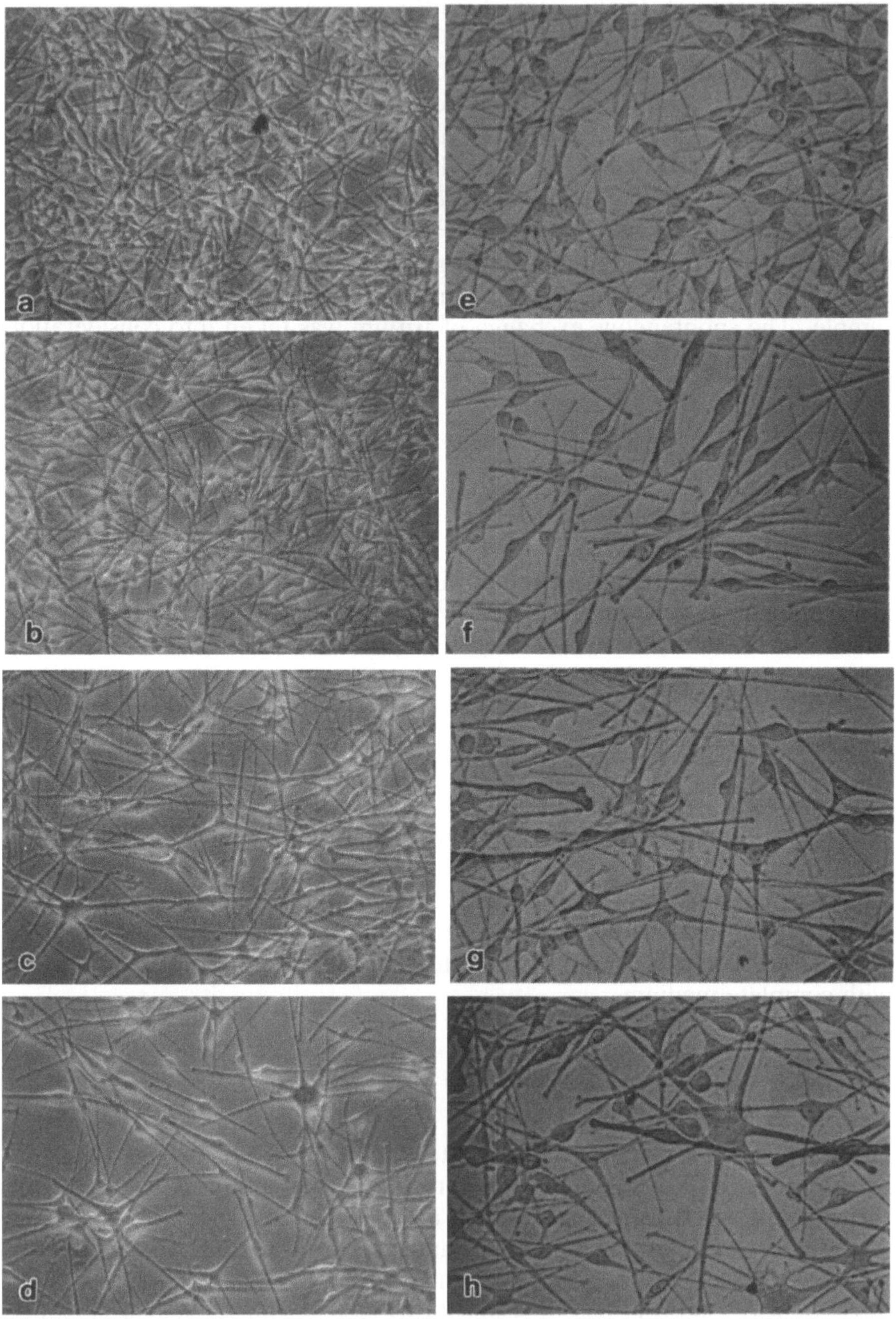

Fig. 1a–h. Effect of UVB on cultures of human melanocytes: **a–d** effect on cell proliferation (light microscopy with phase contrast); **e–h** effect on pigment formation (light microscopy without phase contrast). **a + e**, control; **b + f** 50 mJ/cm^2; **c + g** 100 mJ/cm^2; **d + h** 200 mJ/cm^2

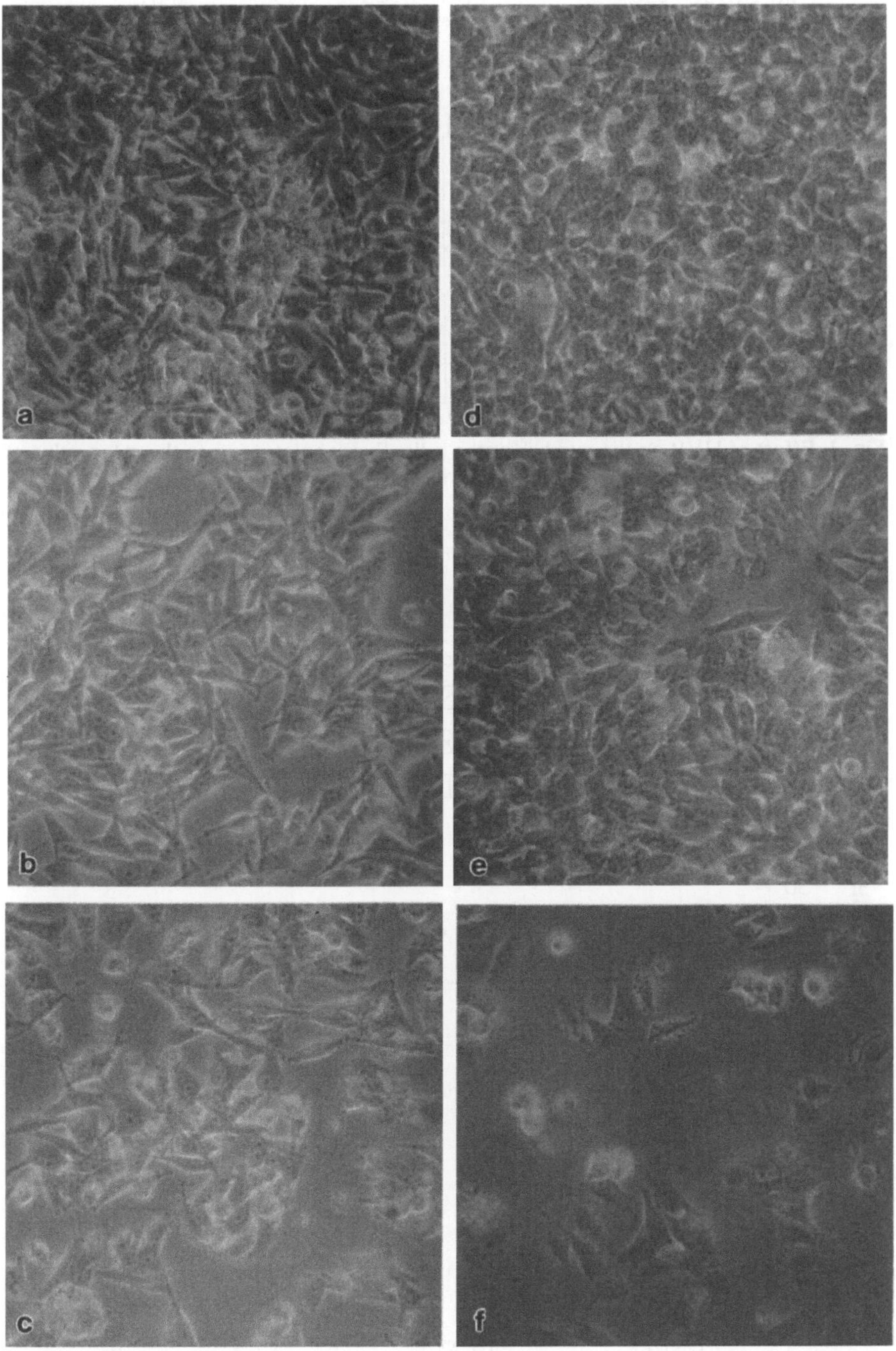

Fig. 2a–f. Effect of UVB on cultures of melanoma cell lines: **a–c** CRL 1424; **d–f** MMLI. **a + d**, control; **b + e**, 50 mJ/cm^2; **c + f**, 100 mJ/cm^2

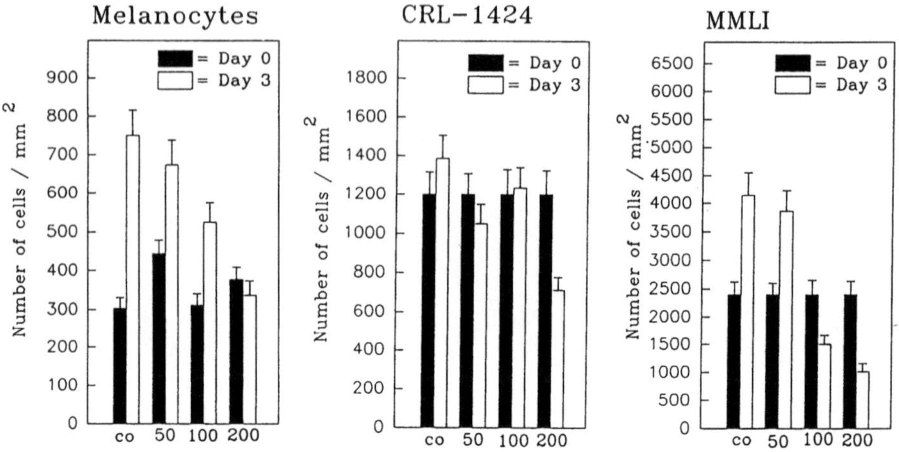

Fig. 3. Effect of UVB on the number of the cells. Every point refers to the mean value of three parallel approaches. The standard deviations are indicated

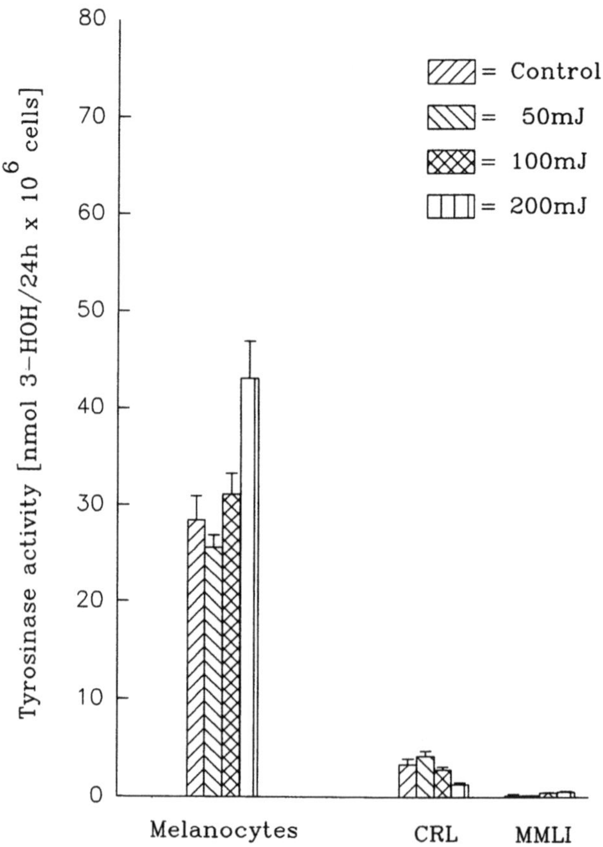

Fig. 4. Effect of UVB on tyrosinase activity of human melanocytes and the human melanoma cell lines CRL 1424 and MMLI. Every point refers to the mean value of three parallel approaches. The standard deviations are indicated

cultures than in nonirradiated controls (Fig. 1 A–D, Fig. 3). The irradiated cells (100–200 mJ/cm^2) changed their morphology to a dendritic shape (Fig. 1 C, D). Also in cultures of irratiated melanoma cells (CRL 1424 and MMLI) the cell number decreased dramatically (Fig. 2 A–F; Fig. 3).

In the same cultures we tested the tyrosinase activity and the melanin content before and after irradiation. The tyrosinase activity of human melanocytes reached maximal values 3 days after UV irradiation with 200 mJ/cm^2. The tyrosinase activity of melanoma cells was not changed after irradiation (Fig. 4). To test whether tyrosinase activity correlated with the melanin content of the cells, melanization was followed using light microscopy (Fig. 1 E–H) and the melanin concentration was determined photometrically (Fig. 5). Synthetic melanin was used as a reference. The extinction at a wavelength of 400 nm correlated with the melanin concentration (regression coefficient r^2 = 0.99888).

Three days after irradiation with 200 mJ/cm^2, the melanin content of human melanocytes was more than twice that of controls (Fig. 5). At lower doses of irradiation no significant changes in melanin content were

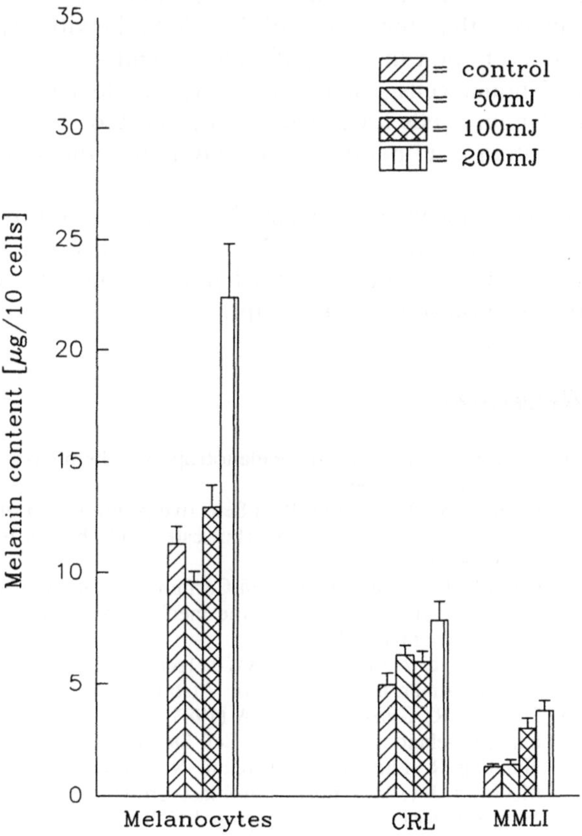

Fig. 5. Effect of UVB on melanin content of human melanocytes and the human melanoma cell lines CRL 1424 and MMLI. Every point refers to the mean value of three parallel approaches. The standard deviations are indicated

observed. Also in the melanoma lines the amount of melanin was not influenced by irradiation (Fig. 5).

Discussion

Melanocytes of malignant melanoma metastases rapidly form colonies, take on elongated and epithelioid shapes, quickly form continuous cell lines, and have a low tyrosinase activity [7]. The melanoma cell lines used in our study also showed very low tyrosinase activities compared with normal melanocytes.

In the primary culture most human melanocytes have a dendritic shape. Under the influence of TPA the morphology changes and they become bipolar. After being exposed to UV the cells become more dendritic again, show more tyrosinase activity, and produce more melanin [10].

Only with the relatively high dose of 200 mJ/cm^2 did we observe a stimulation of the melanin synthesis in vitro. Melanin synthesis is a specific property of differentiated melanocytes [13]. The differentiated state is also characterized by the lack of mitotic activity. UV irradiation induced both an increase of melanin synthesis and a decrease of proliferation. Therefore, we suppose that the effect of UV light in vitro resembles the induction of differentiation in vivo [10]. This would be in accord with the findings that melanoma cells remain relatively unaffected by this treatment. The two melanoma cells lines investigated reacted to UV only with a reduced proliferation rate; parameters of differentiation were not changed.

Acknowledgements. This study was supported by the Alexander von Humboldt Foundation and by the *Landesversicherungsanstalt Hessen.* We thank Prof. Dr. Bereiter-Hahn for helpful discussions and Mrs. E. Kleber for the preparation of the manuscript.

References

1. Eberle AN (1988) The melanotropins. Chemistry, physiology and mechanism of action. Karger, Basel
2. Eisinger M, Marko O (1982) Selective proliferation of normal human melanocytes in vitro in the presence of phorbol ester and cholera toxin. Proc Natl Acad Sci USA 79:2018–2022
3. Galindo J, Pedreno E, Garcia-Carmona F, Garcia Cánovas F, Solano F, Lozano JA (1983) Steady-state study of the mechanisms of dopa oxidase activity of tyrosinase. Int J Biochem 15:1455–1461
4. Halaban R, Alfano FD (1984) Selective elimination of fibroblasts from cultures of normal human melanocytes. In Vitro Cell Dev Biol 20:447–450
5. Halaban R, Ghosh S, Barid A (1983) bFGF is the putative natural growth factor for human melanocytes. In Vitro Cell Dev Biol 23:47–52
6. Halaban R, Langdon R, Birchall N, Cuono C, Baisd A, Scott G, Moellmann G, McGuire J (1988) Basic fibroblast growth factor from human keratinocytes is a natural mitogen for melanocytes. J Cell Biol 107:1611–1619

7. Herlyn M, Herlyn D, Elder DE, Bondi E, LaRossa D, Hamilton R, Sears HF, Balaban G, Clark WH, Kaprowski H (1983) Phenotypic characteristics of cells derived from precursors of human melanoma. Cancer Res 43:5502–5507
8. Jara JR, Solano F, Lozano JA (1988) Assays for mammalian tyrosinase: a comparative study. Pigment Cell Res 1:332–339
9. Lerner AB, Halaban R, Sidney NK, Moellmann G (1987) Transplantation of human melanocytes. J Invest Dermatol 89:219–224
10. Ortonne JP (1990) The effects of ultraviolet exposure on skin melanin pigmentation. J Int Med Res 18:8c–17c
11. Pittelkow MR, Shipley GD (1989) Serum-free culture of normal human melanocytes: growth kinetics and growth factor requirements. J Cell Physiol 140:565–576
12. Pomerantz SH (1964) Tyrosine hydroxlyation catalyzed by mammalian tyrosinase: an improved method of assay. Biochem Biophys Res Commun 16:188–192
13. Slominski A, Moellmann G, Kuklinska E, Bomirski A, Pawelek J (1988) Positive regulation of melanin pigmentation by two key substrates of the melanogenic pathway, L-tyrosine and L-dopa. J Cell Sci 89:287–296
14. Tilgen W, Dzarlieva RT, Breitkreutz D, Hennes B, Engstner M, Matzku S, Fusenig NE (1985) Heterogeneity of human malignant melanoma cells in vivo and in vitro: role of experimental systems. In: Bagnara J, Klaus SN, Paul E, Schartl M (eds) Pigment cell. University of Tokyo Press, Tokyo, pp 435–447
15. Wilkins L, Gilchrest BA, Szabo G, Weinstein R, Maciag T (1985) The stimulation of normal human melanocyte proliferation in vitro by melanocyte growth factor from bovine brain. J Cell Physiol 122:350–361

4. Models in Wound Healing

Re-epithelialization of Incision and Burn Wounds in Culture

J. Bereiter-Hahn, and *M. Vöth*

Abstract

Sheets of epidermal cells emigrated from tadpoles of *Xenopus laevis* retain their in situ relationship of basal and surface cells. These cellular sheets have been used to compare the process of primary closure of wounds made either by incision with a scalpel or by UV irradiation. In the latter case, no response of the cells surrounding the killed epidermis is observed. After mechanical injury, retraction of the wound margins followed by extension of lamellae and subsequent migration of the marginal cells immediately initiates wound closure. We conclude that mechanical interactions of the intact cells with those being killed by irradiation prevents the sequence of reactions taking place in incision wounds.

Introduction

The primary roles of the integument are to maintain the internal homeostasis of the body and to prevent the entry of pathogens from the outside. Thus the integument forms a mechanical and chemical barrier which still allows controlled exchange of matter and information between an organism and its environment. On injury, the passive defense must be restored by the reorganization of the basic structure, and the active defense mechanisms are stimulated. Formation of a blood coagulation clot, re-epithelialization, and inflammatory responses resulting in the development of granulation tissue are the central events which prevent a vertebrate from being invaded by microbial organisms. Re-epithelialization is the process which most easily can be modelled in cell culture systems [2, 5, 6, 8]. In a few cases in situ investigations of the processes involved in epithelial coverage of small wounds have also been possible, including the epidermis of amphibian tadpoles [16] and cornea of mammalian eyes [4, 13, 18]. A flat surface supporting the migrating epithelial sheets is common to the systems used for the in situ investigation of epithelial sheet migration as well as to the in vitro (cell culture) models. These systems allowed the observation of shape changes of cells, of their cytoskeletal patterns, and of the influence of a variety of extracellular matrix components on migration and wound repair.

A. Bernd, J. Bereiter-Hahn,
F. Hevert, H. Holzmann (Eds.)
Cell and Tissue Culture Models
in Dermatological Research

One of the most fascinating questions concerning the re-formation of epithelial continuity is how the tissues "realize" that a wound is present and when the wound is closed. Epithelial continuity results from spreading of epidermal cells from the wound margins toward the center of the wound and – later – from proliferation of these epidermal cells. Small wounds may primarily be covered by migration of the adjacent cells; larger defects require additional proliferative activity. In any case, proliferation is required to restore the original organization of the epidermal layers. Thus, wound healing is closely related to the control of tissue homeostasis, and this emphasizes the general significance of the question raised above on sensing the disruption of epithelia. Loss of mechanical restriction is one important factor which has been proposed to provide the crucial stimulus for wound closure by epithelial sheet migration [2]. In addition, the mechanical stimulus exerted on the leading fronts of migrating cells may be the starting point of a cascade resulting in cessation of cellular motility [12].

The experiments reported in this contribution were performed to determine whether a free surface caused by wounding provides the stimulus for starting migration. For this purpose, small groups of cells in epidermal cell cultures were killed by UV irradiation,and the behavior of the surrounding cells was compared with the behavior of cells lining incision wounds.

Material and Methods

Cell Culture

Tails of tadpoles (stage 49–52 according to [15] of the clawed toad Xenopus laevis Daudin were excised and cultured in Rose chambers, according to the procedure described earlier [1]. Amphibian culture medium devised by Wolfe and Quimby [19] was used for culturing and for the experiments. Bilayered epidermal sheets grow out from these tadpoles (for details see [17]. Strohmeier and Bereiter-Hahn this volume).

Sheets which exhibited a width of at least 20 cell diameters were used for the wounding experiments. Incision wounds were made by scratching with a hypodermic needle under microscopic control. Burn wounds were produced by focusing the beam of a high-pressure mercury arc (HB0 50) via a 100× objective lens onto the cell culture. The field of illumination was limited by a field aperture of 2–6 cell diameters. Irradiation was performed with the UV and the whole visible spectrum of the mercury arc lamp for 60–120 s. The technical setup consisted of a Diavert stand (E. Leitz, Wetzlar, FRG) equipped with the reflection contrast system according to Ploem, which was also used to follow the cellular reactions on wounding. However, for reflection-interference contrast (RIC) microscopy only the 579-nm line was used, which was separated by introducing an edge filter K 570 (Schott, Mainz, FRG).

Healing was recorded using either a Panasonic time-lapse videorecorder or 35-mm film camera.

Staining of Cell Cultures

Cell cultures were stained with TRITC-phalloidin [7] after formalin fixation (10%, 20 min, rom temperature) to visualize their actin pattern. In some cases, exposure of the unfixed cultures to 0.1 trypan blue was used to assess the death of the irradiated cells.

Results

Healing of Incision Wounds

Immediately after a cut through the bilayered sheet of epidermal cells, the wound margins retract as a consequence of elastic recoil. The resulting smooth margin is delineated by actin fibers which persist during the following stages of wound closure, forming a continuous band from one marginal cell to the other via intercellular contact zones. This actin fibrillar zone separates the cell body from the thin and large lamellae which extend with in a few minutes towards the free edges of the marginal cells (Fig. 1). Then these marginal cells of the basal layer start migrating in the direction of the lamella, whereby they elongate and loosen the contacts with their neighbors (Fig. 1, 16 min) which may either be lost totally or restricted to some connecting cytoplasmic strands anchored at desmosomes (Fig. 1, 16 min, phase contrast). In small wounds the lamellae developed at opposite margins meet each other very soon (Fig. 1, 16 min). At the points of contact the lamellae become immobilized, further locomotion ceases, and the lamellae are withdrawn by extension of the cell body. This primary wound closure is finished with the differentiation of new contact zones between the opposing cells (Fig. 1, 51 min). Details of the rearrangements of cytoskeletal elements have been described previously [2]. In larger wounds, cells apart from the wound margin may be involved in active migration and form lamellae after loosing contact with the emigrating marginal cells.

Wounds Made by Irradiation

UV (lambda > 320 nm) irradiation of a field of epidermis emigrated on a slide via an objective lens (100×) kills the exposed cells in approxinately 1 min. Injury can be followed in phase contrast by the appearance of granular cytoplasm and pyknotic nuclei in the field of irradiation (Fig. 2). Also in this case, the cells surrounding the irradiated area retract slightly, forming a smooth margin delineating the wound borders (Figs. 2, 3). The necrotic area attracts some neutrophil granulocytes present in the culture. In RIC microscopy almost no reaction is visible. In particular, no lamellae protrude from the healthy cells at the wound margins. Only in some rounded cells may accumulation of F-actin be observed; in all other cells the poorly developed

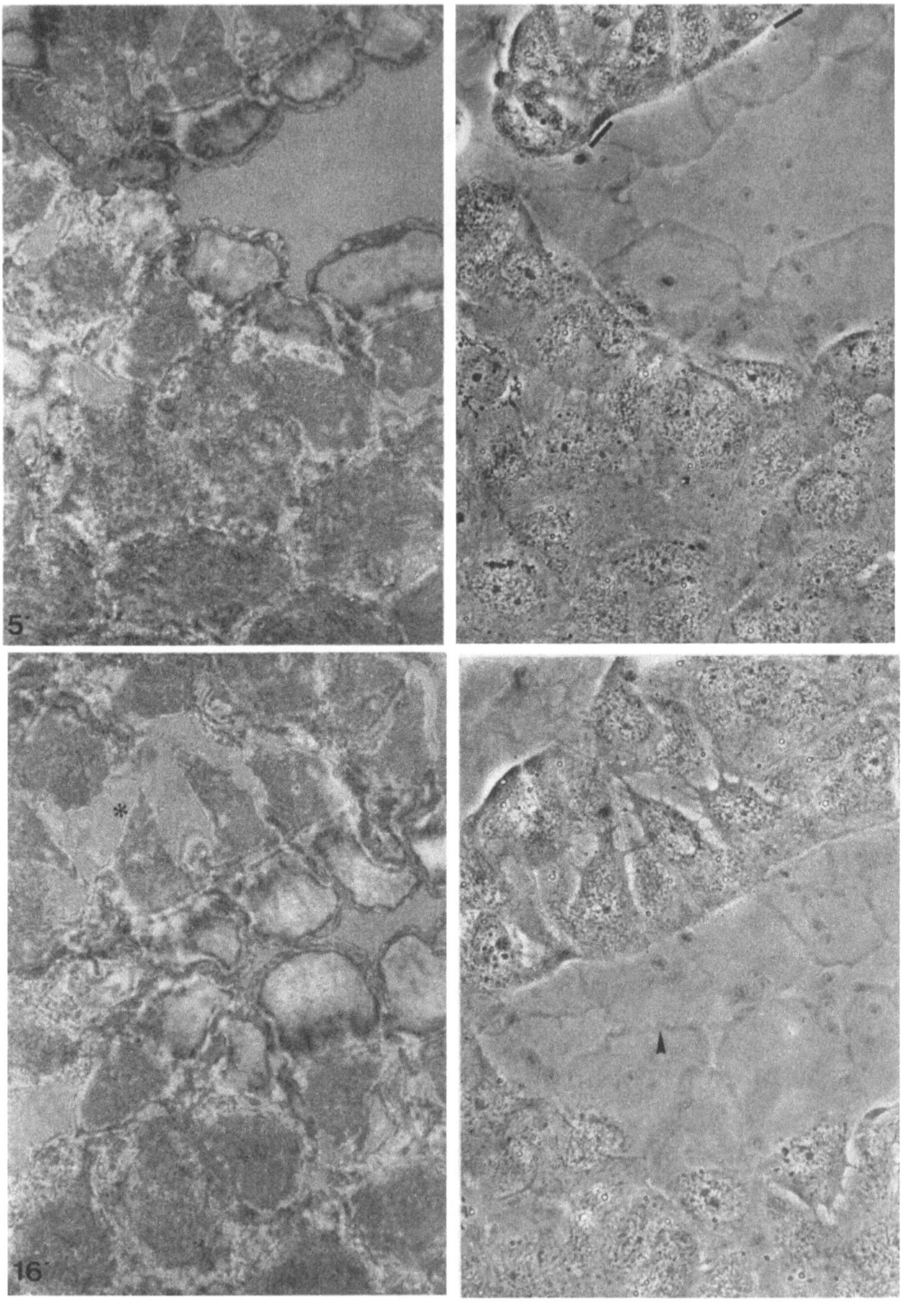

Fig. 1. Healing of a wound made by incision into an epidermal cell sheet in culture. The same areas are shown by phase-contrast microscopy (*right side*) and by reflection – interference contrast microscopy (RIC) (*left side*). RIC provides more detailed insights into the structure and extension of the lamellae formed toward the cell-free wound area. All the marginal cells adhere to each other laong the cell body region were the lamella

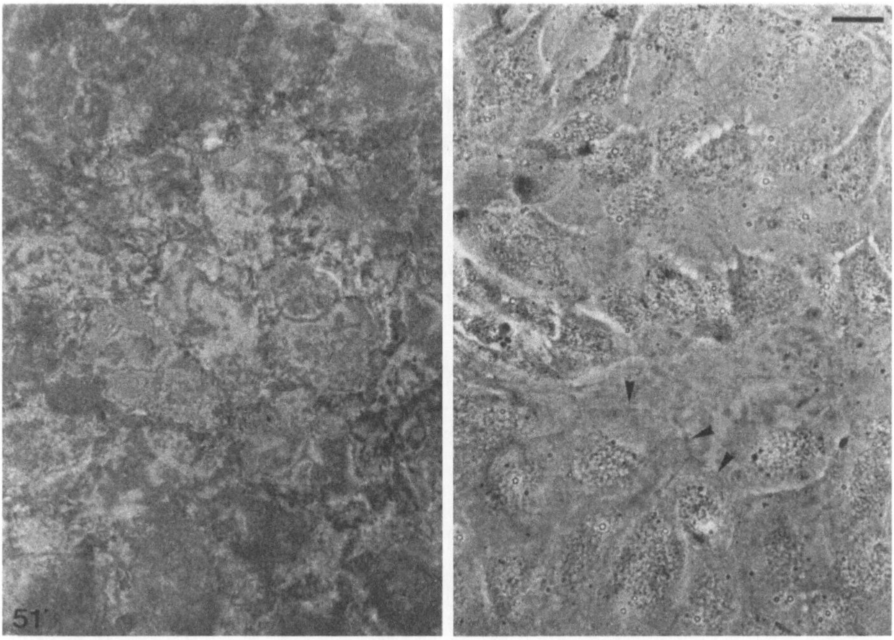

starts (apparent in the phase micrographs at *5* and *16 min*), giving rise to tension along this line, which is marked by the streaks in the phase-contrast image at *5 min*. At *16 min* some cells at the wound margins lost contact with their neighbors in the basal parts (*asterisk* in the *RIC image*), while in the upper parts the cells remain connected by several cytoplasmic strands. At sites where opposing lambellae touch each other they do not protrude further; rather, they broaden the line of contact (*arrowheads* at *16 min*), and later on the distinction between the lamella area and the cell body vanishes (*RIC* at *51 min*) and the contact zones become more distinct (*arrowheads* at *51 min, phase contrast*). *Bar:* 10 μm

F-actin pattern remains unaltered (Fig. 4). Even after several hours the wounds are not closed and the dead cells stick to the glass, surrounded by healthy tissue (Fig. 2). Only after mechanical detachment of the debris (e.g., with a stream of culture medium) do the marginal cells realize the presence of a discontinuity in the cell sheet and close the hole as described above for an incision wound.

Discussion

The observations described above favor the interpretation that the free surface caused by wounding represents an adequate stimulus for the initiation of re-epithelialization by migration which may be sufficient for primary wound closure [2, 9]. In wounds made by irradiation no real free surface is formed, the healthy cells at the wound margins "do not realize" that a

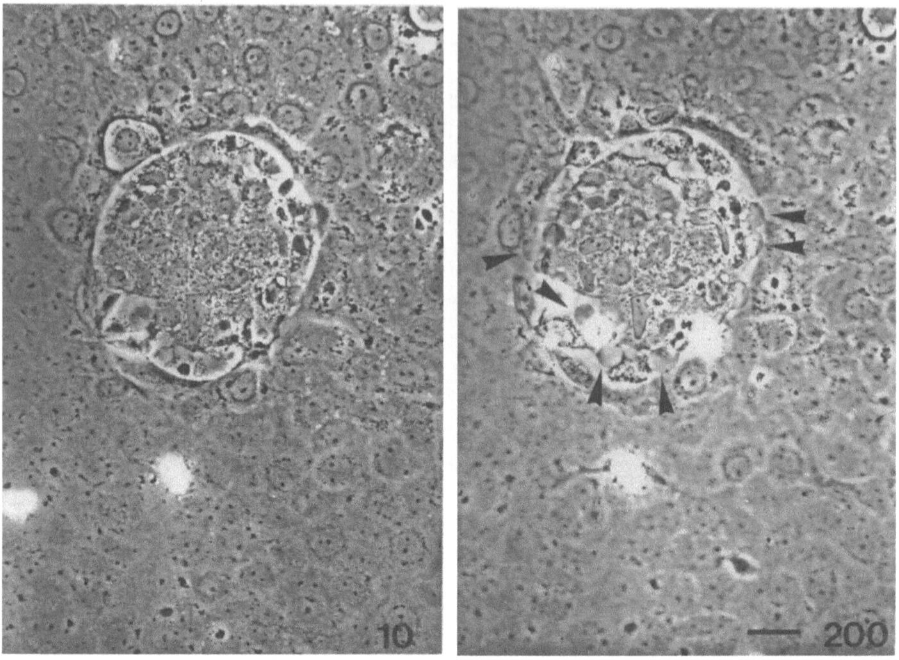

Fig. 2. Phase-contrast micrograph of a bilayered sheet of *Xenopus* epidermal cells. In the central part an approximately circular area was irradiated with unfiltered light from a mecury arc lamp (HBO 50) for 1 min. The burn wound induced by this treatment remained, almost the same for more than 3 h (*10 min* and *200 min* after irradiation). The cells in the wound area are shrunken, more granular than the healthy cells. In the *200 min image* bleb formaton is obvious (*arrowheads*). *Bar:* 10 μm

wound has been made, and the dead cells are not discriminated from living ones. This behavior is similar to that of chick embryo fibroblasts, which show contact inhibition of movement when colliding with living cells as well as with glutaraldehyde-fixed cells [11].

Due to the close relationship between cell motility, energy metabolism, and proliferation [3, 10], no proliferation should be initiated in such burn wounds, a consequence which was not proven in the present study. The results obtained with the cell culture model are in agreement with clinical observations, insofar as healing of burn wounds is severely retarded without débridement. The main differences between the model and the in situ situation are the following:

- Only the behavior of epithelial cells can be investigated: no inflammatory responses and fibroblast activations take place which are specifically altered in burn wounds [20] and are involved in scar formation [14].
- Under in situ conditions the pressure of body fluids disrupts the adhesion of the burnt epidermal layers to the basal lamina; blebs and burn edema

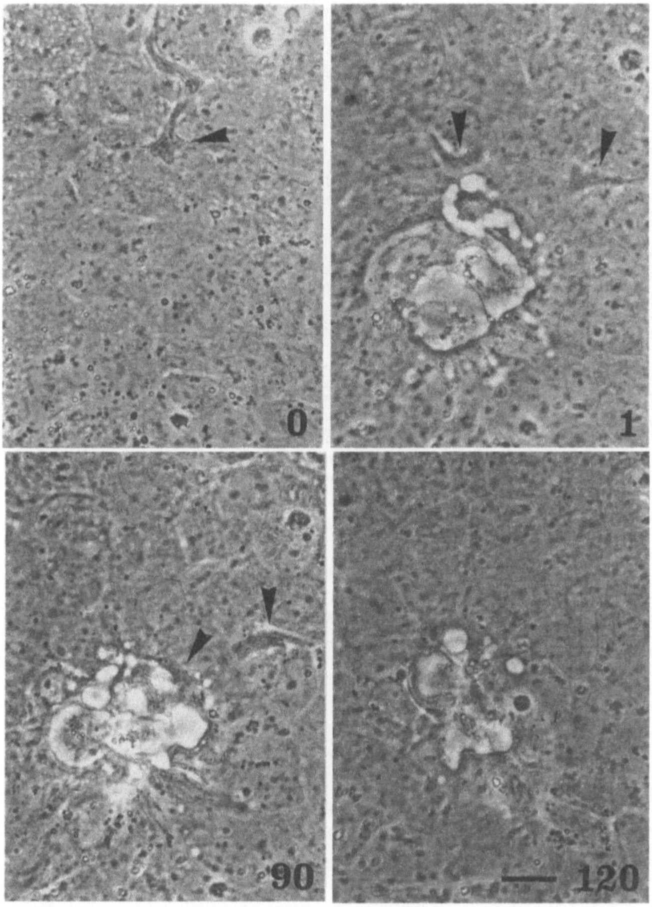

Fig. 3. Phase-contrast micrograph of a bilayered sheet of *Xenopus* epidermal cells before irradiation *(0)*, after 1 min irradiation *(1)*, and *90* and *120* min after irradiation. In this very small wound (diameter 30 μm) the cells swell, become rounded, and pile up. These reactions reduce the wound area. Some lymphocytes are marked by *arrowheads*. *Bar:* 10 μm

are formed, or – in more severe cases – the epithelial coherence is destroyed.

Thus, using cultured epithelial layers with an organization similar to that in situ, investigations of epithelial cell interactions are made possible which may well be of clinical significance. However, the limitations due to the simplification introduced by any model have to be considered.

Acknowledgement. We are grateful for support from the *Institut für den Wissenschaftlichen Film* in Göttingen concerning 35-mm time-lapse filming.

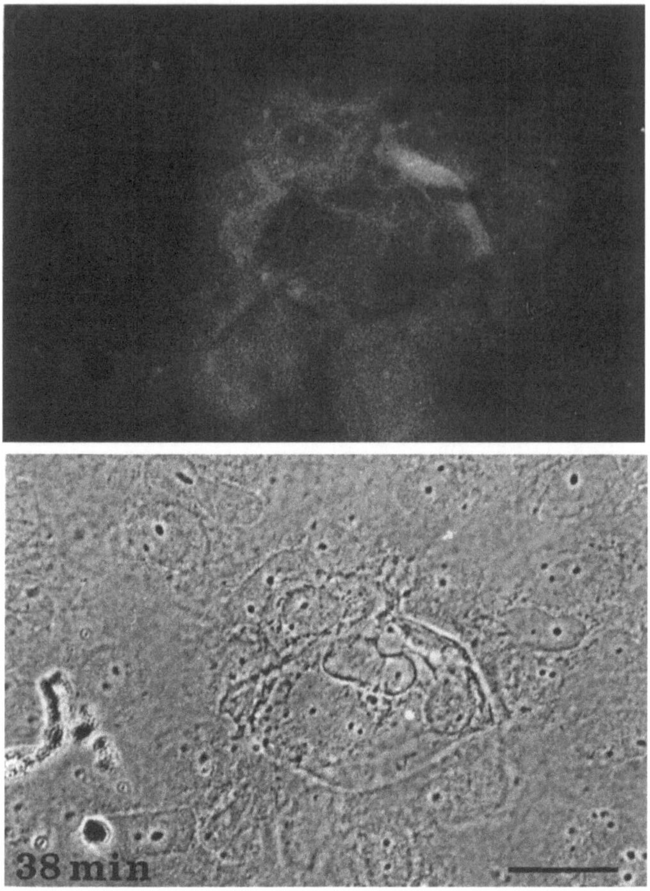

Fig. 4 a, b. A burn wound corresponding to those shown in Figs. 2 and 3, *38 min* after irradiation shown in phase contrast *(a)* and the same site after staining with TRITC-phalloidin for the demonstration of F-actin *(b)*. In this old culture the highly differentiated epidermis cells contain only a very small amount of F-action. Only the marginal cells at the periphery of the wound are clearly F-actin positive. This staining would easily reveal any locomotory organelles if formed. *Bar:* 10 μm

References

1. Bereiter-Hahn J (1967) Dissoziation und Reaggregation von Epidermiszellen der Larven von *Xenopus laevis* (Daudin) in vitro. Z Zellforsch 79:118–156
2. Bereiter-Hahn J (1986) Epidermal cell migration and wound repair. In: Bereiter-Hahn J, Matoltsy AG, Richards S (eds) Biology of the integument, vol. 2. Springer, Berlin Heidelberg New York, pp 443–471
3. Bereiter-Hahn J (1988) Involvement of microcompartmentation in the regulation of cell proliferation. In: Jones DP (ed) Microcompartmentation. CRC, Boca Raton, pp 56–69

4. Chan KY, Patton DL, Cosgrove YT (1989) Time-lapse videomicroscopic study of in vitro wound closure in rabbit corneal cells. Invest Ophthalmol Vis Sci 30:2488–2498
5. Coomber BL, Gotlieb AI (1990) In vitro endothelial wound repair – interaction of cell migration and proliferation arteriosclerosis. Arterisclerosis 10:215–222
6. Donaldson DJ, Mahan JT (1983) Fibrinogen and fibronectin as substrates for epidermal cell migration during wound closure. J Cell Sci 62:117–127
7. Faulstich H, Trischmann H, Mayer D (1983) Preparation of tetramethylrhodaminyl-phalloidin and uptake of the toxin into short-term cultured hepatocytes by endocytosis. Exp Cell Res 144:73–82
8. Grinnell F, Toda K-I, Lamke-Seymour C (1987) Reconstitution of human epidermis in vitro is accompanied by transient activation of basal keratinocytes. Exp Cell Res 172:439–449
9. Klein-Soyer C, Beretz A, Cazenave J-P, Driot F, Maffrand J-P (1990) Behavior of confluent endothelial cells after irradiation. Biol Cell 68:231–238
10. Korohoda W (1971) Interrelations of motile and metabolic activities in tissue culture cells. I. Respiration in chicken embryo fibroblasts actively locomoting, contact-inhibited, and suspended in fluid medium. Folia Biol (Praha) 19:41–51
11. Korohoda W, Kajstura J, Reiss K (1988) Chick embryo fibroblasts show contact inhibition of movement when colliding with glutaraldehyde-fixed cells. Cell Biol Int Rep 12:477–482
12. Korohoda W, Vöth M, Bereiter-Hahn J (1992) Two-phase responses of human polymorphonuclear leucoytes and keratinocytes (epitheliocytes) from *Xenopus laevis*. Protoplasma 167:169–174
13. Matsuda M, Sawa M, Edelhauser HF, Bartels SP, Neufeld AH, Kenyon KR (1985) Cellular migration and morphology in corneal endothelial wound repair. Invest Ophthalmol Sci 26:443–449
14. Montandon D, Gabbiani G (1985) Contractile phenomena in burn wounds. In: May JF, Dogo J (eds), Care of the burn wound. Karger, Basel, pp 111–116
15. Nieuwkoop PD, Faber J (1956) Normal tables of *Xenopus laevis* (DAUDIN). North-Holland, Amsterdam
16. Radice GP (1980) Locomotion and cell-substratum contacts of *Xenopus* epidermal cells in vitro and in situ. J Cell Sci 44:201–223
17. Strohmeier R, Bereiter-Hahn J (1991) Outgrowth of bilayered epidermal cell sheets from tadpole tails (*Xenopus laevis*). Cell Tissue Res 266:615–621
18. Takeuchi S (1972) Wound healing of the cornea in the chick embryo. I. Factors effecting the migration of corneal epithelium. J Fac Sci Univ Tokyo 12:439–447
19. Wolfe K, Quimby MC (1962) Synthetic media. Amphibian culture medium. Excerpta Med 16 (9):
20. Deitch EA, Dobke M, Baxter CR (1985) Failure of local immunity. Arch Surg 120:78–84

Primary Culture of Bilayered Epidermal Cell Sheets

R. Strohmeier, and *J. Bereiter-Hahn*

Introduction

The epidermis separates and protects the organism from the environment. Unfortunately, the attachment of the epidermis to the organism poses technical problems for experimental investigation, especially for direct, microscopic observation. Many experimental culture systems have been established, including the use of organ culture (e.g., rabbit or mouse ear) and epithelial cell sheets reaggregated from isolated cells [2], to overcome these problems.

An ideal model for the study of epithelial cell movement, both as sheets or individually, is the bilayered tadpole epidemis of the clawed toad *Xenopus laevis.* It migrates from the tadpole tail on a coverglass without undergoing significant morphological changes.

This system has several advantages including
a) excellent optical conditions,
b) no loss of normal tissue architecture
c) no need for enzyme treatment to obtain the cell sheet, and
d) no necessity of defined matrices for the attachment of the cells.

Xenopus tadpole tail preparations have yielded information regarding epidermal cell movement during wound closure both in situ [9] and in cell culture [2], as well as on mechanisms of migration of single cells [4, 1, 11, 12].

Preparation of Tadpole Tails

Tadpoles of *Xenopus laevis* Daudin can be generated from hormone – induced mating of adult frogs. The most appropriate embryonic age for cell sheet preparation is stage 49–51 (staged according to [8]). Prior to stage 49 yolk clots hinder optimal observations, and after stage 51 the tails are too big to manipulate.

The tadpole is anesthetized and its tail severed posterior to the anus with a sterile scalpel. The tail is then subsequently washed in two changes of amphibian-Ringer's solution by transfer in a wide-mouth pipette to remove

A. Bernd, J. Bereiter-Hahn,
F. Hevert, H. Holzmann (Eds.)
Cell and Tissue Culture Models
in Dermatological Research

blood cells and debris. The explant is then placed on an ethanol-cleaned, sterilized coverglass and secured by stretching a strip of dialysis membrane across it. The coverglass is placed on the lower aluminium sheet (A in Fig. 1) of the Rose chamber [10]. The dialysis membrane must be pretreated by boilling in distilled water for 10 min and rinsing in amphibian-Ringer's solution before use. A well is formed around the explant on the coverglass using a sterile rubber ring assembly in which two syringe needles are inserted, one for media supply and the other for air relief. The ring is covered by a second coverglass, and the chamber is completed by attaching the upper aluminium plate with four screws. Finally, approximately 1.5 ml of culture medium is added to the chamber using a syringe. After 2–3 days at room temperature (20°–23° C) bilayered epidermal sheets migrate out onto the coverglass.

The upper aluminum plate and the coverglass have to be removed and the medium aspirated prior to experimentation. After removal of the dialysis membrane, the Rose chamber can be re-assembled without the upper coverglass. Using the open chamber, different substances can be added, or experimental manipulations such as wounding can be performed under the microscope.

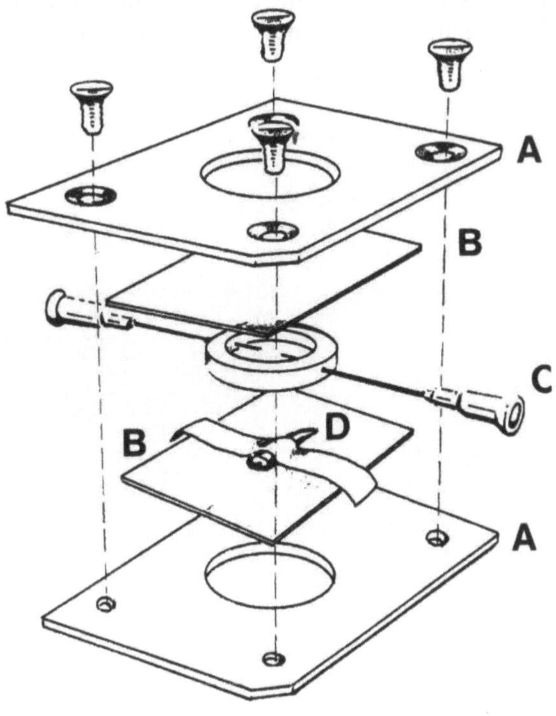

Fig. 1. Rose chamber. *A,* Aluminum plate; *B,* coverglass; *C,* rubber ring with syringe needles for medium supply; *D,* explant, fixed by a strip of dialysis membrane

Structure of the Epithelium In Situ and In Vitro

The epidermis of *Xenopus* tadpoles at stages 49–51 consists of an epithelium two cell-layers thick (Fig. 2). The outer layer is formed by surface cells, which are distinguished from basal cells by mucous vesicles and numerous microridges. At the surface, the entire perimeter of the epithelial cell is bordered by a small ridge at all points of contact with adjacent cells. The cell membrane between adjacent surface cells below this ridge is tightly apposed, forming an occluding and an adhering junction, followed by desmosomes [5]. Basal cells are joined to surface cells almost exclusively by desmosomes, and to the basal lamina by hemidesmosomes (Fig. 2). Basal cells adhere to each other by regions of close contact or, infrequently, by desmosomes.

The bilayered arrangement of basal and surface cells of the tadpole epidermis is largely preserved in the cell sheet; that is, surface cells face the liquid medium side and basal cells line the coverglass (Fig. 3). In comparison to the in situ appearance of the epidermal epithelium, the cell sheet is flatter and the cells are more spread out in vitro. Surface cells remain connected by junctional ridges at the apical cell border; however, they do not form microridges aligning the junctions, as seen in situ. Morphologically differentiated cell contacts between surface and basal cells are rarely development in the cell sheet. Most contacts among basal cells are formed by "interdigitation". Lamellae protrude from surface cells into the space be-

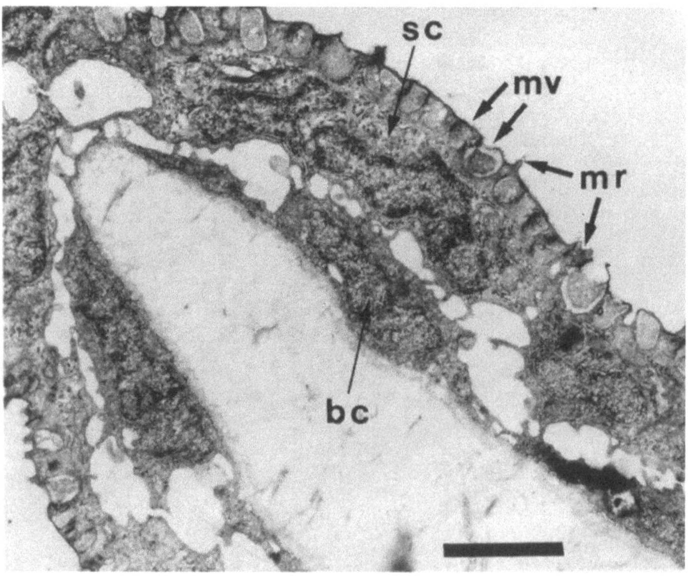

Fig. 2. Transmission electron micrography showing the bilayered epidermis of the tadpole tail. The outer layer is formed by surface cells (*sc*), which are distinguished from basal cells (*bc*) by mucous vesicles (*mv*) and microridges (*mr*). *Bar*, 5 µm

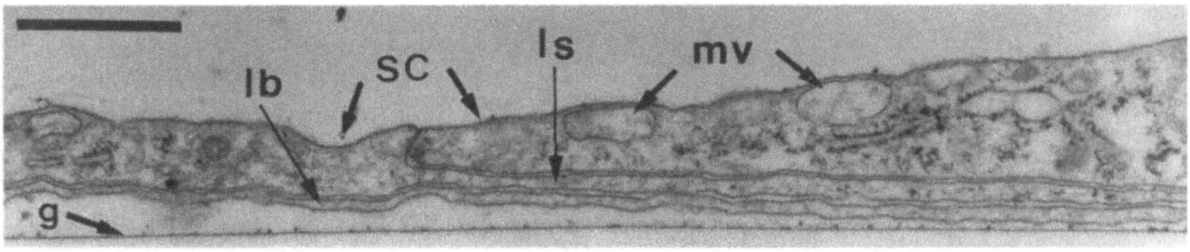

Fig. 3. Transmission electron micrography showing the bilayered epidermis of the cell sheet on the overglass (*g*). Surface cells (*sc*) and basal cells (*bc*) are stretched, lamellae protrude from the surface cells (*ls*) and basal cells (*lb*). Surface cells are still characterized by mucous vesicles (*mv*). *Bar,* 0.5 μm

tween basal and surface cells, whereas lamellae of basal cells protrude under the traılling portion of the cell in front of them, but they do not pass beneath them. In stationary sheets lamellae are small and surround almost the entire cell, so that the cells are no longer polarized.

Migration of Epidermal Cell Sheets

Time-lapse cinemicrography combined with reflection-contrast microscopy [1, 3, 6] allows the movements of cells to be observed close to the coverglass and even beneath the explant. Frame-by-frame analysis shows that the formation of a cell sheet, consisting of basal cells supporting the surface cells, starts with the extension of a lamella of a single basal cell across the surface cell layer; this is followed by the cell body (Fig. 4). The cell extends towards areas devoid of other cells. After about 45 min in culture, the cell is completely attached to the coverlgass and moves toward the fin margin. Cells immediately adjacent to the first migrating cell also migrate in the same direction (Figs. 4 and 5) as the first cell. The lamellae of the moving cells extend in all direction but retract upon contact with adjacent cells. As migrating cells detach their traılling end while extending their lamellae forwards, a free glass surface becomes available behind the cell for the lamellae of following cells. This behaviour leads to a movement of all cells in the same direction. The entire cell sheet moves by active locomotion of the basal cells carrying the surface cells passively. Despite this migration, in no case was the complete epidermis removed from the explant. The movement of the cell sheet ceases when all epidermal cells from the coverglass-facing side of the explant are removed and a continuity is achieved between the epidermis on the surface side of the explant and the cell sheet (Fig. 6).

Conclusions

The type primary culture of amphibian epidermis described in this contribution results in an in vitro system which is closely related to the in situ

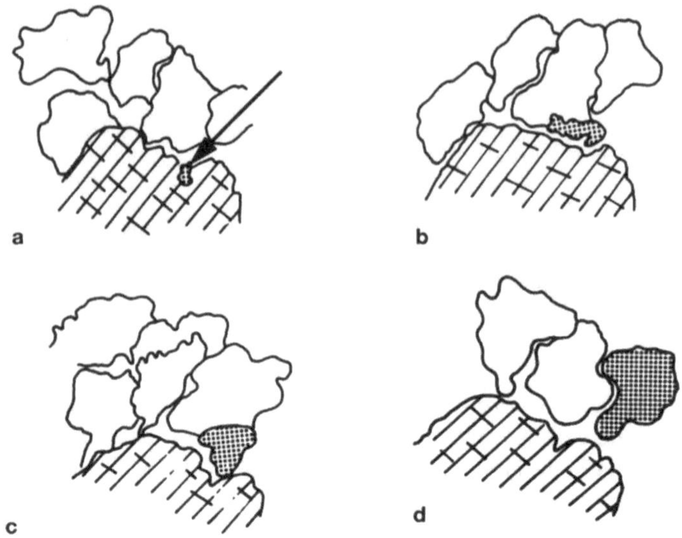

Fig. 4 a–d. Frame-by-frame analysis of a time-lapse cinemicrograph shows the lamella appearing on the coverglass (**a**, *dotted area*) and its enlargement (**b**). After about 45 min the cell body (**c**) can be seen and the cell moves in the same direction as the other cells of the cell sheet (**d**). [From 13]

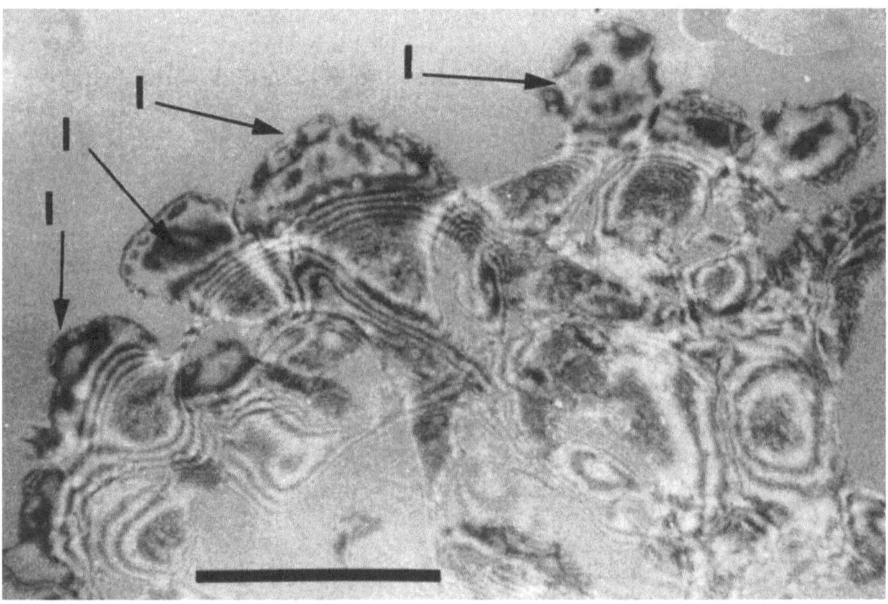

Fig. 5. Reflection-contrast image of a just-formed cell sheet. Lamellae (*l*) are stretched in the direction which is free of other cells. *Bar*, 50 μm [From 13]

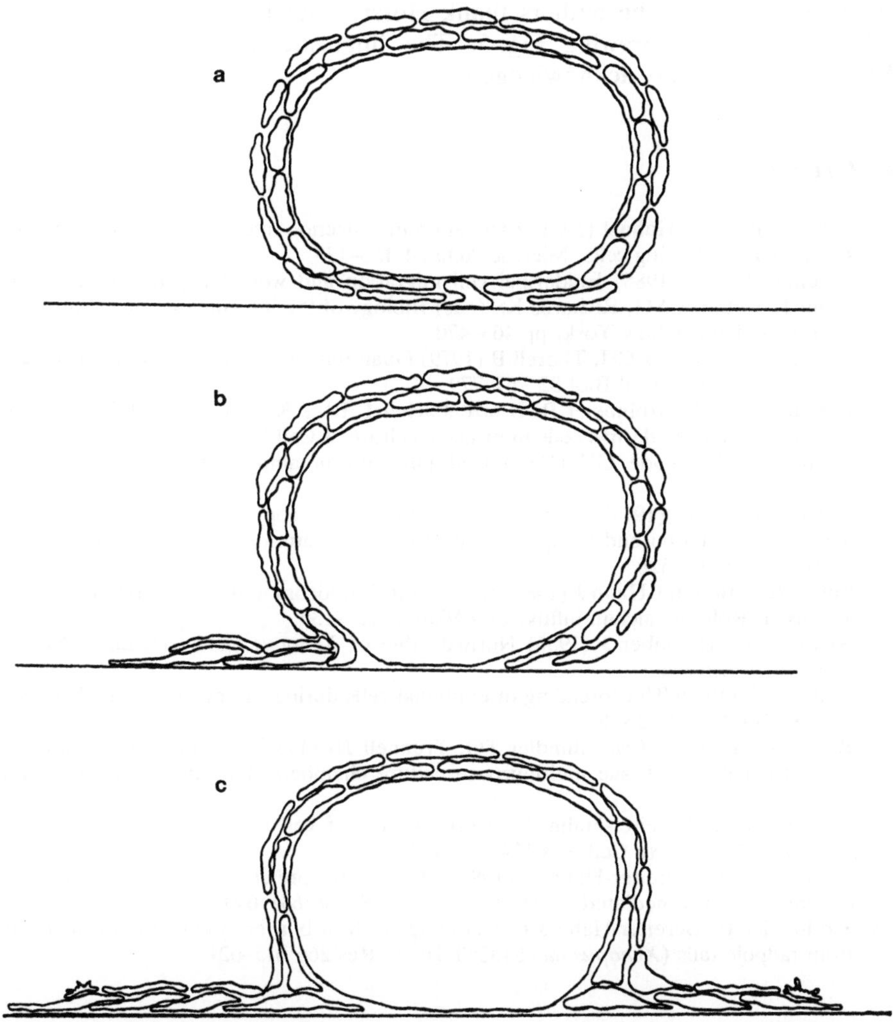

Fig. 6 a–c. Schematic drawing, reconstructed from transmission electron micrographs and reflection-contrast microscopy, of the formation of a cell sheet. **a** A first basal cell protrudes its lamella onto the coverglass. **b** Basal cells carry surface cells while migrating on the coverglass. **c** Migration of the cell sheet ceases when all coverglass-facing cells are removed [From 13]

conditions. The original relations among the cell types participating in the formation of a continuous cell layer are maintained, as are the principle morphological characteristics. The main differences are a consequence of the tension stretching the cells in the emigrating cell sheet. Indicators for this mechanical influence are the loss or flattening of microridges on the surface cells, the larger distance between the mucous granules beneath the apical membranes of surface cells, and the reduction of differentiated structures connecting the surface cells with the basal cells.

Acknowledgement. The authers thank Horst Sauer for drawing Fig. 1 and Ulrike Weich for drawing Fig. 5. The linguistic assistance of Dr. E. Eldredge is gratefully acknowledged.

References

1. Beck K, Bereiter-Hahn J (1981) Evaluation of reflection interference contrast microscope images of living cells. Microsc Acta 84:153–178
2. Bereiter-Hahn J (1986) Epidermal cell migration and wound repair. In: Bereiter-Hahn J, Matoltsy AG, Richards KS (eds) Biology of the integument vol 2. Springer, Berlin Heidelberg New York, pp 463–470
3. Bereiter-Hahn J, Fox CH, Thorell B (1979) Quantitative reflection contrast microscopy of living cells. J Cell Biol 82:767–779
4. Bereiter-Hahn J, Strohmeier R, Kunzenbacher I, Beck K, Voeth M (1981) Locomotion of *Xenopus* epidermis cells in primary culture. J Cell Sci 52:289–311
5. Farquahar M, Palade GE (1965) Cell junctions in amphibian skin. J Cell Biol 26:263–291
6. Gingell D, Todd I (1979) Interference reflection microscopy: a quantiative theory for image interpretation and its application to cell-substratum separation measurement. Biophys J 26:507–526
7. Mittal AK, Bereiter-Hahn J (1985) Ionic control of locomotion and shape of epithelial cells. I. Role of calcium influx. Cell Motil 5:123–136
8. Nieuwkoop PD, Faber J (1956) Normal table of *Xenopus laevis* (Daudin). North-Holland, Amsterdam
9. Radice GP (1980) The spreading of epithelial cells during wound closure in *Xenopus* larvae. Dev Biol 76:26–46
10. Rose GG, Pomerat CM, Shindler TO, Trunnell JB (1958) A cellophane-strip technique for culturing tissue multipurpose culture chambers. J Biophy Biochem Cytol 4:761–764
11. Strohmeier R, Bereiter-Hahn J (1984) Control of cell shape and locomotion by external calcium. Exp Cell Res 154:412–420
12. Strohmeier R, Bereiter-Hahn J (1987) Hydrostatic pressure in epidermal cells is dependent on Ca-mediated contractions. J Cell Sci 88:631–640
13. Strohmeier R, Bereiter-Hahn J (1991) Outgrowth of bilayered epidermal cell sheets from tadpole tails (*Xenopus laevis*).Cell Tissue Res 266:615–621

Plasminogen Activator Contained in Secretions of a Keratinocyte Cell Line and in the Fluid of Intraepidermal Pemphigus vulgaris *Blisters Is Activated by Sodium dodecyl sulfate*

J. Reinartz, P. Boukamp, N.E. Fusenig, and *M.D. Kramer*

Abstract

We have studied the influence of sodium dodecyl sulfate (SDS) on plasminogen activator (PA) activity in culture supernatants derived from a human keratinocyte cell line (HaCaT) and in fluid aspirated from intraepithelial blisters of pemphigus vulgaris skin lesions. Addition of SDS to these specimens resulted in a marked increase of PA activity, as tested by chromogenic substrate assays. Stimulation of PA activity in the high-molecular-weight range after molecular sieve chromatography of HaCaT secretions or pemphigus vulgaris blister fluid suggested that SDS may activate PA activity from high-molecular-weight PA (PA/PAI?) complexes and that this mechanism could also be operative in vivo.

Introduction

The plasminogen activator system plays an important role in extracellular proteolysis (for review see [2, 5]). Plasminogen activators (PAs) can convert the proenzyme plasminogen, which is present in plasma and the interstitial fluid, into the broadly specific trypsin-like serine proteinase plasmin. Two types of plasminogen activators are known: the urokinase-type PA (uPA) and the tissue-type PA (tPA). The active enzymes are regulated by specific inhibitors, i.e., plasminogen activator inhibitor (PAI)-1 and PAI-2 [8] or α_2-antiplasmin [5], respectively. Keratinocytes have been shown to produce PAs of both types [4, 6], as well as PAI [3].

Activation of the PA system is observed under diverse conditions of cutaneous inflammation in vivo. Active plasmin is thought to possess proinflammatory activity [7]. We have obtained evidence that the PA system is locally activated in vivo at the site of sodium dodecyl sulfate (SDS)-induced skin irritation (Kramer et al., this volume). After penetration of the horny layer, topically applied skin irritants interact with keratinocytes as the primary and major cellular target. It was our aim to analyze the influence of SDS on components of the PA system excreted by keratinocytes into the extracellular space. Therefore, we analyzed SDS effects on PA activity in culture supernatants derived from a human keratinocyte cell line (HaCaT

A. Bernd, J. Bereiter-Hahn,
F. Hevert, H. Holzmann (Eds.)
Cell and Tissue Culture Models
in Dermatological Research

[1]) grown under serum-free conditions in vitro as well as in the fluid aspirated from intraepidermal blisters of pemphigus vulgaris skin lesions. PA activity in the fluid of these blisters is thought to be keratinocyte-derived and to be crucially involved in acantholytic blister formtion observed in this disease.

Addition of SDS to these specimens resulted in a marked increase of PA activity as tested by chromogenic peptide substrate assays. When HaCaT culture supernatants were separated by molecular sieve chromatography we found that SDS stimulated PA activity in the high, intermediate-, and low-molecular-weight (mol. wt.) range. In fractionated fluid of pemphigus vulgaris blisters a stimulatory effect was observed in high-mol-wt. fractions only, whereas PA activity in the intermediate- mol.-wt. range was inhibited by SDS.

In conclusion, our data suggest that SDS might activate the PA system via different mechanisms, and that reactivation of PA activity from high-molo-wt. PA (PA/PAI?) complexes may be one of these. The fact that SDS-mediated PA stimulation is also observed in ex vivo-derived intraepidermal blister fuid of pemphigus vulgaris lesions points to the possibility that this mechanism may also be operative in vivo. Therefore, in future studies SDS effects on the PA system components of normal, ex vivo-derived epidermis will have to be explored.

Results and Discussion

Immunoblot Analysis of uPA, tPA, PAI-1, and PAI-2 in Secretions of the Human Keratinocyte Cell Line HaCaT

As shown in Fig. 1, serum-free culture supernatants of the HaCaT cell line contain proteins reacting with antibodies specific for uPA, tPA, PAI-1, or PAI-2. uPA- or tPA-specific antibodies (lanes D or E, respectively) stained various low- and high-mol.-wt. bands: The low-mol.-wt. bands correspond to the heavy and light chains of the heterodimeric active molecules, which is in line with our previous observations that active uPA and tPA are contained in HaCaT-derived secretions [6]. For uPA a strong band corresponding to the single chain molecule pro-uPA (54 kD; arrow) was stained, whereas for tPA only a minor band was found at the position of single chain tPA (68 kD; arrow). The tPA-specific antibodies also stained one pronounced (filled triangle) and several weaker high-mol.-wt. bands. For PAI-1 (50 kD, lane B) and PAI-2 (48 kD, lane C) [8], weaker bands in the position of the free inhibitors (arrows) were observed. PAI-1-specific antibodies stained several additional compounds in the mol. wt. range above free PAI-1 (filled triangles). Both anti-PAI antibodies stained lower-mol.-wt. bands (open triangles), which possibly are cleavage products of the inhibitor molecules: Nonimmune IgG did not yield a positive staining result (lane A).

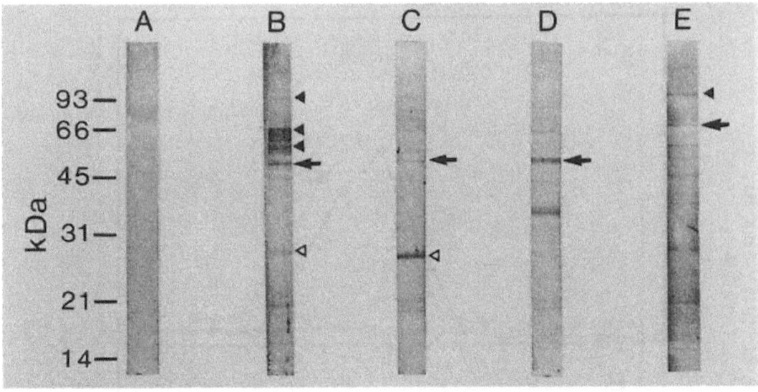

Fig. 1. Immunoblot analysis of HaCaT-derived secretions. Serum-free cell culture supernatants of HaCaT cells grown for 48 h in HL-1 medium (Paesel, Frankfurt, FRG) were 10 × concentrated by centricon centrifugation, and 10 μl/lane of this concentrate was separated by SDS PAGE using 12% separating gels. Separated proteins were electrophoretically transferred onto nitrocellulose and analyzed using nonimmune IgG (*lane A*) and specific IgG against the following proteins: PAI-1 (*lane B*), PAI-2 (*lane* Cm), uPA (*lane D*), and tPA (*lane E*) (American Diagnostica via Ortho, Neckargemünd, FRG). Marks are explained in the text under Results and Discussion

Taken together, these data are evidence for the production of both types of PAs and PAIs by the keratinocyte cell line HaCaT. Moreover, the data indicate that HaCaT-derived extracellular secretions contain a mixture of free, complexed, and degraded compounds of the PA system. Although the immunoblot results do not allow quantitative analysis, it appears that tPA- and PAI-1-containing complexes prevail. However, the final molecular characterization of the hig-mol.-wt. complexes must await further experimentation.

SDS-mediated Activation of the PA System in Culture Supernatants of the Keratinocyte Cell Line HaCaT and in the Fluid of Intraepidermal pemphigus vulgaris Blisters

Using a chromogenic peptide substrate assay, we tested PA activity in HaCaT-derived culture supernatants and in blister fluid from pemphigus vulgaris lesions in the presence of varying concentrations of SDS. As shown in Fig. 2A and B, SDS caused a dose-dependent increase of PA activity in both specimens which reached stimulation to 3000% above the basal activity in HaCaT-derived culture supernatants and about 4000% above basal activity in pemphigus vulgaris blister fluids. However, the dose-response curves obtained for these two specimens were different: In culture supernatants maximum stimulation was observed at 1.25×10^{-3}% SDS (w/v) with a decrease of stimulatory activity at higher and lower concentrations. In blister fluid increasing stimulation was observed from 7.8×10^{-4} to 2×10^{-2}%.

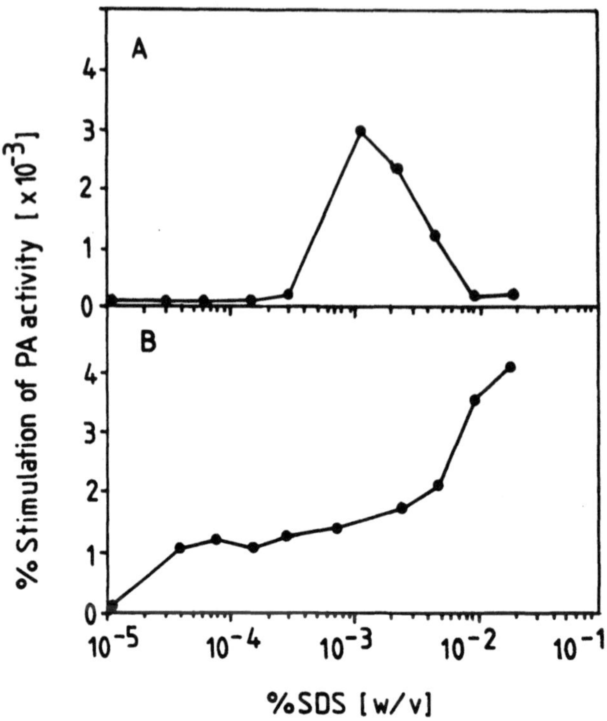

Fig. 2. Activation of the PA system by SDS. Either serum-free culture supernatants of HaCaT cells (60 μl per setup, *A*) or fluid from pemphigus vulgaris blisters (2 μl per setup, *B*) were tested for PA activity in the presence of graded concentrations of SDS [7.8×10^{-4} – 2×10^{-2}% (w/v) final concentration]. PA activity was determined in an indirect chromogenic assay using human plasminogen and the chromogenic plasmin substrate H-D-Val-Leu-Lys-pNA [9]. The enzyme reaction was run for 2 h at 37°C and the absorption at 405 nm was afterwards read directly from the 96-well flat-bottomed microtiter plates. The data are plotted as % stimulation over the activity in the absence of SDS (= 100%)

Higher concentrations of SDS were not testable, since these caused turbidity of the reaction mixture. However, the results indicate that SDS can strongly activate the PA system in secretions derived from keratinocytes in vitro and in ex vivo-derived fluid of intraepidermal blisters.

Activation of the PA System by SDS in Fractionated HaCaT Secretions and Pemphigus vulgaris Blister Fluid

To explore whether stimulation of PA activity is mediated by interference with free PAs or with high-mol.-wt. PA (putative PA/PAI) complexes we studied SDS effects in fractionated specimens: serum-free culture supernatants of the HaCaT cell line and pemphigus vulgaris blister fluid were separated by molecular sieve chromatography and analyzed for PA activity in the absence or presence of SDS. As shown in Fig. 1A and B, SDS has a slight stimulating effect on purified tPA [elution volume (Ve) 13–15 ml] and uPA (Ve 11–14 ml). As shown in Fig. 3C for HaCaT culture supernatants, SDS caused increased activity of high- (Ve 10–15 ml), intermediate- (Ve 15–23 ml) and low- (Ve 25–28 ml)-mol.-wt. PAs. Stimulating activity was most pronounced in the intermediate- and low-mol.-wt. fractions. In contrast, PA activity in blister fluid (Fig. 3D) was stimulated only in high-mol.-wt. fractions (Ve 9–13 ml). PA activity in intermediate-mol.-wt. fractions (Ve 15–19 ml) was inhibited by SDS.

From these data we conclude that the stimulatory "net effect" of SDS observed in either culture supernatants or in blister fluid is due to interference, either positive or negative, with different mol.-wt. PAs and PA (PA/PAI?) complexes. Moreover, it is obvious that different modes of PA activity are operative in these two specimens, and that activation of high-mol.-wt. PA complexes is the major pathway in ex vivo-derived intraepidermal blister fluid. These data extend our knowledge about the influence of SDS on keratinocyte-derived PA activity, and they agree with our finding

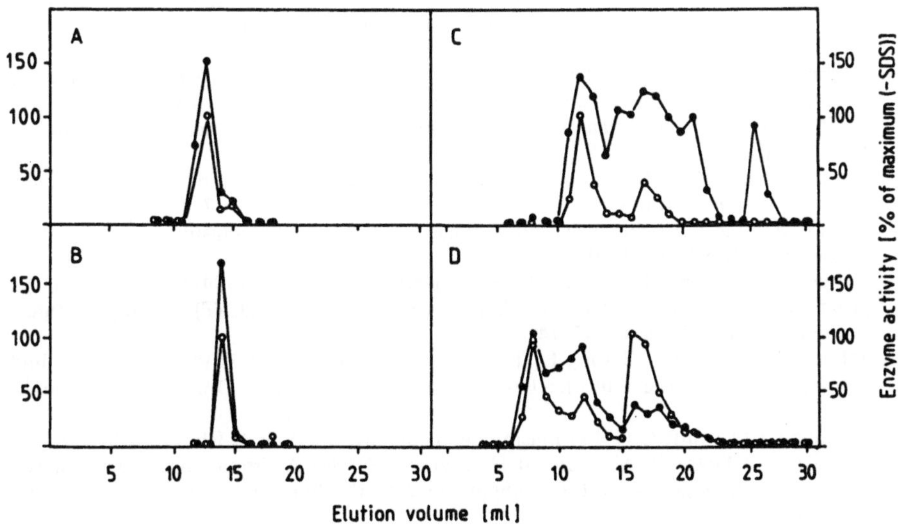

Fig. 3. SDS effect on PA activity of uPA, tPA, and HaCaT secretions and pemphigus vulgaris blister fluid after separation by molecular sieve chromatography. Purified uPA (40 μg/200 μl, *A*) and tPA (40 ng/200 μ, *B*), HaCaT supernatant (200 μl of a 50 × concentrate, *C*) and blister fluid of a Pemphigus vulgaris patient (200 μl, *D*), were separated by gel filtration using a Superose 6 column connected to an FPLC separation system (Pharmacia, Freiburg, FRG). Sixty microliters of the resulting fractions were tested for PA activity in the absence or presence of SDS (2.5×10^{-3}% w/v). The data are plotted as % of maximum activity compared with the highest activity in the absence of SDS (= 100%) for the respective PA-containing sample

that SDS activates tPA/PAI-1 complexes in cutaneous suction blister fluid derived experimentally from human skin (Kramer et al., this volume). Our present studies are addressed to the following questions: (a) Are there additional (dermal?) factors present in intraepidermal pemphigus vulgaris blister fluid that are responsible for the different mode of SDS activity observed for HaCaT secretions and blister fluid? (b) What is the exact molecular composition and possible relation of the high-mol.-wt. (Ve 10–15 ml) PA complexes observed in HaCaT secretions and blister fluid? (c) How does SDS influence the epidermal PA system contained in normal human skin?

Acknowledgements. This work was supported in part by the *Deutsche Forschungsgemeinschaft* (DFG Kr 931/2-2) and the *Bundesministerium für Forschung und Technologie* (01 KG 8808). The authors are indebted to Ms. U. Schirmer and Ms. S. Jobstmann for their expert technical assistance.

References

1. Boukamp P, Petrusevska RT, Breitkreuz J, Hornung J, Markham A, Fusenig NE (1988) Normal keratininization in a spontaneously immortalized aneuploid human keratinocyte cell line. J Cell Biol 106:761–771
2. Dano K, Andreasen PA, Grondahl-Hansen J, Kristensen P, Nielsen LS, Skriver L (1985) Plasminogen activators, tissue degradation and cancer. Adv Cancer Res 44:139–266
3. Hashimoto K, Wun T-C, Baird J, Lazarus GS, Jensen P (1989) Characterization of keratinocyte plasminogen activator inhibitors and demonstration of the prevention of pemphigus IgG-induced acantholysis by a purified plasminogen activator inhibitor. J Invest Dermatol 92:310–315
4. Jensen PJ, Baird J, Morioka S, Lessin S, Lazarus GS (1988) Epidermal plasminogen activator is abnormal in cutaneous lesions. J Invest Dermatol 90:777–782
5. Mayer M (1990) Biochemical and biological aspects of the plasminogen activation system. Clin Biochem 23:197–211
6. Reinartz J, Boukamp P, Schickel E, Fusenig NE, Kramer MD (1991) Activation of the plasminogen activator system in a keratinocyte cell line (HaCaT) by alkyl sulfates. Biomed Biochim Acta 50:743–748
7. Salonen EM, Tervo T, Törmä E, Takkanen A, Vaheri A (1987) Plasmin in tear fluid of patients with corneal ulcers: basis for new therapy. Acta Ophthalmol (Copenh) 65:1–12
8. Sprengers ED, Kluft C (1987) Plasminogen activator inhibitors. Blood 69:381–387
9. Zaoui D, Le Fevre B, Magdalenat H, Bieth JE (1984) A simple spectrophotometric assay of plasminogen activator: comparison with the fibrinolytic method. Clin Chim Acta 141:101–109

Are Rat Granulation Tissue Fibroblasts Regulated by Cytokines in the Same Manner as Skin Fibroblasts?

C. Maus, S. Gebauer, M. Bruns, K. Herrmann,
and *U.-F. Haustein*

Abstract

The aim of the present study was to compare the collagen synthesis and proliferation of fibroblasts from human and rat skin with those from granulation tissue before and after incubation with transforming growth factor β1 (TGFβ) and interferon-gamma (IFNg). TGFβ and IFNg are well-characterized cytokines that are able to regulate fibroblast metabolism. We showed that rat skin and human skin fibroblasts did not differ in proliferation or in their capacity to synthesize collagen. However, rat granulation tissue fibroblasts seemed to be in an activated state. Their proliferation and their collagen synthesis were significantly increased compared with normal rat skin and human skin fibroblasts. This effect was significant in 3-day-old granulation tissue fibroblasts. The effects of the cytokines TGFβ and IFNg on collagen synthesis and proliferation differed quantitatively between skin and granulation tissue fibroblasts. In reaction to the cytokines used, day-3 granulation tissue fibroblasts produced the highest levels of collagen, followed by day-6 granulation tissue fibroblasts. Our findings support the idea of a time-dependent activation of granulation tissue fibroblasts.

Introduction

In the complex regulation of wound healing different cell types are involved, which produce a variety of soluble mediators. Among these mediators, the cytokines have been intensively studied with regard to their regulatory activity in the wound healing process.

Fibroblasts belong to the most important cell types in wound healing. Disturbances of their function may cause severe impairment in the healing process, i.e., remodeling of extracellular matrix and connective tissue formation. During wound healing, fibroblasts exhibit distinct functions in different stages. They seem to adapt themselves to the requirement of the healing process by an unknown mechanism. Our aim was to examine whether fibroblasts in different ages of granulation exhibit differences in collagen synthesis and proliferation as compared with fibroblasts of intact skin. It has been shown that cytokines are able to modify the fibroblast metabolism,

A. Bernd, J. Bereiter-Hahn,
F. Hevert, H. Holzmann (Eds.)
Cell and Tissue Culture Models
in Dermatological Research

e.g., the syntheses of extracellular matrix proteins (collagen, fibronectin, glycosaminoglycans), collagenase, and PGE_2. In addition, they can modulate proliferation, migration, and protease release by fibroblasts.

We studied the effect of the well-characterized cytokines transforming growth factor β (TGFβ) and interferon gamma (IFNg) on the proliferation and collagen synthesis of human skin fibroblasts, rat skin fibroblasts, and rat granulation tissue fibroblasts. TGFβ is known to promote, IFNg to inhibit wound healing. We wanted to find out whether they influence granulation tissue fibroblasts in a different way than they do skin fibroblasts.

The Action of TGFβ and IFNg in Wound Healing

TGFβ

TGFβ is produced by a wide variety of cell types. The TGFβ family consists of four peptides: TGFβ1, -β2, β3, -β4. In the present study our interest was focused on TGFβ1, which is a homodimeric peptide with a molecular mass of 25000 Daltons [1]. TGFβ alone can induce the formation of granulation tissue, including the chemotactic attraction of monocytes and fibroblasts [1, 2]. It stimulates fibroblast proliferation and secretion of extracellular matrix components such as collagen and fibronectin, as well as the production of protease inhibitors [3]. Neovascularization is stimulated subsequently to the organization of endothelial cells into capillary tubes [1]. It has been demonstrated that both an acceleration of healing and an increase in tensile strength of the scar follow the exogenous TGFβ [2, 4], a time-dependent increase in the levels of endogenous TGFβ into wound fluid has also been shown [5]. Wound healing also requires immunologic responses. In this regard, TGFβ seems to play a pleiotropic role, acting both as a chemoattractant and an activator of monocytes [6] and as an inhibitor of lymphocyte proliferation and differentiation [7].

On the molecular level, TGFβ is known to activate the expression of mRNA for various connective tissue components, including collagen types I and III, fibronectin, and the β-subunit of the fibronectin receptor on the transcriptional level [8].

TGFβ increases the synthesis of protease inhibitors (plasminogen-activator-inhibitor, tissue inhibitor of metalloproteinases) and decreases the synthesis of proteases (serine protease, plasminogen activator, a thiol protease, collagenase, transin/promyelysin) [9, 10].

In fibroblast cultures TGFβ induces the expression of fibronectin receptors which are members of the β1 integrin family and are necessary for an undisturbed cell-matrix interaction [3]. TGFβ may also stimulate migrating epidermal cells to produce fibronectin and fibronectin receptors as well as other integrins, and thereby promote translocation over the surface of the wound [3].

IFNg

Interferons are a group of polypeptides which exhibit immunoregulatory, antiviral, and antiproliferative effects and have been shown to demonstrate some antitumor effects in vivo. They are a multigene family whose protein products (15–40 kDa) have been divided into three main types: interferon alpha (IFNα), interferon β (IFNβ), interferon gamma (IFNg). IFNa was originally derived from peripheral blood mononuclear cells, IFNβ from fibroblasts, and IFNg from activated T lymphocytes and natural killer cells. They all suppress collagen synthesis in dermal fibroblasts [11]. Furthermore, IFNg inhibits the elevated collagen synthesis of fibroblasts derived from scleroderma patients [12] and suppresses collagen synthesis of myofibroblasts [13]. IFNg was found to decrease angiogenesis by inhibiting the proliferation and migration of endothelial cells [13]. It has been reported that tumor necrosis factor alpha and IFNg are synergistic in their inhibition of collagen synthesis of human fibroblasts [14] and rat myofibroblasts. The inhibitory effect of IFNg on the synthesis of types I and- III collagen and the alpha chain of type-VI collagen is associated with decreased levels of the corresponding mRNAs [15, 16].

Furthermore, IFNg inhibits fibroblasts chemotaxis [17]. In contrast to IFNα and β, IFNg increases the production of glycosaminoglycans [11]. IFNg also enhances the transcriptional rate of fibronectin [11, 15] and activates collagenase [11].

Material and Methods

Fibroblast Cultures

Human fibroblast cultures were established by outgrowth from skin biopsies of healthy volunteers. Rat skin and rat granulation tissue fibroblasts were derived from 400 g Wistar rats. Rat granulation tissue was taken from 3- and 6-day-old wounds. Biopsy material was cut into fragments of 1 mm^3 and transferred into 25-cm^2 tissue culture flasks (Falcon). Cells were maintained in Dulbecco's minimal essential medium (DMEM, Difco), supplemented with 10% fetal calf serum, 400 U/ml penicillin, 50 μg/ml streptomycin, 300 μg/ml glutamine, and 50 μg/ml Na ascorbic acid in a humified atmosphere of 5% CO_2 in air at 37°C. Fibroblast cultures were subcultivated by trypsinization and were used in the fourth to fifth passage.

Proliferation

Subconfluent monolayer cultures were incubated with 10μCi ^{3}H-thymidine for 24 h. Cultures were rinsed twice with PBS buffer, extracted with ice-cold 5% (w/v) trichloroacetic acid, and washed twice with distilled water. The cell layer was solubilized by incubation for 2 h at 37° C in 0.5 *N* NaOH. Radioactivity was determined in a liquid scintillation counter (LKB Pharmacia).

Collagen Synthesis

Confluent monolayer cultures were incubated with L-(2,3-^{3}H).proline (10 μCi/ml) for 24 h. Cells were harvested, homogenized, and pooled with the medium, dialysed for 12 h in 1 *M* $CaCl_2$ (pH 7.0), then washed four times in 0.5% acetic acid to remove free ^{3}H-proline. Labeled collagen was quantified after digestion with highly purified collagenase type VII, free of any proteolytic activity. Probes were added to 10 ml scintillation fluid and radioactivity was determined in a liquid scintillation counter.

Cytokines

Hu-r TGFβ1 (Boehringer, Mannheim, FRG) was used in the concentrations 0.5, 1.0, and 5.0 μg/ml medium. Hu-r IFNg was purchased from the Dr. Rentschler Arzneimittel GmbH & Co., FRG, and used in concentrations of 100, 1000, and 10000 IE/ml medium.

Each experiment was done in triplicate using 10^5 cells/dish. Averaged results are expressed as ^{3}H-proline and ^{3}H-thymidine incorporation, respectively, in DPM ± standard deviation.

Results

Proliferation and Collagen Synthesis of Rat and Human Skin Fibroblasts

There was no significant difference in the proliferation and collagen synthesis between rat and human skin fibroblasts, as shown in Figs. 1 and 2.

Proliferation and Collagen Synthesis in Granulation Tissue Fibroblasts

Rat granulation tissue fibroblasts showed a two- to threefold increase in proliferation and collagen synthesis compared with skin fibroblasts. This effect was more distinct in the cultures of 3-day-old granulation tissue fibroblasts than in those of the 6-day-old fibroblasts (Figs. 1 and 2).

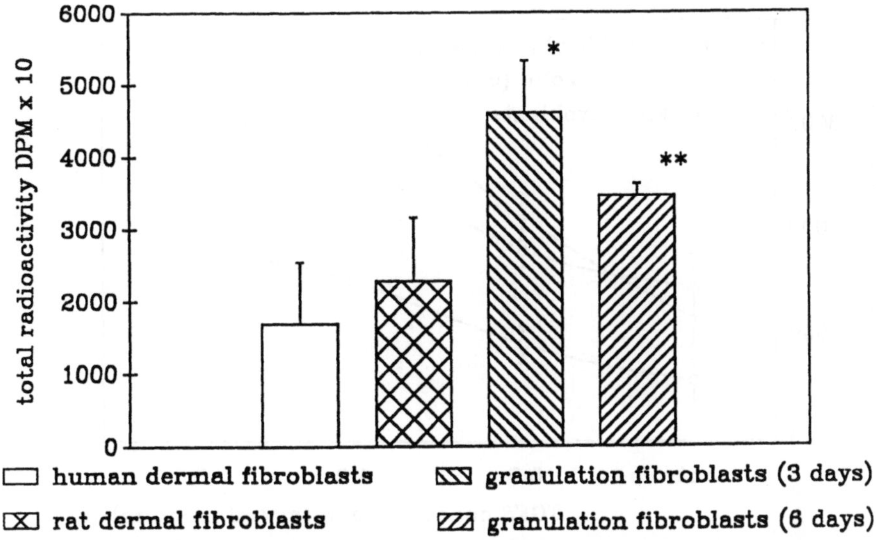

Fig. 1. Proliferation of rat granulation tissue fibroblasts, rat skin fibroblasts, and human skin fibroblasts. The data are expressed as averages ± standard deviation of three experiments in triplicate (10^5 cells/dish). * $p \leq 0.001$ (Wilcoxon test), ** $p \leq 0.001$ (Wilcoxon test)

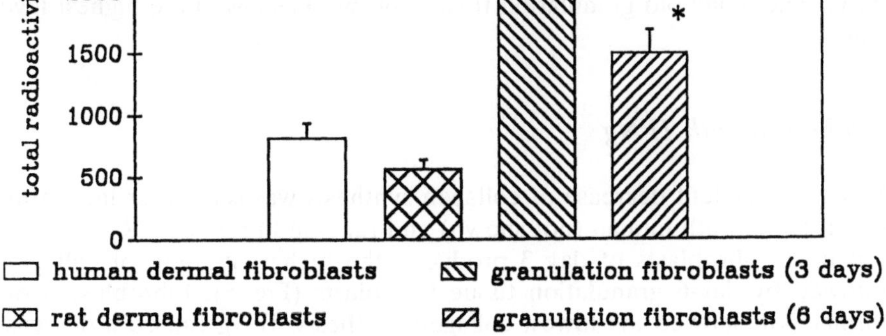

Fig. 2. Collagen synthesis of rat granulation tissue fibroblasts, rat skin fibroblasts and human skin fibroblasts. The data are expressed as averages ± standard deviation of three experiments in triplicate (10^5 cells/dish). * $p \leq 0.001$ (Wilcoxon test)

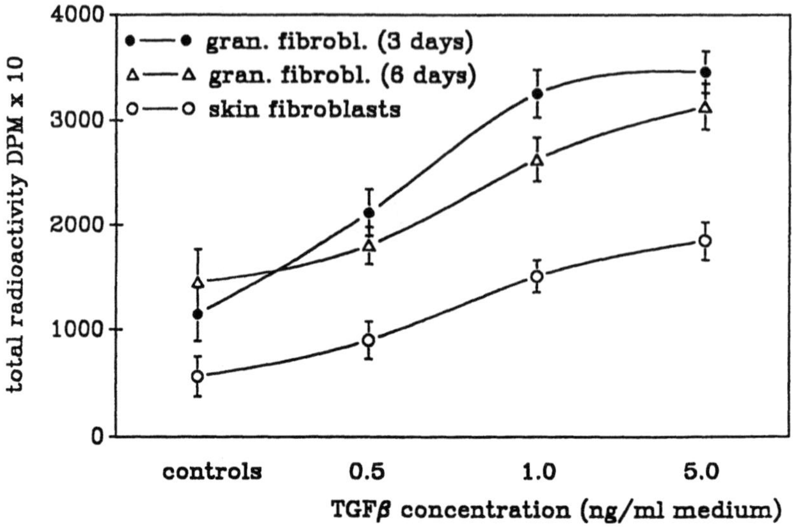

Fig. 3. Influence of TGFβ on the collagen synthesis by rat fibroblasts. The data are expressed as averages ± standard deviation of three experiments in triplicate (10^5 cells/dish)

Incubation with TGFβ

Collagen synthesis and proliferation of human monolayer fibroblasts were significantly enhanced after incubation with TGFβ in a dose-dependent manner (not shown). The same effect was seen in rat skin and rat granulation tissue fibroblasts. Day-3 granulation tissue fibroblasts showed the highest amount of collagen production, followed by day-6 granulation tissue fibroblasts, in comparison with the skin fibroblasts (Fig. 3). Proliferation of the granulation tissue fibroblasts in the presence of TGFβ (shown in the concentration of 1 ng/ml) was clearly elevated compared with skin fibroblasts. The 3-day-old granulation tissue fibroblasts showed the highest level (Fig. 4).

Incubation with IFNg

A dose-dependent decrease of collagen synthesis was seen after incubation with IFNg in both human (not shown) and rat skin fibroblasts. Rat granulation tissue fibroblasts of day 3 produced the highest amount of collagen, followed by day-6 granulation tissue fibroblasts (Fig. 5). Fibroblasts from normal skin showed the lowest collagen synthesis. IFNg caused a decrease of proliferation in a dose-dependent manner. The proliferation rate of 3-day- old granulation tissue fibroblasts was higher than that of 6-day-old fibroblasts, followed by skin fibroblasts (Fig. 6).

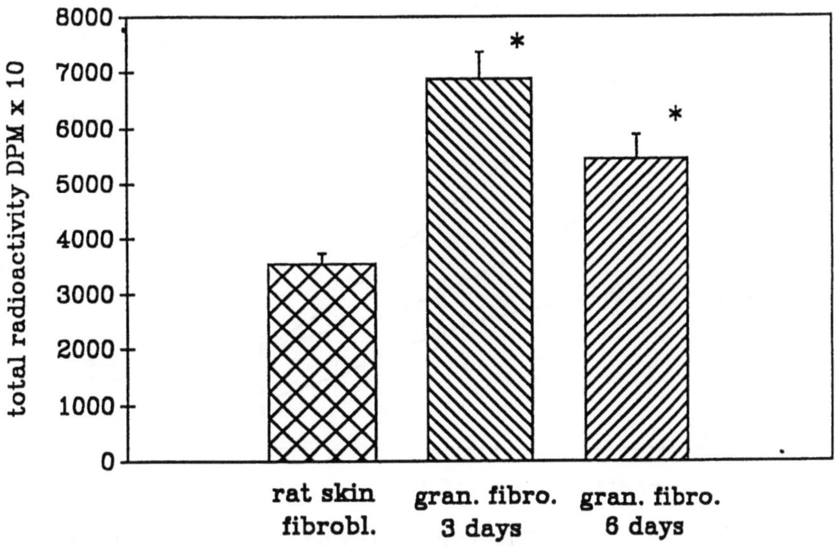

Fig. 4. Influence of TGFβ on the proliferation (thymidine incorporation) of rat fibroblasts. The data are expressed as averages ± standard deviation of three experiments in triplicate (10^5 cells/dish). * $p \leq 0.001$ (Wilcoxon test)

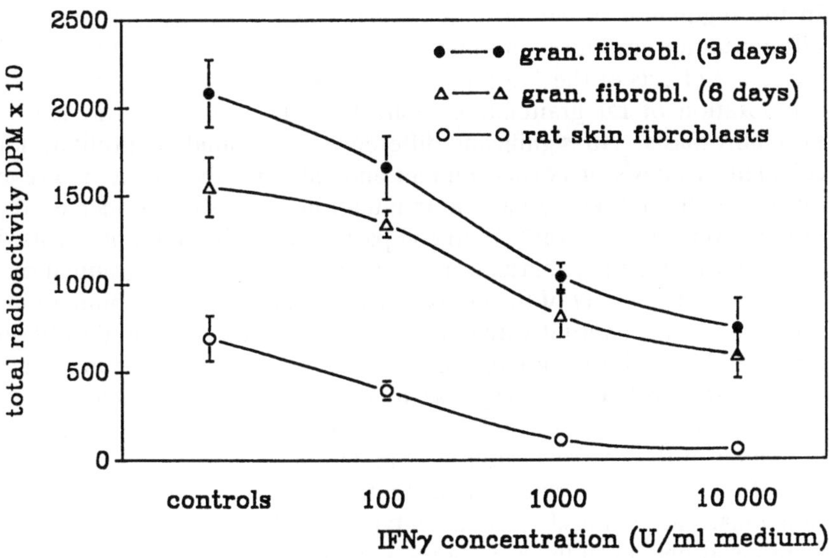

Fig. 5. Influence of IFNg on the collagen synthesis of rat fibroblasts. The data are expressed as averages ± standard deviation of three experiments in triplicate (10^5 cells/dish)

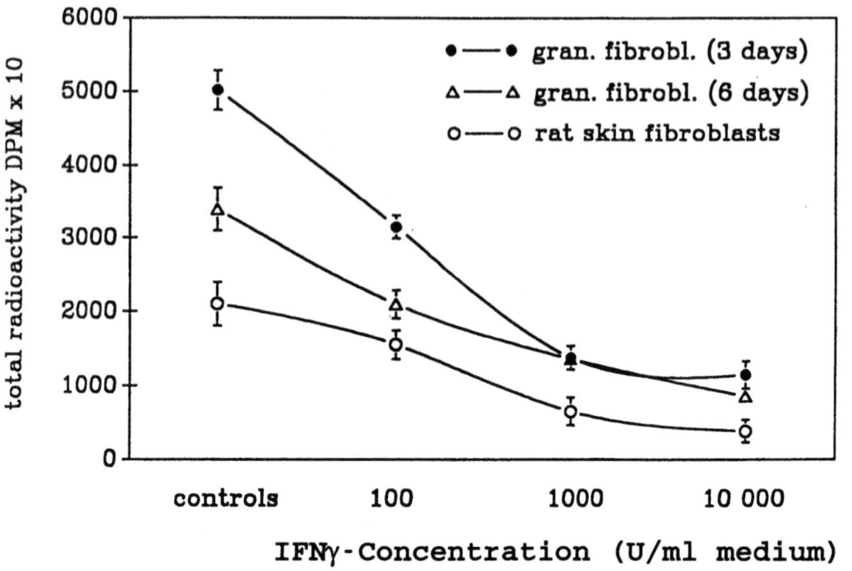

Fig. 6. Influence of IFNg on the proliferation (thymidine incorporation) of rat fibroblasts. The data are expressed as averages ± standard deviation of three experiments in triplicate (10^5 cells/dish)

Discussion

In the regulation of wound healing cytokines exert distinct effects on cell differentiation, motility, and matrix protein synthesis. In our study we compared the effects of the cytokines TGFβ and IFNg on collagen synthesis and proliferation of rat granulation tissue fibroblasts, normal rat, and human skin fibroblasts. No significant difference was found in proliferation and collagen synthesis of normal human and rat skin fibroblasts. However, rat fibroblasts from 3- and 6-day-old granulation tissue showed increased proliferation and collagen synthesis in comparison with the skin fibroblasts. Day-3 granulation tissue fibroblasts were more potent than day-6 granulation tissue fibroblasts. The effect of the cytokines TGFβ and IFNg on granulation tissue fibroblasts was quantitatively dinstinct from that on the skin fibroblasts. We observed higher levels of collagen synthesis and proliferation in the granulation tissue fibroblasts in response to the cytokines investigated. Our data demonstrate that granulation tissue fibroblasts seem to be activated, in particular in the earlier period of granulation.

In the process of wound healing fibroblasts are able to change their functions. The synthesis of collagen types I and III is strongly induced at the pretranslational level in early stages of wound healing following injury to the skin and then followed by sharply reduced down-regulation [18]. As Scharffetter et al. [19] demonstrated by in situ hybridization of collagen alpha 1 (I) chain mRNA, collagen type-I gene expression is activated as early as 16 h after injury. At that time, collagen type-I-producing fibroblasts were localized in the deep layer of granulation tissue and at the wound edges. After 6 days, most of the fibroblasts in the granulation tissue were seen to produce collagen type I. After 8 days, collagen type-I mRNA was detectable directly underneath the epidermis, and hardly any collagen type-I-producing cell was seen after 26 days.

In addition, Kurkinen et al. [20] demonstrated that fibronectin appeared very early with invading fibroblasts. Fibronectin receptors have been shown to be expressed by fibroblasts just prior to wound contraction [21].

The mechanism of fibroblasts regulation in wound healing is not clearly understood.

Furthermore, it remains open whether activated fibroblasts immigrate into the wound area or if fibroblasts in the wound site adapt to the requirement of increased matrix protein synthesis in healing wounds.

It has been supposed that fibroblast function in wound healing is regulated by cell-cell, cell-matrix interaction and by cytokines released from platelets, endothelial cells, and inflammatory cells [18]. Interestingly, we observed in our study that granulation tissue fibroblasts remained in an activated state despite isolation and cultivation in monolayers. Similar observations have been made with fibroblasts obtained from patients with systemic scleroderma (SSc). Cultured SSc fibroblasts retained their capability to produce high levels of collagen up to the tenth to fifteenth passage [22], (own observations). The existence of high collagen-producing subsets of fibroblasts in SSc has been suggested, which become expanded by exposure to cytokines [22]. Such highly productive fibroblast subsets may play a role in the wound healing process, too. They many invade the granulation tissue at different stages of wound healing.

On the other hand, the observations of Bayreuther et al. argue for a differentiation of fibroblasts at the wound site. Fibroblasts of distinct differentiation states can be distinguished by their different proliferative properties and by the amount and type of collagen they synthesize [23, 24]. Further investigations have to be done to solve this question.

The present study has shown that granulation tissue fibroblasts produced higher amounts of collagen and exhibited a higher proliferation rate than skin fibroblasts even though they were cultured in monolayers. The influence of the cytokines TGFβ and IFNg on granulation tissue fibroblasts did not differ from skin fibroblasts in principle. There was, however, a quantitative difference in the investigated parameters, such as proliferation and collagen synthesis, between granulation and skin fibroblasts.

References

1. Rodland K, Muldoon LL, Magun BE (1990) Cellular mechanism of TGFβ action. J Invest Dermatol 94:33S–40S

2. Pierce GF, Mustoe GA, Lingelbach J, Masakowski VR, Griffin GL, Senior RM, Deuel TF (1989) Platelet-derived growth factor and transforming growth factor-β enhance tissue repair activities by unique mechanisms. J Cell Biol 109:429–440
3. Clark RAF (1991) Cutaneous wound repair: a review with emphasis on integrin receptor expression. In: Janssen H, Rooman R, Robertson JIS (eds) Wound healing. Wirghtson:
4. Cromack DT, Sporn MB, Roberts AB, Merino MJ, Dart LL, Norton JA (1987) Transforming growth factor beta levels in rat wound chambers. J Surg Res 42:622–628
5. Mustoe TA, Pierce GF, Thomason A, Gramates P, Sporn MB, Deuel TF (1987) Accelerated healing of incisional wounds in rats induced by transforming growth factor-beta. Science 237:1333–1336
6. Wahl SM, Hunt DA, Wakefield LM, Mc-Cartney-Francis N, Wahl LM, Roberts AB, Sporn MB (1987) Transforming growth factor type beta induces monocyte chemotaxis and growth factor production. Proc Natl Acad Sci USA 84:5788–5792
7. Rook AH, Kehrl JH, Wakefield LM, Roberts AB, Sporn MB, Burlington DB, Lane HC, Fauci AS (1986) Effects of transforming growth factor beta on the functions of natural killer cells: depressed cytolytic activity and blunting of interferon responsiveness. J Immunol 136:3916–3920
8. Raghow R, Postlethwaite AE, Keski-Oja J, Moses HL, Kang AG (1987) Transforming growth factor-β increases steady-state levels of type-I procollagen and fibronectin messenger RNAs posttranscriptionally in cultured human dermal fibroblasts. J Clin Invest 79:1285–1288
9. Roberts AB, Sporn MB (1988) Transforming growth factor β. Adv Cancer Res 51:107–145
10. Overall C, Wrana JL, Sodek J (1989) Transforming growth factor-β regulation of collagenase, 72 kDa-progelatinase, TIMP and PAI-1 expression in rat bone cell populations and human fibroblasts. Connect Tiss Res 20:289–294
11. Duncan MR, Berman B (1989) Differential regulation of glycosaminoglycan, fibronectin and collagenase production in cultured human dermal fibroblasts by interferon-alpha, -beta, and -gamma. Arch Dermatol Res 281:11–18
12. Duncan MR, Berman B (1987) Persistence of a reduced-collagen-producing phenotype in cultured scleroderma fibroblasts after short-term exposure to interferons. J Clin Invest 79:1318–1324
13. Sato N, Nariuchi T, Tsuruoka N, Nishihara T, Beitz JG, Calabresi P, Frackelton AR jr (1990) Actions of TNF and IFNg on angiogenesis in vitro. J Invest Dermatol 95:85S–89S
14. Scharffetter K, Heckmann M, Hatamochi A, Mauch C, Stein B, Riethmüller G, Ziegler-Heitbrock H-W, Krieg T (1989) Synergistic effect of tumor necrosis factor alpha and interferon-gamma on collagen synthesis of human skin fibroblasts in vitro. Exp Cell Res 181:409–419
15. Czaja MJ, Weiner FR, Eghbali ME, Giambrone M-A, Eghbali MA, Zern MA (1987) Differential effects of g-interferon on collagen and fibronectin gene expression. J Biol Chem 27:13348–13351
16. Heckmann M, Aumailley M, Hatamochi A, Chu ML, Timpl R, Krieg T (1989) Down-regulation of alpha3 (VI) chain expression by interferon gamma decreases synthesis and deposition of collagen.VI. Eur J Biochem 182:719–726
17. Adelmann-Grill BC, Hein R, Wach F, Krieg T (1987) Inhibition of fibroblasts chemotaxis by recombinant human interferon gamma and interferon alpha. J Cell Physiol 130:270–275
18. Kulozik M, Heckmann M, Mauch C, Scharffetter K, Krieg T (1991) Cytokine regulation of collagen metabolism during wound healing in vitro and in vivo. In: Janssen H, Rooman R, Robertson JIS (eds) Wound healing. Wirghtson
19. Scharffetter K, Kulozik M, Stolz W, Lankat-Buttgereit B, Hatamochi A, Söhnchen R, Krieg T (1989) Localization of collagen al (I) gene expression during wound healing by in situ hybridization. J Invest Dermatol 93:405–412

20. Kurkinen M, Vaheri A, Roberts P, Stenman S (1980) Sequential appearance of fibronectin and collagen in experimental granulation tissue. Lab Invest 43:47–51
21. Clark RAF (1990) Fibronectin matrix deposition and fibronectin receptor expression in healing and normal skin. J Invest Dermatol 94:128S–134S
22. White-Needleman B, Ordonez JV, Taramelli D, Alms W, Gayer K, Choi J (1990) In vitro identification of a subpopulation of fibroblasts that produces high levels of collagen in scleroderma patients. Arthritis Rheum 33:842–849
23. Mollenhauer J, Bayreuther K (1986) Donor-age-related changes in the morphology, growth potential, and collagen biosynthesis in rat fibroblast subpopulations in vitro. Differentiation 32:165–172
24. Bayreuther K. Rodemann HP, Francz PI, Maier K (1988) Differentiation of fibroblasts stem cells. J Cell Sci [Suppl] 10:115–130

Cultured Allografts for Wound Conditioning and Enhanced Epithelialization of Chronic Venous Ulcers

K. Ramme, Z. Ruszczak, and *R. Stadler*

Abstract

Cultured epidermal allografts were prepared from living HBsAG- and HIV-negative skin donors, using the "serum-free medium technique" to treat ten patients with chronic venous ulcers. The subsequent dressing with allografts led to clean granulation tissue, and within 4–6 weeks epitheliazation was achieved. Under this continuous grafting of cultured allogenic keratinocyte sheets we noticed clinically that the bacterial contaminiation of the wound bed was remarkably reduced without internal or external use of antibiotics. Our data suggest that the treatment of chronic venous ulcers with cultured allografts is a simple, readily available, and effective method with some clinical benefit in healing standard therapy-resistant chronic venous ulcers.

Introduction

Chronic nonhealing ulcers of the extremities are a major health problem. In industrialized countries 0.5%–1.5% of the population suffers from chronic leg ulcers. In 1979, 128000 patients in the USA were treated primarily for stasis ulceration, with an average length of hospitalization of 11.4 days, which corresponds to a total amount of $ 300 million in health-care support.

A treatment which could reduce wound healing time would be of benefit for both the patient and the community. Previous studies have demonstrated the effectiveness of cultured allografts in healing chronic ulcers [19]. We treated ten patients (Table 1) who suffered from chronic therapy-resistant venous or arteriovenous ulcers with subsequent dressings of cultured allogenic keratinocytes.

Material and Methods

Skin samples were obtained from living HBsAG- and HIV-negative skin donors, using neonatal human foreskin or adult split skin. Skin specimens were floated in 0.25% Trypsin for 1–2 h at 37° C to split the skin through the

A. Bernd, J. Bereiter-Hahn,
F. Hevert, H. Holzmann (Eds.)
Cell and Tissue Culture Models
in Dermatological Research

Table 1. Grafting of ten patients with chronic ulcers with cultured allogenic keratinocytes

Patient	Age (years)	Sex	Underlying diseases	Mean duration (years)
L. H.	66	f	Venous stasis ulcer	17
G. K.	76	f	Venous stasis ulcer	2.5
S. D.	90	f	Venous stasis ulcer	37
S. J.	60	f	Venous stasis ulcer	10
B. J.	66	f	Venous stasis ulcer	0.5
F. E.	78	f	Arteriovenous stasis ulcer	31
M. J.	59	m	Arteriovenous stasis ulcer	3
P. H.	68	m	Arteriovenous stasis ulcer	10
E. F.	78	f	Arteriovenous stasis ulcer	30
B. J.	16	m	Venous stasis ulcer/neurotrophic ulcer	3

dermoepidermal junction. This procedure is selective for keratinocytes, since fibroblasts require collagenase for optimal isolation [15].

The epidermis was separated from the dermis, and a single cell suspension was prepared from epidermal fragments. Then the keratinocytes were seeded in a serum-free medium [keratinocyte growth medium (KGM) at a density of 2×10^5 cells/cm^2; KGM was purchased from Clonetics Corporation, San Diego, CA].

The basal medium is complete MCDB 153 (3) with amino acid concentrations as described by Pittelkow and Scott [16]. MCDB is supplemented with epidermal growth factor (0.1 μg/ml), insulin (5 μg/ml), hydrocortisone (0.5 *mM*), bovine pituitary extract (0.4%), penicillin streptomycin (100 μg/ml), fungizone (2.5 mg/ml), and a variety of trace metals [15]. The cells were first grown in KGM low calcium (0.1 m*M*) at 37° C, 5% CO_2, 95% air.

A first passage (subcultivation) was possible on days 7–14. The cells were grown to confluence. Postconfluent cells were then cultured in KGM hig calium (1.2 m*M*) to allow the cultures to differentiate and to stratify for a further 7–10 days.

At the end of this period, stratified sheets 7–8 cells thick were obtained. Immediately prior to clinical use the cultures were detached from the Petri dishes after dispase treatment (Boehringer Mannheim, FRG), 1.2 units/ml in DMEM plus 10% fetal calf serum and mounted on Surfasoft (Mediprof, The Netherlands). The grafts showed a translucent layer which was easily applied to the ulcer bed with basal cells next to the ulcer after short preparation by débridement to produce a clean granulation tissue.

After application of the cultured sheets, tulle gras and elastic bandages were applied. Then the patients were confined to bed for 2 days. After 4–5 days the wound was inspected, and subsequent dressings were applied at 4–5 day intervals.

Clinical Results

Seven of ten patients who were subsequently grafted with cultured keratinocyte sheets showed a complete epithelialization after 4–6 weeks. After each application of allografts the healing process of the ulcer wound beds was markedly accelerated, starting with epithelialization of the wound edges or disseminated epithelial islands in the wound bed. The cosmetic results were excellent (Figs. 1–5).

Immediately following application of the allografts all patients noted a marked relief of pain.

There was no evidence of acute rejection or any other adverse effects after healing. All patients showed a remarkable reduction of bacterial contamination. The ulcers were rapidly filled with markedly healthy granulation tissue.

Due to an unfavorable mechanical localization or lack of compliance, three of the ten patients achieved only incomplete epithelialization after 4 weeks.

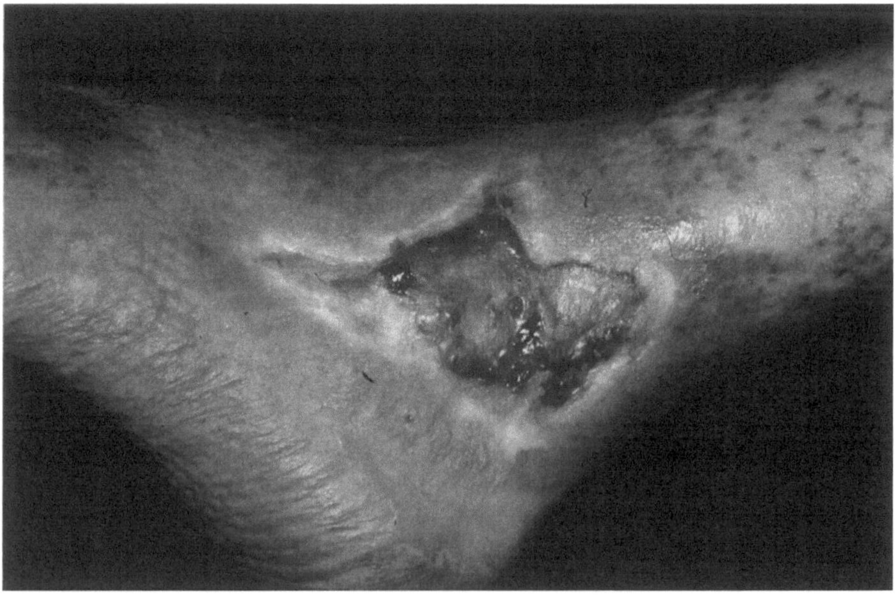

Fig. 1. Chronic end-stage stasis ulceration above the left ankle before grafting

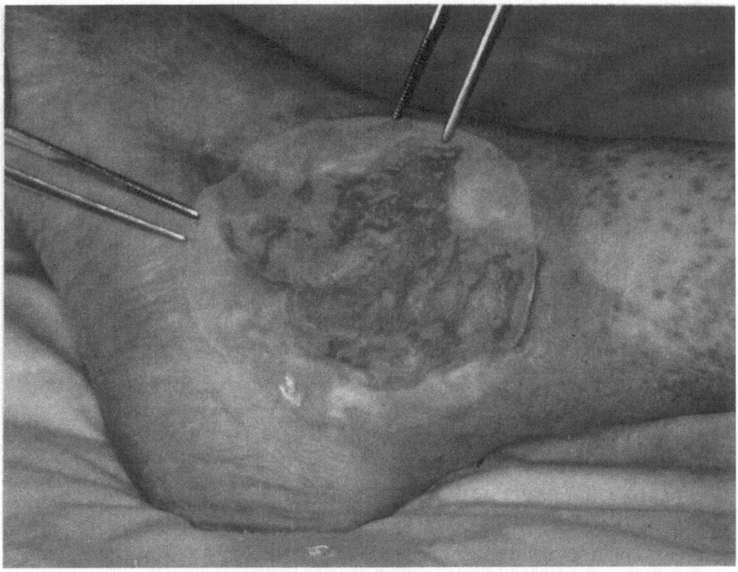

Fig. 2. Translucent sheet of cultured keratinocytes attached to a nonadherent dressing

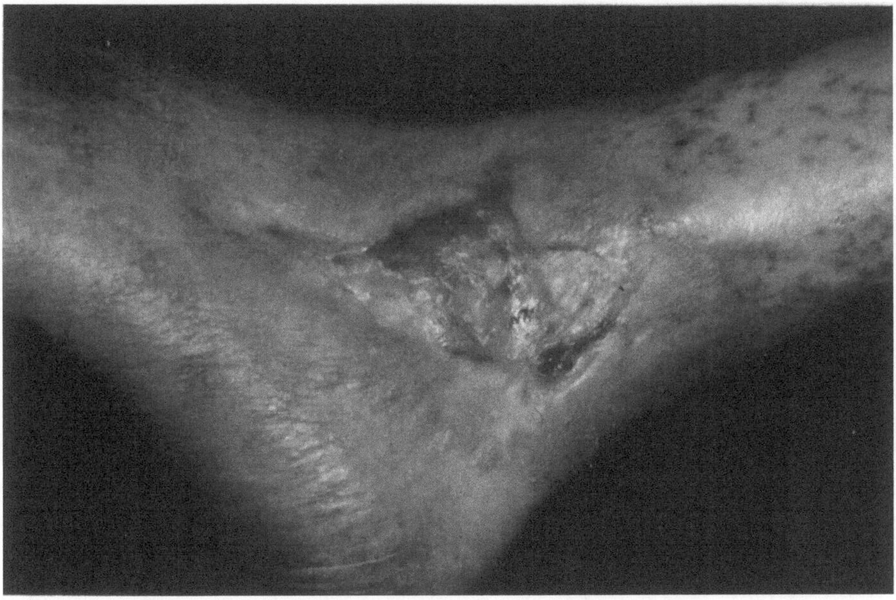

Fig. 3. Excellent cosmetic results after healing (5.5 weeks after subsequent graftings)

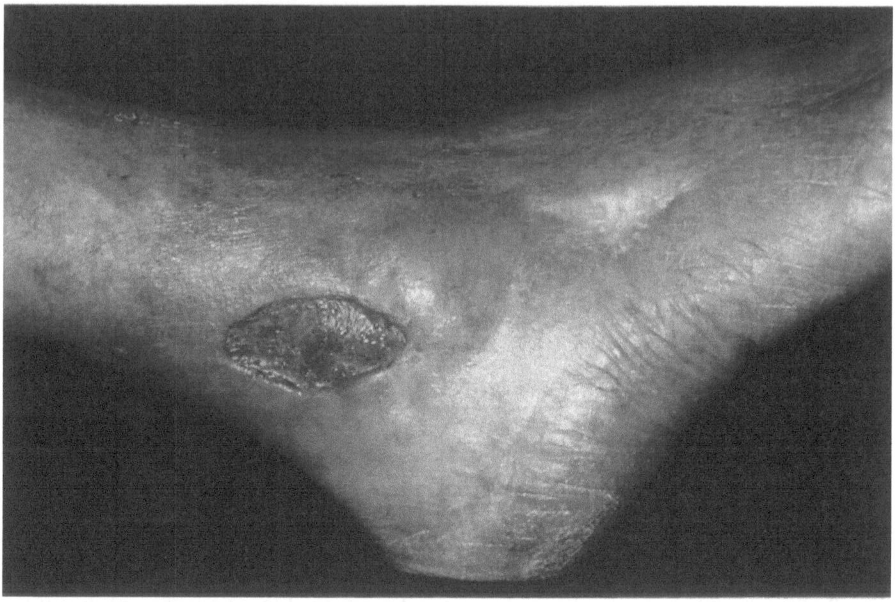

Fig. 4. Chronic arteriovenous stasis ulcer above the right ankle before graftinga00

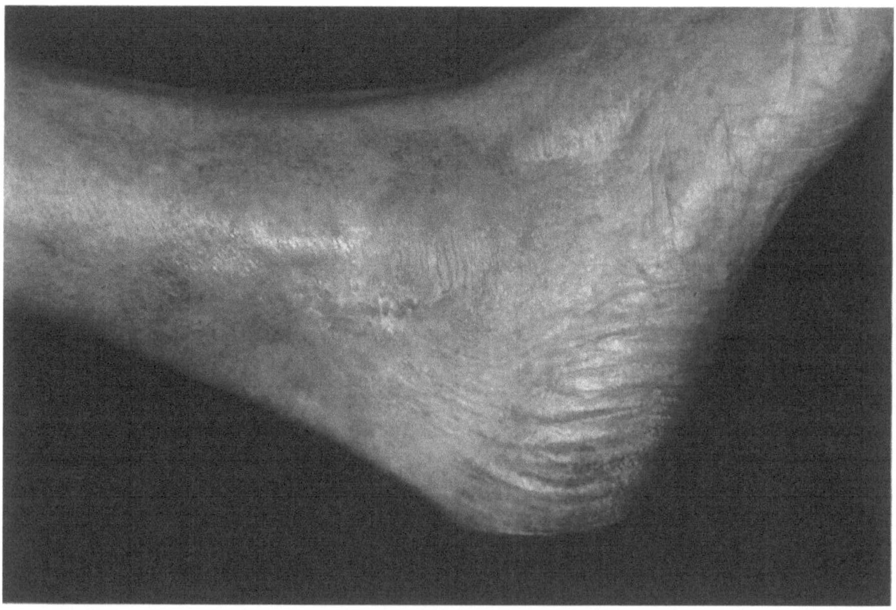

Fig. 5. Excellent cosmetic result after wound healing time of 3 weeks

Table 2. History of keratinocyte grafting

Year	Reference	Method
1869	Reverdin [17]	Transplantation of free epidermis
1950	Billingham and Medawar [2]	Enzymatic separation of dermis and epidermis in vivo
1966	Karasek [12]	Cultivation of human keratinoytes in vitro
1975	Rheinwald and Green [18]	Subcultivation/expansion of human keratinocytes
1979	Eisinger [8]	Growing keratinocytes in culture without dermal matrix
1981	O'Connor et al. [11]	Transplantation with cultured human autografts
1990	Teepe et al. [14]	Transplantation with cryopreserved cultured human allografts

Discussion

The first transplantation of epidermis was described by Reverdin in 1869 (Table 2). More than one century later the ability to grow keratinocytes in culture opened up new perspectives for skin transplantation [20]. Developments have rapidly progressed since 1975, when Rheinwald and Green first described a method for subcultivation and thus expansion of human keratinocytes.

New culture systems with improved culture and growth conditions have been developed, using altered culture media, ph, substrate, and growth factors [13]. In addition, keratinocyte cultures were established on a thick collagen gel with additional fibroblasts, either as "organotypical cultures" with separate layers or as "skin equivalents" with meshed fibroblasts in collagen gel.

Cultured keratinocytes can be used for skin grafting of extensive burns [14], excised giant congenital nevi [7], degloving skin injuries [6], junctional epidermolysis bullosa [4], and chronic skin ulcers [11, 13, 19]. We treated ten patients suffering from therapy-resistant chronic venous ulcers with cultured allografts, since recent studies revealed that cultured epidermis is devoid of class-II MHC antigens and may be used as permanent allografts. Our present results confirmed shortening of the healing time with good cosmetic results. Most recent studies have demonstrated that cultured allografts are not permanently accepted but progressively replaced by the recipient cells [10], although graft survival times is slightly prolonged [1].

The cultured keratinocyte sheet apparently acts as a biological wound dressing, producing epidermal-cell-derived factors (Table 3). These factors seem to stimulate the migration and proliferation of keratinocytes from the wound edges and epidermal appendages, such as the outer root sheath of human hair follicles [19] and the efferent ducts of sweat glands. In addition, fibroblast-induced collagen contraction is inhibited [9]. Further investiga-

Table 3. Epidermal-cel-derived factors produced by cultured keratinocytes

Il1:	Regulates various aspects of T- and B-lymphocyte development and functional activation, including the maturation of thymic T-cell and B-cell precursors. Induction of lymphokine receptor synthesis and expression; induction of lymphocyte proliferation; action on inflammatory cells. Acts as pyrogen
Il3:	Stimulates histamine-liberating cells. Has influence on hematopoietic growth and differentiation
Il6:	Influences the growth and differentiation of T- and B-lymphocytes
Il8:	Activation of neutrophilic granulocytes and chemotoxis
ENKAF:	Stimulation of natural killer cells
KTGF:	(≙ GM-CSF) keratinocyte T-cell growth factor
bFGF in vitro:	Affects cell morphology, proliferation, and differentiation

tions have demonstrated that fibronectin, which is also synthesized by keratinocytes, may provide a scaffold for epithelial and endothelial cells, as well as for fibroblast migration during wound repair [5]. Our findings may have clinical relevance for patient care, as cultured allgrafts are easily available as a biological "off-the-shelf" product and can be transferred rather easily. The grafting requires neither special skill nor anesthesia.

Due to the keratinocyte expansion small skin samples can serve to graft extensive wound areas. Immediately after grafting, pain is remarkably reduced and cosmetic results are good, comparable to those of thin split-skin graftings. On the other hand, keratinocyte culturing initially requires some personal and material effort, with a certain time of training in growing keratinocytes in culture and adequate cell culture equipment. Compared with split-skin grafts in vitro epidermis is more vulnerable, at least during the first 2 weeks, since in vitro epidermis has no stratum corneum, which is the most important barrier against loss of moisture, electrolytes, and proteins, and against invasion of micro-organisms.

At present, we are conducting further experimental studies to substantiate our clinical observation that the application of keratinocyte sheets remarkably reduce the bacterial contamination of wound beds.

References

1. Auböck J, Irschick E, Romani N et al. (1988) Rejection after a slightly prolonged survital time, of Langerhans cell-free allogeneic cultured epidermis used for wound coverage in humans. Transplantation 45:730–737
2. Billingham RE, Medawar P (1950) Technique of free skin grafting in mammals. J Exp Biol 28:385–402
3. Boyce ST, Ham RG (1983) Calcium-regulated differentiation of normal human epidermal keratinocytes in chemically defined clonal cultures and serum-free serial cultures. J Invest Dermatol 81 [Suppl]:33–40
4. Carter DM, Lin AN, Varghese MD et al. (1987) Treatment of junctional epidermolysis bullosa with epidermal autografts. J Am Acad Dermatol 17:246–250

5. Clark RAF, Winn HJ, Dvorak HF, Colvin RB (1983) Fibronection beneath reepithelializing epidermis in vivo. Sources and significance. J Invest Dermatol 80 [Suppl]:265–305
6. Clarke JA, Burt AM, Eldack A, Gusterson BA (1986) Cultured skin for burn injury. Lancet 2:809
7. Dubertret L (1988) Cultured keratinocytes and synthetic skin. In: Orfanos CE, Stadler R, Gollnick H (eds) Proceedings of the 17th World Congress of Dermatology. Springer, Berlin Heidelberg New York, p 731
8. Eisinger M, Lese JS, Hefton JM, Derzynkiewicz JW, Chiao JW, De Harven E (1979) Human epidermal cell cultures: growth and differentiation in the absence of dermal components or medium supplements. Proc Natl Acad Sci USA 76:5340–5344
9. Eisinger M, Sadan S, Silver IA, Flick RB (1988) Growth regulation of skin cells by epidermal cell-derived factors. implications for wound healing. Proc Natl Acad Sci USA 85:1937–1941
10. Gielen V, Faure M, Mauduit G, Thiovolet J (1987) Progressive replacement of human cultured epithelial allografts by recipient cells as evidenced by HLA class-I antigen expression. Dermatologica 175:166–170
11. Hefton JM, Caldwell D, Biozes DG et al. (1986) Grafting of skin ulcers with cultured autologous epidermal cells. J Am Acad Dermatol 14:399–405
12. Karasek M (1966) In vitro culture of human skin epithelial cells. J Invest Dermatol 47:533–540
13. Leigh JM, Purkis PE, Navsaria HA, Phillips TJ (1987) Treatment of chronic venous ulcers with sheets of cultured allogenic keratinocytes. Br J Dermatol 117:591–597
14. O'Connor NE, Mulliken JB, Banks-Schlegel S et al. (1981) Grafting of burns with cultured epithelium prepared from autologous epidermal cells. Lancet 1:75–78
15. Pillai S, Bikle DD, Hincenbergs M, Elias PM (1988) Biochemical and morphological characterization of growth and differentiation of normal human neonatal keratinocytes in a serum-free medium. J Cell Physiol 134:229–237
16. Pittelkow MR, Scott RE (1986) New techniques for the in vitro culture of human skin keratinocytes and perspectives on their use for grafting of patients with extensive burns. Mayo Clin Proc 61:771–777
17. Reverdin SL (1869) Greffe epidermigue. Bull Soc Imp (Paris) 10:511
18. Rheinwald JG, Green H (1975) Serial cultivation of strains of human epidermal keratinocytes: the formation of keratinizing colonies from single cells. Cell 6:331–344
19. Teepe RGC, Koebrugge EJ, Ponec M, Vermeer BJ (1990) Fresh versus cryopreserved cultured allografts for the treatment of chronic skin ulcers. Br J Dermatol 122:81–89
20. Thivolet J, Faure M, Denudem A et al. (1986) Long-term survival and immunological tolerance of human epidermal allografts produced in culture. Transplantation 42:274

Collagen Type-I Is Chemoattractive for the Human Keratinocyte Cell Line (HaCaT)

K. Scharffetter-Kochanek, G. Heinen, S. Lange, K. Kirchberg, G. Goerz, N. E. Fusenig, and *G. Plewig*

Abstract

The migratory response of the human keratinocyte cell line (HaCaT) to collagen type I was investigated in vitro. Beside a minor haptotactic and random migratory response, a major chemotactic component was observed. Denatured random coil formation was less stimulatory compared with the native triple helical conformation of the collagen molecule. The migratory process appears to be mediated by an RGD-independent receptor, which is different from that of fibronectin. Since collagen is degraded and synthesized at sites of tissue injury and inflammation, it is tempting to speculate that collagen type I might function as a migratory stimulus for epidermal cells in vivo, thus enhancing epidermal restoration during cutaneous wound healing.

Introduction

Cutaneous wound healing involves a complexly regulated sequence of biochemical and cellular events which result in complete or partial reconstitution of the skin. A subtle coordination of epidermal and dermal wound healing is a prerequisite for preventing hypertrophic and keloid scar formation, wound dehiscence, and atrophy. During cutaneous restoration newly synthesized dermal matrix proteins are thought to play a central role in the regulation of dermal-epidermal interactions. Fibronectin has been shown to stimulate epithelial cell migration [6, 8], thus promoting reepithelialization. In contrast, laminin specifically inhibits keratinocyte migration and therefore may signal keratinocytes to stop migration when reepitelialization has been completed [25]. Collagen type I has been also found to be chemotactic for fibroblasts and monocytes [13, 14]. Collagen is the major newly synthesized structural protein during cutaneous wound healing, and following dissolution of the basement membrane it is directly exposed to epidermal cells. Its stimulatory effect on epidermal cell migration has been discussed previously [8, 25]; however, it has not been characterized in detail. We therefore studied the effect of collagen type I on HaCaT cell migration using a modified blindwell-type Boyden chamber. Our results suggest that this

A. Bernd, J. Bereiter-Hahn,
F. Hevert, H. Holzmann (Eds.)
Cell and Tissue Culture Models
in Dermatological Research

epidermal cell line is specifically attracted to collagen type I in a dose-dependent fashion involving both directed and undirected migration.

Material and Methods

Cell Culture

HaCaT cells are a spontaneously transformed human keratinocyte cell line with a differentiation pattern similar to that of normal keratinocytes [5, 18]. HaCaT cells were cultivated in DMEM supplemented with penicillin (400 U/ml), streptomycin (50 μg/ml), glutamine (300 μg/ml), and 10% FCS in a humified atmosphere of 5% CO_2 and 95% air at 37°C. Cells were detached for passaging or for the assay with 0.1% trypsin, 0.04% EDTA in PBS.

Proteins

BSA, RGD peptides, and fibronectin were purchased from Boehringer Mannheim (Mannheim, FRG), gelatin from Bio Rad (Munich), and collagen type I from Biomol (Hamburg, FRG). Purity was verified by PAGE.

Cell Motility Assays

Determination of Chemotaxis

Chemotactic cellular migration was assessed using modified Boyden chambers as described elsewhere [12]. Briefly, increasing concentrations of collagen were diluted in DMEM and added to the lower compartment, and polycarbonate filters (13-mm diameter, 8- μm pore size, Nucleopore Corp., Pleasonton, CA, USA) were placed above. After adding 2×10^5 cells to the upper compartment, chambers were incubated at 37°C for a period of 6 h. At the end of the assay, filters were removed, fixed, and stained with Diff-Quick (Merz and Dade, Düdingen, Switzerland) and then mounted on glass slides. Cells that had migrated were quantitated by light microscopy under 200 × magnification. For each replicate, the number of cell in five randomly chosen microscope fields was determined and counts were averaged. Random migration was evaluated by checkerboard analysis as previously described [26].

Determination of Haptotaxis

Haptotactic cellular migration was assessed in triplicates using modified Boyden chambers as described above. Filters were pretreated with gelatin as published elsewhere [1] and then coated with collagen, either on the lower side or on both sides of the filter [11]. Negative control filters were floated on DMEM alone. Chambers were assembled with DMEM in the lower compartment. All filters were washed five times with DMEM and placed in the chamber so that the side of the filter with the higher concentration of protein faced the lower compartment. Following addition of 2×10^5 cells to the upper compartment, the assay was processed and migration was quantitated as for chemotaxis.

Determination of Collagen on Filters

The binding of ^{125}I-labeled collagen was quantitated on gelatin-coated or uncoated filters following incubation in the chambers using a gamma liquid scintillation counter. The amount of collagen bound to the filter was calculated by the following formula:

$$\frac{\text{CPM } ^{125}\text{I on filter} \times \text{amount protein in well} = \text{amount protein bound to the filter}}{\text{CPM total available } ^{3}\text{H.}}$$

Diffusion of Collagen Through Filter

Experiments were performed using ^{125}I-labeled collagen to determine if the filters used for migration maintained a concentration gradient of this attractant. Radioactive collagen was added at a concentration of 100 μg/ml to the lower compartment of the Boyden chamber. After assembling of the chamber, medium was filled into the upper compartment. Samples were removed from the upper compartment sequentially during a 6-h period. Radioactive collagen diffusing to the upper compartment was quantitated in a gamma liquid scintillation counter.

Digestion of Collagen Type I with Bacterial Collagenase

Native unlabeled collagen at concentrations of 100 μg/ml and 5 μg/ml was digested with clostridial collagenase (Wortington, CLSPA, Biochrom, Berlin, FRG) as described previously [1]. Digestion was stopped after various periods of time. The digests of collagen at initial concentrations of 100 μg/ml were used for migration experiments. The digests of collagen at initial concentrations of 5 μg/ml were subjected to SDS/PAGE [10] using an 8% polyacrylamide gel. Gels were stained with Coomassie-blue and photographed.

Preparation of Denatured Collagen

Denatured collagen was prepared by heating pepsin-digested native collagen in 0.5% acetic acid for 30 min at 45° C and subsequently dialyzing in the cold, overnight, against DMEM without supplement as described previously [1].

Effect of RGD Peptides on HaCaT Migration to Collagen

HaCaT cells were preincubated with RGD peptides at concentrations of 25 μg/ml) and 250 μg/ml for 1 h at 25°C before assays. Chemotaxis to soluble collagen (100 μg/ml) or to soluble fibronectin (25 μg/ml) was evaluated in the presence of RGD peptide in the upper compartment of the Boyden chamber.

Results

Chemotactic Response of HaCaT Cells to Collagen Type I

HaCaT cells migrated in a concentration-dependent manner to soluble collagen. In Fig. 1, representative concentration-response curves of stimulated chemotactic motility are shown. Migration declined at concentrations higher than 100 μg/ml, which elicited a maximal response, resulting in a bimodal curve. Conversion of native triple helical to the random coil conformation of the constituent alpha chains of collagen results in a reduced but not totally abolished chemoattraction (Fig. 1).

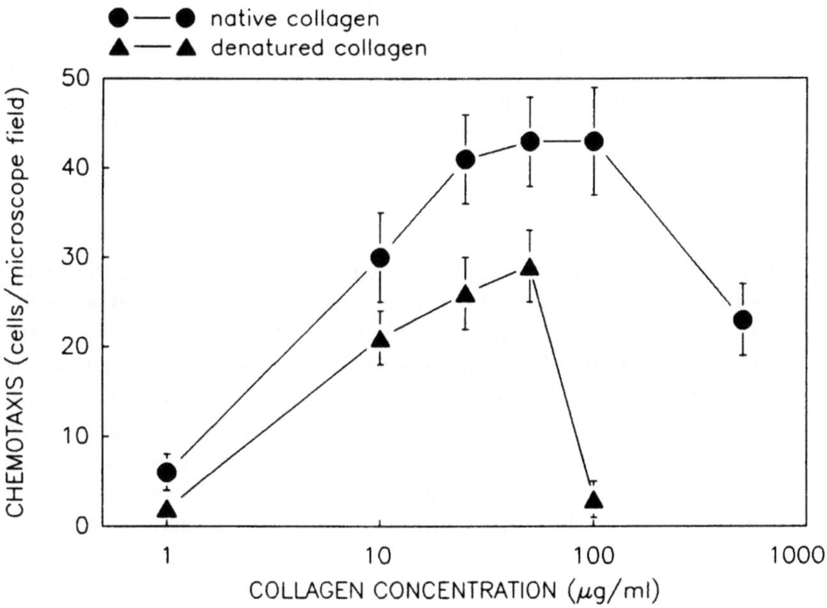

Fig. 1. Dose-dependent effect of native triple helical and denatured collagen type I on HaCaT cell chemotaxis. Chemotaxis was carried out as detailed in "Material and Methods". Number of migrated cells/microscope field are shown as a function of various concentrations of collagen type I. Three independent experiments were performed, each in triplicate. Determinations are expressed as the mean number of migrated cells per microscope field plus or minus the standard error of the mean ($\bar{x} \pm$ SEM)

lower compartment \ upper compartment	0	5	50	250
0	0	1,4 ± 1	2 ± 1	3 ± 1
5	16 ± 1	11 ± 2	9 ± 2	3 ± 2
50	65 ± 2	45 ± 2	35 ± 3	8 ± 1
250	51 ± 1	39 ± 4	38 ± 2	6 ± 2

Fig. 2. Checkerboard analysis of collagen-induced HaCaT cell migration. Varying concentrations of soluble collagen were added to the upper compartment with the cells or to the lower compartment, as indicated. *Boldface vertical* data indicate migration of HaCaT cells to collagen in the lower wells (chemotaxis). Data along the *diagonal* indicate HaCaT cell migration in the presence of equal concentrations of collagen above and below the filter (random migration). Motility response is expressed as the mean of triplicate determinations of migrated cells per microsope field ($\bar{x}$ ± SEM)

Checkerboard analysis (Fig. 2) was carried out with collagen in solution to assess the random or directed (chemotactic) nature of HaCaT cell migration. Chambers were assembled with different ratios of collagen above and below the filter. Data above the diagonal represent migratory responses to a negative gradient, while those below the diagonal represent responses to a positive gradient (chemotaxis). The values along the diagonal reflect random migration to equal concentrations of collagen added to both sides of the chamber. Checkerboard analysis showed that collagen induced both random and directed motility with a major chemotactic response (values below the diagonal). Diffusion of ^{125}I-labeled collagen across the filter during a 6-h incubation period was about 15% (Fig. 3), indicating that the concentration gradient was maintained. To confirm that collagen remained in solution under the assay conditions, binding of ^{125}I-labeled collagen was assessed. At the end of the incubation only small amounts of the radioactivity initially added to the lower well were associated with the filter (data not shown).

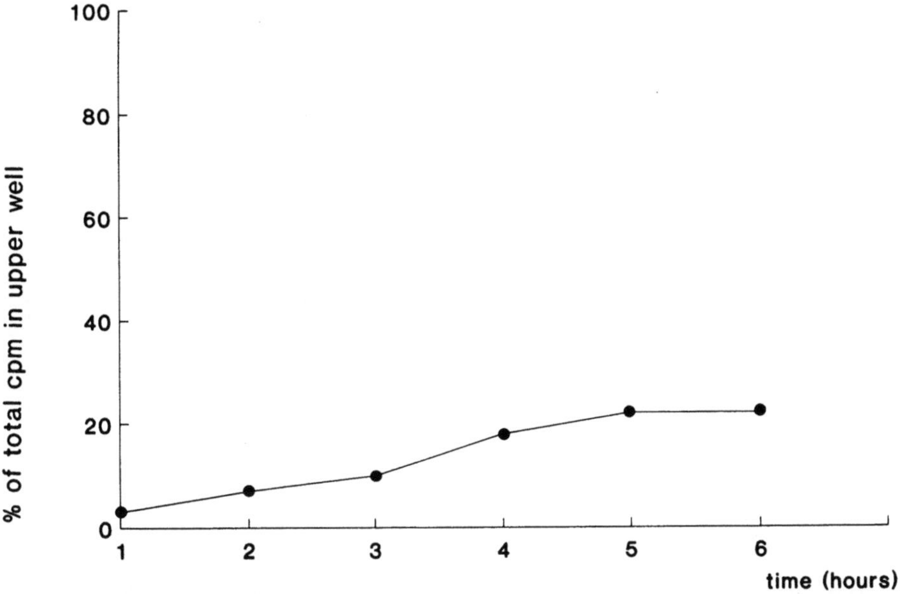

Fig. 3. Diffusion of ^{125}I-collagen across the filter during a 6-h incubation period. Aliquots of ^{125}I-collagen were added to the lower well of the Boyden chamber. The chamber was assembled with an 8.0 µm filter and medium was taken from the upper well at times indicated and analyzed for the presence of radioacitivity. Data represent the mean of duplicate determinations. Standard deviations were $\bar{x} \leq 10\%$

Haptotactic Migratory Response of HaCaT Cells to Collagen Type I

To better characterize directed motility of HaCaT cells to collagen type I, migration over a step gradient of filter-bound collagen in the absence of soluble collagen was tested. Prior to coating filters with collagen type I, filters have to be pretreated with gelatin. Gelatin-pretreated filters are able to bind high amounts of collagen (30%) as opposed to the low binding capacity of filters which have not been treated with gelatin prior to collagen coating (0.03%). Significant migration occurred when the filter was coated on the distal side in the absence of soluble collagen. The migratory response was reduced by 30% when the same concentration of collagen was used to coat both the upper and the lower side of the filter (data not shown).

Migratory Response of HaCaT Cells to Collagenase or Digested Collagen

When collagen type I was digested with bacterial collagenase, which is known to completely digest the triple helix of the collagen molecule, cell

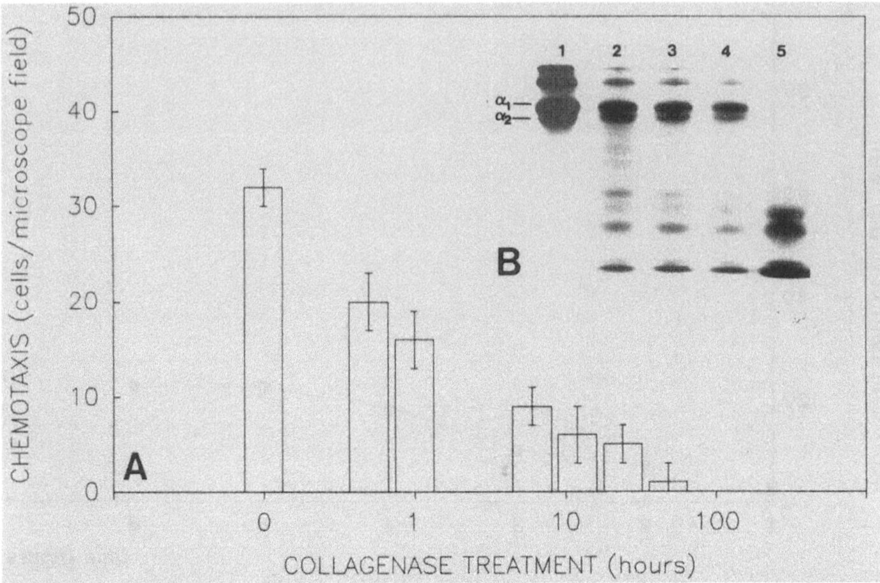

Fig. 4. A Chemotactic response of HaCaT cells to collagen type I (100 μg/ml) after collagenase treatment as a function of incubation time. Native collagen was digested with clostridial collagenase as described in "Material and Methods". Digests were then evaluated for their chemoattractive properties using the Boyden chamber assay. Data represent the mean number of migrated cells per microscope field ($\bar{x} \pm$ SEM). Determinations were done in triplicate. **B** After digestion of collagen with collagenase for various periods of time, digests were subjected to polyacrylamide gel electrophoresis (*lanes 2–5*). *Lane 1,* native collagen (5 mg/ml); *lane 2,* native collagen digested with collagenase for 0.5 h; *lane 3,* native collagen digested with collagenase for 1 h; *lane 4,* native collagen digested with collagenase for 12 h; *lane 5,* native collagen digested with collagenase for 24 h

migration was totally abrogated. Collagenase was not chemotactic itself (data not shown). As shown in Fig. 4A, loss of chemoattraction of collagen type I was clearly dependent on the duration of collagenase digestion. These data corresponded to the results of PAGE, showing a time-dependent decrease of the specific α_1, 2_2, β- and components of collagen type 1 (Fig. 4B).

Preincubation of HaCaT Cells with RGD Tripeptides Prior to Chemotaxis to Collagen Type I and Fibromectin

Following preincubation with RGD tripeptides no inhibition of the chemotaxis to collagen type I was apparent, whereas chemotaxis to fibronectin was found to be almost completely inhibited (Fig. 5).

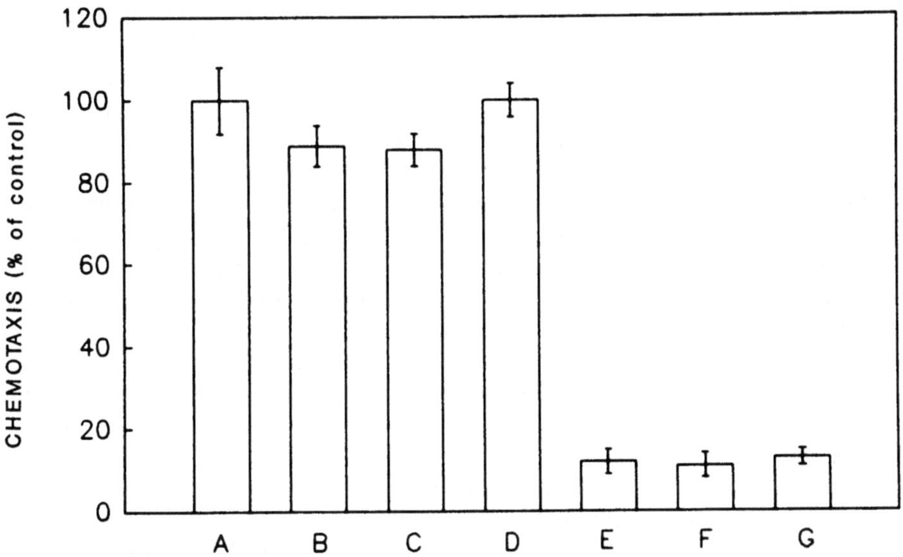

Fig. 5. Effect of RGD peptides on the migratory response of HaCaT cells to collagen and to fibronectin. HaCaT cells were pretreated with RGD peptides at the indicated concentrations for 1 h at 25° C before migration assays. Migration assays were carried out in the presence of RGD peptides in the upper compartment of the Boyden chamber. Data represent the mean of triplicate determinations of migrated cells per microscope field plus or minus SEM. *A,* No pretreatment, 100 μg/ml collagen in the lower compartment of the Boyden chamber; *B,* pretreatment with 25 μg/ml RGD, 100 μg/ml collagen in the lower compartment of the Boyden chamber; *C,* pretreatment with 250 μg/ml RGD, 100 μg/ml collagen in the lower compartment of the Boyden chamber; *D,* no pretreatment, 25 μg/ml fibronectin in the lower compartment of the Boyden chamber; *E,* pretreatment with 25 μg/ml RGD, 25 μg/ml fibronectin in the lower compartment of the Boyden chamber; *F,* pretreatment with 250 μg/ml RGD, 25 μg/ml fibronectin in the lower compartment of the Boyden chamber; *G,* pretreatment with 250 μg/ml RGD, 250 μg/ml RGD in the lower compartment of the Boyden chamber

Discussion

The results of this study indicate that the migratory response of the human keratinocyte cell line (HaCaT) to collagen is complex, including a major chemotactic and a minor haptotactic as well as a random migratory response. Experiments were performed to distinguish chemotaxis involving the directed migration of cells in response to soluble concentration gradients of an attractant from haptotaxis, which is due to substratum-bound attractant [7]. A dose-dependent chemotactic response was observed at low concentrations (up to 100 μg/ml) whereas at higher concentrations of collagen chemotaxis was significantly decreased. This bimodal response has been reported for other chemoattractants and other cells [1, 20, 2]. Only a small amount of ^{125}I-collagen (0.3%) bount to the untreated filters under the assay conditions used for chemotaxis.

In addition, as postulated for chemotaxis, a concentration gradient of collagen was maintained even after a 6-h incubation period. Checkerboard analyses were performed to determine whether or not the migration was exclusively directional. These results indicate that maximal migration of HaCaT cells to collagen occurred in response to positive gradients of collagen (chemotaxis). However, to a lesser extent, migration also occurred in response to challenge with increasing levels of collagen added simultaneously on both sides of the filter (random migration). Thus, migration of HaCaT cells was due to increased chemotaxis as well as to accelerated random migration. Furthermore, there is some evidence that haptotaxis is also involved in HaCaT cell migration to collagen.

The specific effect of collagen type I on HaCaT cell migration was further substantiated by time-dependent inhibition of epidermal cell migration following disintegration of the triple helix into small fragments by bacterial collagenase.

Transition of the triple helical to the random coil conformation of the alpha-chains was associated with some, but not total loss of the chemoattractive activity. This suggests that the ordered triple helical configuration may contribute significantly to efficient contact between the collagen molecule and its corresponding receptors on the surface of HaCaT cells.

Attractant-receptor interactions are thought to play a central role in mediating cellular migration [24]. Our results provide evidence that the putative receptor involved in migration of the human keratinocyte cell line HaCaT to collagen is different from that of fibronectin. This has been postulated by Donaldson et al. [8]. However, the migration assay applied by these authors cannot distinguish conclusively proliferation from migratory effects. The fibronectin receptor involved in migration has earlier been characterized as a 140-kD glycoprotein which interacts with the RGD peptide (Arg-Gly-Asp) [6, 17] of the fibronectin molecule [2,3]. Identical peptides have been found in the triple helical domain of the collagen molecule [4]. For this reason, we investigated whether or not the RGD peptides within the collagen molecule are involved in mediating chemotaxis to collagen. Inhibition studies with RGD peptides did not result in a decreased migration to collagen, whereas migration of HaCaT cells to fibronectin was almost abolished by pretreatment of the cells with RGD peptides. These results indicate that the receptor mediating migration to collagen is different from the receptor responsible for migration to fibronectin. Recently, an RGD-independent receptor for collagen belonging to the integrin family [19, 22] was characterized. However, it remains to be established whether or not this integrin receptor is involved in the chemotactic attraction of epidermal cells to collagen.

The biological significance of the finding that collagen type I is chemotactic for the human keratinocyte cell line HaCaT can only be speculated upon, and it remains to be elucidated whether human primary keratinocytes reveal a similar response. It is, however, tempting to envisage a minimal sequence of events. During cutaneous restoration, epidermal cells – physiologically separated from the dermis by the basement membrane – are in direct contact with the newly synthesized collagen type I [23]. Since epidermal reconstitution by migration parallels de novo synthesis of collagen type I and other extracellular matrix proteins, it might be reasonable to assume extracellular matrix proteins to be directly involved in chemotaxis of epidermal cells, as has already been shown for fibronectin [6]. Interstitial collagen types I and III have been found to be sequentially synthesized following initial production of fibronectin [9] during cutaneous restoration, thus possibly perpetuating migration required for epidermal reconstitution.

Acknowledgement. We wish to thank W. Neuse for excellent phototechnical work and V. Sahnen for patiently typing the manuscript. This work was supported by the *Deutsche Forschungsgemeinschaft* (Scha 411/1–2).

References

1. Albini A, Adelmann-Grill BC (1985) Collagenolytic cleavage products of collagen type I as chemoattractants for human dermal fibroblasts . Eur J Cell Biol 36:104–107
2. Albini A, Richter H, Pontz BF (1983) Localization of the chemotactic domain in fibronectin. FEBS Lett 156:222–226
3. Albini A, Allavena G, Melchiori A, Giancotti F, Richter H, Comoglio PM, Parodi S, Martin GR, Tarone G (1987) Chemotaxis of 3T3 and SV 3T3 cells to fibronectin is mediated through the cell attachment site in fibronectin and a fibronectin cell surface receptor. J Cell Biol 105:1867–1872
4. Bernard MP, Myers JC, Chu M-L, Ramirez F, Eikenberry EF, Prockop D (1983) Structure of a cDNA for the pro 2 chain of human type I procollagen. Comparison with chick cDNA for pro 2 (I) identifies structurally conserved features of the protein and the gene. Biochemistry 22:1139–1145
5. Boukamp P, Petrusseva RT, Breitkreuz D, Hornung I, Markham A, Fusenig NE (1988) Normal keratinization in a spontaneously immortalized aneuploid human keratinocyte cell line. J Cell Biol 106:761–771
6. Boxerson J, Sergente N (1982) Chemotactic response of epithelial cells in response to fibronectin. Cancer Res 42:2547–2551
7. Carter SB (1987) Haptotaxis and the mechanism of cell motility. Nature 213:256–260
8. Donaldson DJ, Mahan JT, Smith GN (1988) Newt epidermal cell migration over collagen and fibronectin involves different mechanisms. J Cell Sci 90:325–333
9. Kurkinen M, Vaheri A, Roberts PJ, Stenman S (1980) Sequential appearance of fibronectin and collagen in experimental granulation tissue. Lab Invest 43:–47–51
10. Laemmli UK (1970) Cleavage of structural proteins during the assembly of the head of bacteriophage T4. Nature 227:680–685
11. McCarthy JB, Palm SL, Furcht LT (1983) Migration by haptotaxis of a Schwann cell tumor line to the basement membrane glycoprotein laminin. J Cell Biol 97:772–777
12. Postlethwaite AE, Synderman R, Kang AH (1976) The chemotactic attraction of human fibroblasts to a lymphocyte-derived factor. J Exp Med 144:1188–1203
13. Postlethwaite AE, Synderman R, Kang AH (1976) Collagen and collagen peptide-induced chemotaxis of human blood monocytes. J Exp Med 144:1188–1202
14. Postlethwaite AE, Seyen JM, Kang AH (1978) Chemotactic attraction of human fibroblasts to type I, II and III collagens and collagen-derived peptides. Proc Natl Acad Sci USA 75:871–875
15. Postlethwaite AE, Kesky-Oja J, Balian G, Kang AH (1980) Induction of fibroblast chemotaxis by fibronectin: localization of the chemotactic region to a 140000 molecular weight non-gelatin-binding fragment. J Exp Med 153:494–499

16. Ruoslahti E, Pierschbacher MD (1986) Arg-Gly-Asp: a versatile cell recognition signal. Cell 44:517–518
17. Ruoslahti E, Pierschbacher MD (1987) New perspectives in cell adhesion – RGD and integrins. Science 238:491–497
18. Ryle CM, Breitkreuz D, Stark H-J, Leight IM, Steinert PM, Roop D, Fusenig NE (1989) Density-dependent modulation of synthesis of keratins 1 and 10 in the human keratinocyte line HaCaT and in *ras*-transfected tumorigenic clones. Differentiation 40:42–54
19. Santoro A (1986) Identification of a 160000-Dalton platelet membrane protein that mediates the initial divalent cation-dependent adhesion of platelets to collagen. Cell 46:913–920
20. Senior RM, Griffin GL, Mecham RP (1982) Chemotactic responses of fibroblasts to tropoelastin and elastin-derived peptides. J Clin Invest 70:614–618
21. Senior RM, Griffin GL, Huang DA, Walz A, Deuel T (1983) Chemotactic activity of platelet alpha granule proteins for fibroblasts. J Cell Biol 96:382–385
22. Staatz WD, Walsh JJ, Pexton T, Santoro S (1990) The $\alpha_2\beta_1$ integrin cell surface collagen receptor binds to the α_1(I)-CB3 peptide of collagen. J Biol Chem 265:4778–4781
23. Stanley JR, Alvarez DM, Berke EW, Eaglestein WH, Katz SI (1981) Detection of basement membrane zone antigens during epidermal wound healing in pigs. J Invest Dermatol 77:240–243
24. Synderman R, Goetzel EJ (1981) Molecular and cellular mechanisms of leukocyte chemotaxis. Science 213:830–837
25. Woodley DT, Bachmann PM, O'Keefe EJ (1988) Laminin inhibits human keratinocyte migration. J Cell Physiol 136:140–146
26. Zigmond SH, Hirsch GJ (1973) Leukocyte locomotion and chemotaxis. new methods for evaluation and demonstration of a cell-derived chemotactic factor. J Exp Med 137:387–410

5. *Models in Fibroblast Function and Collagen Metabolism*

*The Molecular Biology of the Terminal Differentiation of Dermal Fibroblasts in the Fibroblast Stem Cell System In Vivo and In Vitro**

K. Bayreuther, P. I. Francz, and *H.-G. Meinrath*

Abstract

Fibroblast populations, that earlier were dealth with as homogeneous non-differentiating cell populations, have now been demonstrated to be developing cell populations of the fibroblast stem cell systems, when analyzed with stem cell system-specific concepts and methods. Such undefined fibroblast populations have been the predominant cell systems for qualitative and/or quantitative studies in general cell biology, biochemistry, and virology, and more specialized issues like differentiation, aging, apoptosis, and transformation. Many data resulting from such investigations are inconsistent, to such an extent that many scientists consider the fibroblast cell system to be unsuited for studies of the molecular mechanisms of developing fibroblast populations. Defined fibroblast populations with up to 10^8 cells of one cell type or a constant composition of several cell types can be made available. Cell-biological and biochemical studies of properly defined fibroblast populations that are pools of cell populations with a verified constant composition of fibroblast cell types, and thus with identical biochemical parameters, have provided reproducible results in our hands. Evidence is accumulating that makes it very likely that all cell systems are stem cell systems with a design resembling that of the hematopoietic and the fibroblast stem cell system. It will be of utmost importance for future progress in molecular cell biology when the nature of the cell populations under study is clarified. In case cell populations under study are part of a differentiating stem cell system, it would be unavoidable to apply stem cell system-specific concepts and methods if future studies should yield reproducible and meaningful results.

Introduction

Higher eukaryotic animal and human organisms are made up of organs which in turn are composed of tissues, which are built up of cell systems. A

* Supported by grants from the *Deutsche Forschungsgemeinschaft* and the *Bundesministerium für Forschung und Technologie.*

A. Bernd, J. Bereiter-Hahn,
F. Hevert, H. Holzmann (Eds.)
Cell and Tissue Culture Models
in Dermatological Research

human body is made up of about 200 cell systems [4], the majority of which are cell systems with specialized functions, e.g., keratinoblasts/keratinocytes. In addition there are two cell systems with helper or accessory functions. These are the glia cell system of the brain and the fibroblast cell system of the connective tissues. It is very likely that all cell systems are stem cell systems with a design resembling that of the hematopoietic stem cell system [9, 10].

We have presented evidence that the primary fibroblast cell system isolated from connective tissues of prenatal and postnatal skin of Valo-chicken (unpublished work), of prenatal and postnatal skin of C3H-mice (unpublished work), of prenatal and postnatal skin und lung of BN-rats [8, 11], and of prenatal and postnatal skin of man [3] have a stem cell system nature. In addition, secondary fibroblast populations derived from the connective tissue of prenatal and postnatal skin of Valo-chicken (unpublished work), C3H-mice (unpublsihed work), BN-rats [8], and man [2,3] were also found to develop along a terminal differentiation sequence typical for stem cell systems.

In the primary stem cell compartment of the prenatal connective tissue of Valo-chicken ex vivo-in vitro, a multipotential stem cell S1 with four developmental potentials for fibroblasts and chondroblasts plus fibroblasts and myoblasts proliferates and segragates two types of stem cells S2 with two potential functions for fibroblasts and chondroblasts or fibroblasts and myoblasts. S2 proliferates and throws off four types of stem cells S3. S3 are committed stem cells with one developmental quality for fibroblast or chondroblast or fibroblast or myoblast. S3 cells move over from the stem cell compartment to the progenitor compartment, in which the progenitor cells proliferate and differentiate along the fibroblast, the myoblast, and the chondroblast progenitor-cell lineage (unpublished work).

Primary and secondary fibroblast populations established from postnatal connective tissue of skin of Valo-chicken (unpublished work), C3H-mice (unpublished work), BN-rats [8, 11], and man [2, 3] in vitro proliferate and differentiate along the differentiation sequence of the mitotic fibroblasts (MF) MF I – MF II – MF III in the fibroblast progenitor compartment. When the mitotic renewal capacity of the fibroblast progenitors is exhausted, MF III moves on to the fibroblast maturing compartment, in which the reversibly or irreversibly postmitotic fibroblasts (PMF) develop along the sequence PMF IV – PMF V – PMF VI – PMF VIIa/PMF VIIb. The postmitotic fibroblast PMF VI is the terminally differentiated end cell of the fibroblast stem cell system. It is the only cell that expresses the full functions of the fibroblast stem cell system. All earlier cells in the stem cell compartments, the progenitor compartments, and the maturing compartments are restricted to mitotic and postmitotic renewal functions. After a tissue- and species-specific period of high metabolic activity, PMF VI either degenerates in the fibroblast-degenerating compartment to PMF VIIa by apoptosis or transforms to PMF VIIb = TMF I (transformed mitotic fibroblast) in the fibroblast transforming compartment. Spontaneous transformation has been observed for PMF VI in primary and secondary populations of C3H-mice (unpublished work), BN-rats [8, 11] and secondary populations of the normal human fibroblasts of the cell line WI-38 (Bayreuther and Gogol, this volume).

Results

Primary Human Fibroblast Populations Ex Vivo – In Vitro

Primary human skin fibroblast populations were established ex vivo – in vitro from postnatal abdominal skin of 45 female donors of four age-groups: 1–20, 20–40, 40–60, and 60 90 years, with 19, 7, 8, and 11 donors, respectively. Primary skin fibroblasts were studied in explant cultures, in mass cultures of low density, and in clonal populations ex vivo – in vitro. Primary fibroblast populations were found to be composed of eight fibroblast cell types, the mitotic fibroblasts (MF), MF I, MF II, MF III in the progenitor compartment, the postmitotic fibroblasts (PMF) PMF IV, PMF V, PMF VI in the maturing compartment, PMF VII a in the degenerating compartment, and PMF VII b in the transforming compartment. MF I is a small spindle-shaped cell with a high mitotic capacity, MF II is an epitheloid cell with intermediate proliferative properties, and MF III is the largest epitheloid cell with a low mitotic capacity. The postmitotic fibroblast PMF IV is a large spindle-shaped cell with the morphology of a myofibroblast, PMF V is a large epitheloid cell, and PMF VI is the largest epitheloid cell, the terminally differentiated end cell. PMF VII a is a cell with a heteropyknotic nucleus and a contracted, degenerating cytoplast. PMF VII b is an epitheloid cell with 10–20 nuclear fragments. The nuclear fragments, together with a small cytoplast, bud through the cell membrane and give rise to transformed mitotic fibroblasts TMF I [3] (Bayreuther and Gogol, this volume).

The composition of primary fibroblast populations that migrate from skin explants during the first 4–7 days is a function of the age of the donor. Such populations are always heterogeneous with regard to their combination of cell types. Populations migrating out of the explant of a 7-year-old donor are predominantly made up of mitotic fibroblasts MF II plus 10%–20% of PMFs; those of a 37-year-old donor are composed mainly of MF II and MF III plus 10%–20% PMFs, and those of an 87-year-old donor are made up of MF III, PMF IV, and PMF V, but mainly of PMF VI [3].

Primary low-density mass cultures of skin fibroblasts extracted from skin material of human donors of different age by collagenase/trypsin treatment can be analyzed 4 days after seeding. At this time the fibroblasts have expressed their morphological characteristics best and not yet started to divide. Primary low-density mass cultures were established from the same skin material as that from which the primary explant populations were set up. The age-dependent changes in the frequencies of the primary mitotic and postmitotic fibroblasts were similar or identical to those found for the mitotic and postmitotic fibroblast cell types in primary explant populations [3].

Five different clone types can be identified in primary clone cultures 14 days after seeding. These consist either of uniform populations of MF I in the fibroblast clone type (CTF) CTF I, MF II in CTF II, and MF III in CTF III, or mixtures of MF I and MF II in CTF I/II, and MF II and MF III in clonetype CTF II/III. Fibroblasts in a postmitotic state at the time of seeding into clonal cultures formed plated cells of the types PMF V and PMF VI. The age-dependent changes in primary clone type frequencies are similar or identical to those found for the mitotic fibroblast cell types in primary explants and low-density mass populations [3].

The reproducible appearance of the eight fibroblast cell types in primary fibroblast populations of 45 donors and the reproducible age-related changes in the composition of the primary fibroblast populations makes it very likely that equivalents of the primary fibroblasts exist in vivo and that age-related changes of these equivalents take place in vivo.

Primary mass populations of donors in the age-group 21–40 years that are predominantly made up of MF II and MF III can be experimentally induced by UV [13] or MMC [5] to differentiate to the postmitotic populations of cell type PMF IV after 4 days, PMF V after 8 days, and PMF VI after 21 days. When the [^{35}S]-methionine polypeptide pattern of the pure primary clonal populations of MF I, MF II, MF III and experimentally induced pure primary populations of PMF IV, PMF V, and PMF VI is compared with that of secondary clonal populations of MF I, MF II, MF III and spontaneously arisen or experimentally induced secondary populations of PMF IV, PMF V, and PMF VI of human skin, the cell-type-specific [^{35}S]-methionine marker polypeptide pattern of the primary and secondary cell types are found to be identical [3].

Primary fibroblast populations from skin and lung of BN-rats of different in vivo age were studied in (ex vivo-in vitro) primary explant cultures, low-density mass cultures, and clone cultures. Again, three mitotic fibroblasts (MF) and five postmitotic fibroblasts (PMF) are found in the three primary cell systems [11]. The age-dependent changes in the frequencies of the fibroblast cell types in primary explant populations were similar to those found in low-density mass cultures. The in vivo age-dependent changes in primary clone type frequencies resembled those found for the mitotic fibroblast cell types in primary explant populations and in low-density mass populations [11].

When skin of mature BN-rats was sliced horizontally and fibroblasts were allowed to migrate from these sections in vitro, the reticular layers of the dermis were found to yield predominantly fibroblasts MF I and MF II, while the papillary layer rendered MF III and PMF IV, PMF V, and PMF VI. These findings make it very likely that there is a stratification of the fibroblast cell types in the dermis with the mitotic fibroblasts in the reticular and the postmitotic fibroblasts in the papillary layer, the earlier postmitotic to be placed in the lower part of the papillary layer, with

PMF IV located in the boundary zone undeneath the basal lamina [11] (unpublished work).

Secondary Human Skin Fibroblasts In Vitro

A secondary human dermal fibroblast population was established from abdominal skin of a female donor, 8 years of age. Mitotic populations of this cell line, designated HH8, underwent 53.6 ± 6.0 cumulative population doublings (CPD) in 302 ± 27 days. When the growth capacity of the mitotic fibroblasts was exhausted and when appropriate methods were applied, mitotic fibroblasts did not degenerate in a phase-III phenomenon [6, 7], but shifted spontaneously to postmitotic fibroblasts. The mean life-span of postmitotic HH8 fibroblasts in stationary culture was 418 ± 35 additional days [2].

Mitotic and postmitotic secondary fibroblast populations of HH8 are heterogeneous, with reproducible changes in the proportions of cell stages of mitotic and postmitotic fibroblasts. Secondary fibroblasts have the same forms and sizes as primary fibroblasts. The reproducible changes in the populations of mitotic and postmitotic fibroblasts in the secondary populations make it evident that secondary fibroblasts develop in one direction along the eight-stage terminal-cell lineage MF I – MF II – MF III – PMF IV – PMF V – PMF VI – PMF VII a/PMF VII b.

As described for primary clonal populations, five different types of clones arise in the secondary clonal cultures. These are pure clones CTF I, CTF II, and CTF III, and the heterogeneous clones CTF I/II and CTF II/III. Fibroblasts in a postmitotic state at the time of seeding into clonal cultures formed plated cells of the types PMF V and PMF VI. Changes in the frequencies of the distinct clone types can be observed as a function of the CPD level of the heterogeneous populations. Thus, the changes in cell type and clone type frequencies in secondary low-density mass cultures and clonal populations are a function of the CPD level of the mitotic mass populations or the period of time of postmitotic mass populations in stationary culture, i.e. of the so-called in vitro age of the populations [2, 3]. They are similar or identical to those found for the cell type frequencies in primary explant populations, low-density mass populations, and clonal populations, which are a function of the in vivo age of the donors [3].

The shifts in the frequencies of mitotic and postmitotic fibroblasts in secondary mass populations of HH8 are accompanied by reproducible alterations of biochemical parameters, for example in the [^{35}S]-methionine polypeptide pattern of the developing mass populations. The [^{35}S]-methionine polypeptide pattern of homogeneous subpopulations of MF I, MF II, MF III, PMF IV, PMF V, and PMF VI, isolated from heterogeneous mass populations by the clonering technique, revealed that the six fibroblast morphotypes studied express their cell-type-specific polypeptide pattern in the heterogeneous populations. Thus, the [^{35}S]-methionine polypeptide pattern of the mixed populations of the cell line HH8 is the sum of the polypeptide patterns of the distinct fibroblast cell types that make up the total population at different stages of development [2, 5].

For the analysis of the biological and biochemical changes accompanying the cytodifferentiation of the secondary fibroblasts of HH8 in the cell lineage, methods have been developed for the selective enrichment of the three mitotic and four postmitotic fibroblasts from the heterogeneous mitotic populations [5, 13]. Almost homogeneous mass populations of MF II with 78%–87% purity can routinely be obtained at the CPD level 28–34, those of MF III with 73%–86% purity at CPD level 48–53. Another cell system suited for this purpose is the quantitative clonal system, with MF I in fibroblast clone type CTF I, MF II in CTF II, and MF III in CTF III, with mean sitzes of 4303, 2352, and 302 cells per clone type [13]. In order to achieve a selective enrichment of four postmitotic fibroblast cell types, methods have been worked out for the experimental induction of the differentiation from a lower state (MF II, MF III) to a higher state (PMF IV, PMF V, PMF VI, and PMF VII a) in the lineage of HH8 by physical methods (UV [13] and electromagnetic fields [14] and chemical agents (MMC [5]).

The [^{35}S]-methionine polypeptide pattern of nearly homogeneous sub- or total populations of the mitotic fibroblasts MF I, MF II, and MF III and their corresponding clonal populations CTF I, CTF II, and CTF III are identical [2, 5] (Bayreuther and Gogal, this volume). Likeweise, postmitotic fibroblast cell types PMF IV, PMF V, PMF VI, and PMF VII a – either spontaneously arisen or experimentally induced – show identical [^{35}S]-methionine polypeptide pattern for cell-type-specific marker proteins [5, 13]. In the lineage of MF II → PMF VI, 14 cell-type-specific marker proteins have been identified in the nuclear and cytoplasmic fraction, 24 cell-type-specific marker proteins in the membrane-bound fraction, and 11 cell-type-specific marker proteins in the secreted protein fraction [5]. Degenerating postmitotic fibroblasts PMF VII a express a new set of cell-tye-specific [^{35}S]-methionine polypeptides, indicating that PMF VII a degenerates under the control of an apoptosis program [3].

Macromolecule content and synthesis have been investigated in nearly homogeneous mass populations of 75%–88% purity of MF II, MF III, PMF IV, PMF V, PMF VI, and PMF VII of the cell line HH8, in an attempt to characterize the differentiating fibroblasts by other biochemical parameters. The DNA content, RNA synthesis, protein synthesis, collagen synthesis, and protein content were determined in fully contact regulated mitotic populations at CPD 30 (78%–87% MF II) and at CPF 52 (73%–86% MF III),that of PMF IV, PMF V, PMF VI, and PMF VII a (all at about 75%–85% purity) in experimentally induced postmitotic fibroblasts in stationary cultures. The DNA content increases in the course of cytodifferentiation. MF II has 2C DNA, MF III and PMF IV contain 4C DNA, PMF V 8C DNA, 64% of PMF VI have 8C DNA, and the remaining 36% have 16C DNA. The degenerating PMF VII have DNA contents in the range of 8C–2C. In nearly homogeneous mass populations of MF II, MF III, PMF

IV, PMF V, and PMF VI, RNA synthesis increases by a factor of 9 (MF II = factor 1), protein synthesis increases by a factor of 4.8, collagen synthesis by 6.8, and total protein content by 11.5, when MF II and PMF VI are compared (unpublished work). Differential degradation of intracellular proteins in human skin fibroblasts of mitotic and postmitotic states in vitro has been described for the cell line HH8 [12].

Secondary fibroblast populations of Valo-chicken (unpublished work), C3H-mice (unpublished work), and BN-rats [8] also develop along the same cell-biological and cell-biochemical differentiation sequence in the progenitor, maturing, degenerating, and transforming compartments of the fibroblast stem cell system. Spontaneous transformation has been observed for primary and secondary populations of dermal fibroblasts of C3H-mice (unpublished work) and BN-rats [8, 11] and in secondary fibroblasts of the normal human embryonic cell line WI-38 (Bayreuther and Gogol, this volume). In all cell populations studied, transformation starts with the fragmentation of the nucleus (karyorrhexis) of PMF VI into 10–20 micronuclei. The micronuclei, surrounded by a small cytoplast, bud through the cell membrane and give rise to transformed cell lines with three transformed mitotic fibroblasts (TMF) TMF I, TMF II, and TMF III and four transformed postmitotic fibroblasts (TPMF) TPMF IV, TPMF V, TPMF VI, TPMF VII a, and TPMF VII b. In TPMF VII b, nuclear fragmentation (karyorrhexis) occurs again. After the budding of the micronuclei plus cytoplast through the cell membrane, three neoplastically transformed mitotic fibroblasts (NTMF) – NTMF I, NTMF II, and NTMF III – and five neoplastically transformed postmitotic fibroblasts (NTPMF) – NTPMF IV, NTPMF V, NTPMF VI, NTPMF VII a, and NTPMF VII b – are established [5, 8] (unpublished work). In the transformed cell lines TMF II represents the stem line with 87%–91% of the cells, in neoplastically transformed cell lines, NTMF II makes up the stem line with 85%–89% of the cells in the population.

Discussion

Primary and secondary fibroblast cell systems of Valo-chicken (unpublished work), C3H-mice (unpublished work), BN-rats [8, 11], and man have been studied with stem cell system-specific concepts and methods. The data reveal that all fibroblast cell systems studied are stem cell systems with 11 cells developing in an 11-stage differentiation sequence [1–3, 5, 8, 11–14] (Bayreuther and Gogol, this volume). The development of the cells is under the control of a genetic program, which is expressed by the reproducible, unidirectional, and cell-type-specific changes of the [^{35}S]-methionine polypeptide pattern in the terminal differentiation cell lineage. Normally, the differentiation sequence proceeds spontaneously, but it can be speeded up by a greater number of physical and chemical agents (UV [13], MMC [5]). Spontaneously arisen and experimentally induced fibroblast cell types marker have identical polypeptide patterns, indicating that normal terminal differentiation and experimentally accelerated terminal differentiation follow the same genetically programmed pathway [5, 13, 14].

As a function of the in vivo and in vitro aging of the primary and secondary fibroblast populations studied, fibroblast populations with reproducible age-dependent distinct compositions of the fibroblast cell types can be found [1–3, 5,8, 11–14] (Bayreuther and Gogol, this volume). These findings support the hypothesis that aging in the fibroblast stem cell systems in vivo and in vitro results from the spontaneous or induced unidirectional progression of the fibroblasts along the multistage terminal differentiation sequence in the morphologically and biochemically recognizable cell compartments of the fibroblast stem cell system. At higher in vivo and in vitro age, the mitotic and postmitotic fibroblasts of the renewing progenitor and maturing compartments in the dermal fibroblast stem cell systems of BN-rats or man are completely or partially depleted [3, 11], leading to a modification or breakdown of the helper function of the dermis for the epidermis.

The terminally differentiated end cell PMF VI degenerates to PMF VII a under the control of a genetic program by apoptosis. In spontaneously occurring or experimentally induced apoptosis, three new cell-type PMF VII a-specific [^{35}S]-methionine polypeptides are expressed than turn off the synthesis of all [^{35}S]-methionine polypeptides by a hitherto unknown mechanism (unpublished work). The evolution of a molecular mechanism for the programmed removal of terminally differentiated end cells is additional evidence for the genetically programmed control of the development of the fibroblast stem cell system.

In the stem cell systems of C3H-mice (unpublished work), BN-rats [8, 11], and man (Bayreuther and Gogol, this volume), transformed cells orginate by nuclear fragmentation and budding from PVM VI. In transformed and neoplastically transformed cell lines, three mitotic and five postmitotic cells with cell-biological and cell-biochemical individuality can be identified. In contrast to the normal fibroblasts, the progression along the differentiation lineage of most of the transformed and neoplastically transformed fibroblasts is blocked at differentiation stages TMF II and NTMF II. Because of this, TMF II and NTMF II are the predominant cell types (stem line) with 84.1% and 78.9% cells of the populations studied (unpublished work).

Fibroblast populations, which have been dealt with to date as homogeneous nondifferentiating cell populations [6, 7], have now been demonstrated to be developing cell populations of the fibroblast stem cell systems when analyzed with stem cell system-specific concepts and methods [1–3, 5, 8, 11–14] Bayreuther and Gogol, this volume). Defined fibroblast populations can be made available for qualitative and/or quantitative studies in general cell biology, biochemistry, and virology, and of more specialized issues like differentiation aging, apoptosis, and transformation. Cell-biological and- biochemical studies of properly defined fibroblast populations, which are pools of cell populations with a verified constant composition of fibroblast cell types and thus with identical biochemical parameters, are destined to provide reproducible results.

References

1. Bayreuther K, Francz PI (1990) Aging in the human dermal fibroblast stem cell system in vivo and in vitro. UCLA Symp Mol Cell Biol [New Ser] 123:205–215
2. Bayreuther K, Rodemann HP, Hommel R, Dittman K, Albiez M, Francz PI (1988) Human skin fibroblasts in vitro differentiate along a terminal cell lineage. Proc Natl Acad Sci USA 85:5112–5116
3. Bayreuther K, Rodemann HP, Francz PI, Maier K (1988) Differentiation of fibroblast stem cells. J Cell Sci [Suppl] 10:115–130
4. Bloom W, Fawcett DW (1975) A textbook of histology. 10th ed. Saunders, Philadelphia
5. Francz PI, Bayreuther K, Rodemann HP (1989) Cytoplasmic, nuclear, membrane-bound and secreted [^{35}S]-methionine-labelled polypeptide pattern in differentiating fibroblast stem cells in vitro. J Cell Sci 92:231–239
6. Hayflick L (1981) Cell death in vitro. In: Bowen ID, Lockshin G (eds) Cell death in biology and pathology. Chapman and Hall, London, pp 243–285
7. Hayflick L, Moorhead PS (1961) The serial cultivation of human diploid cell strains. Exp Cell Res 25:585–621
8. Kontermann K, Bayreuther K (1979) The cellular aging of rat fibroblasts in vitro is a differentiation process. Gerontology 25:261–274
9. Metcalf D (1977) Hemopoietic colonies. Recent Results Cancer Res 61
10. Metcalf D (1988) The molecular control of blood cells. Harvard University Press, Cambridge/MA
11. Mollenhauer J, Bayreuther K (1986) Donor-age-related changes in the morphology, growth potential, and collagen biosynthesis in rat fibroblast populations in vitro. Differentiation 32:165–172
12. Rodemann HP (1989) Differential degradation of intracellular proteins in human skin fibroblasts of mitotic and mitomicyn-C (MMC) – induced post-mitotic differentiation states in vitro. Differentiation 42:37–43
13. Rodemann HP, Bayreuther K, Francz P, Dittmann K, Albiez M (1989) Selective enrichment and biochemical characterization of seven human skin fibroblast cell types in vitro. Exp Cell Res 180:84–93
14. Rodemann HP, Pfleiderer G, Bayreuther K (1989) The differentiation of normal and transformed human fibroblasts in vitro is influenced by electromagnetic fields. Ex Cell Res 182:610–621

Cytokine-dependent Regulation of Human Dermal Fibroblasts Cultured in a Three-dimensional Collagen-Gel

K. Herrmann, V. Kunzelmann, M. Bruns, and *U.-F. Haustein*

Abstract

In the present study we cultured human dermal fibroblasts within type-I collagen gels and in a monolayer. We found that the collagen synthesis of fibroblasts cultured in a gel was significantly reduced when compared with conventional monolayer cultures. Furthermore, we studied the regulation of collagen synthesis by the cytokines interleukin-1β (IL-1), interferon-γ-(IFN-γ),and transforming growth factor β (TGFβ) in monolayer and gel culture. While Il-1 showed no significant influence on either of the culture systems, IFNγ led to different results in a dose-dependent manner, i.e, stimulation at lower doses and inhibition at higher ones in the gel. TGFβ markedly stimulated the collagen synthesis in both monolayer and gel cultures. These data make it obvious that the specific properties of the culture system used are to be considered when studying functional parameters of fibroblasts and their regulation.

Introduction

The classical method for studying the fibroblast metabolism is the in vitro culture in a monolayer. There was a great deal of evidence for an artificial influence of monolayer culture conditions on the cells that did not resemble the in vivo situation [1]. This refered mainly to the conversion of procollagen to collagen and its fibril formation [2]. It has been clearly demonstrated that fibroblasts cultured in a monolayer are "activated" according to their collagen synthesis. These cells expressed a markedly reduced synthesis in a three-dimensional gel [3, 4]. On the other hand, the synthetic capacity of the collagen-degrading enzyme collagenase was strongly enhanced under gel growth conditions [3]. Fibroblasts growing in such a gel stopped their cell division and changed their morphology [5]. One of the reasons for such a different regulation of the fibroblast metabolism is the absence of the cell-matrix interaction within the monolayer culture. It has been shown that these interactions mediated by different adhesion molecules (integrins) play an important role in the in vivo regulation of the fibroblasts.

A. Bernd, J. Bereiter-Hahn,
F. Hevert, H. Holzmann (Eds.)
Cell and Tissue Culture Models
in Dermatological Research

Beside this cell-matrix interaction, many soluble mediators, such as cytokines, can also modulate the fibroblast functions [6]. Among these cytokines, TGFβ plays an important role. For example, in scleroderma the fibroblasts exhibited an elevated collagen synthesis, and in sclerodermatous skin the coexpression of collagen type I mRNA and transforming growth factor β (TGFβ) mRNA was shown by in situ hybridization [7]. There is evidence pointing towards an activation as well as an inhibition of fibroblast functions by the addition of various cytokines to the culture [6, 8, 9]. It was therefore the aim of the present study to investigate the influence of the cytokines interleukin-1 (IL-1), interferon-gamma (IFN), and TGFβ on the collagen synthesis of cultured fibroblasts within a three-dimensional collagen gel as compared with the monolayer.

Material and Methods

Cells

Fibroblasts were established from skin biopsies of healthy volunteers by outgrowth and maintained in Dulbecco's modified Eagle medium(DMEM), supplemented with 10% fetal calf serum, 400 U/ml penicillin, 50 μg/ml streptomycin, 300 μg/ml glutamine, and 50 μg/ml Na-ascorbic acid in a humified atmosphere of 5% CO_2 in air at 37° C. Confluent fibroblast monolayers were propagated by trypsinization (0.1% trypsin, 0.02% EDTA) and replaced 1:2. The cells for the experiments were used in passages 4–8.

Collagen gels were prepared according the protocol of Mauch et al. [1]. Briefly, collagen type I was extracted from rat tail tendons as described by Benoit et al. [10]. The lyophylized collagen was sterilized by washing with ethanol and dissolved in sterile 0.1% acetic acid at a concentration of 1.5 μg/ml. For gel preparation, 0.4 ml of collagen solution was neutralized with 1.6 ml of 1.5 × concentrated DMEM containing 10% fetal calf serum and 0.6 ml of 0.1 *N* NaOH, then mixed with 0.4 ml fibroblast suspension (1.5×10^5/ml). The polymerization of the collagen in bacteriological dishes must take place within a few minutes.

Collagen Synthesis

Confluent monolayer cultures and fibroblast growing in a collagen gel were incubated simultaneously with the cytokines and with 10 μCi L-(2,3-^{3}H)-proline/ml medium for 24 h. After labeling, the cells were harvested and pooled with the medium. The proteins were precipitated with 10% (w/v) trichloroacetic acid containing 1% tannic acid. The precipitate was washed four times and the labeled collagen was quantified after digestion with highly purified collagenase type VII, free of any proteolytic activity. The collagen gels with medium were homogenized and precipitated with 10% (w/v) trichloroacetic acid containing 1% tannic acid. The precipitate was treated in the same way as the precipitate of the monolayer culture. The radioactivity was measured in a liquid scintillation counter (LKB Pharmacia).

Each experiment was done in triplicate and averaged results are expressed as percent ^{3}H-proline incorporation (DPM) per 2×10^3 cells ± standard deviation (SD) compared with the controls (100%).

Cytokines

Human recombinant (hu-r) Interleukin 1β and the hu-r transforming growth factor β1 (TGFβ) were obtained from Boehringer, Mannheim, FRG. Hu-r interferon gamma (IFNγ) was a friendly gift from Dr. Rentschler Arzneimittel GmbH & Co., FRG).

Results

A shown in Fig. 1, the collagen synthesis of human dermal fibroblasts is significantly enhanced (fourfold) in the monolayer compared with the three-dimensional collagen gel. These results are in a good agreement with those in the literature [1, 11]. The total protein concentration in the collagen gel was distinctly diminished, suggesting a reduced cell growth (results not shown). In the monolayer, IL-1 significantly enhances the collagen synthesis in a dose-dependent manner (Fig. 2). This stimulatory effect is seen only in subconfluent fibroblast cultures; in confluent cultures we did not observe a similar effect. In the collagen gel we demonstrated only a slight stimulatory influence of IL-1 on the collagen synthesis (Fig. 2). These findings emphasize a dependence of the II-1 effect on the cell-cell (confluence) and cell-matrix (collegen cell) interactions of fibroblasts .

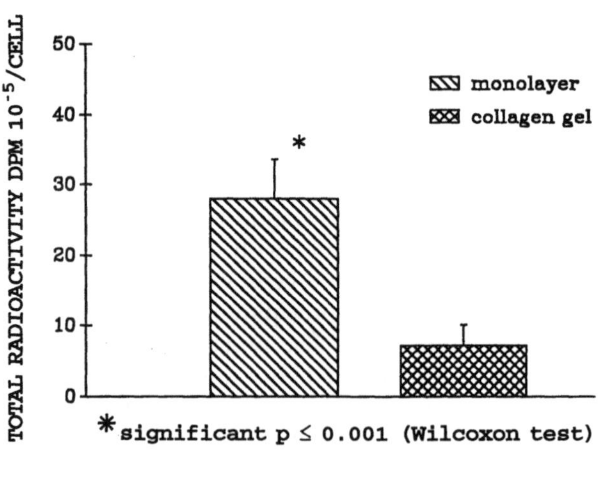

Fig. 1. Collagen synthesis of human dermal fibroblasts cultured in a monolayer and collagen gel. The data are expressed as averages ± standard deviation of six experiments in triplicate

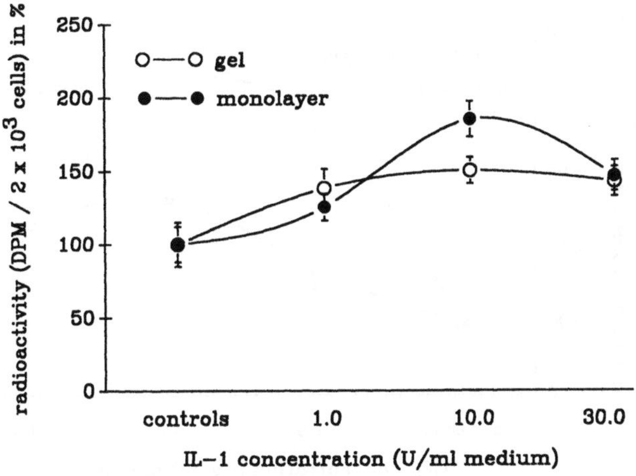

Fig. 2. Influence of interleukin-1 (IL-1) on the collagen synthesis of human dermal fibroblasts cultured in a monolayer and collagen gel. The data dare expressed as averages ± standard deviation of six experiments in triplicate

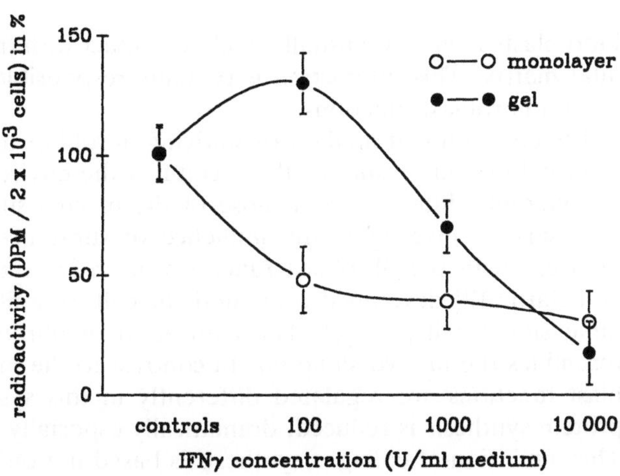

Fig. 3. Influence of interferon-gamma (IFNγ) on the collagen synthesis of human dermal fibroblasts cultured in a monolayer and collagen gel. The data are expressed as averages ± standard deviation of six experiments in triplicate

IFNγ is able to inhibit significantly the collagen synthesis of fibroblasts in the monolayer as well as in the collagen gel (Fig. 3). However, when we used a low IFNγ concentration in the collagen gel, a stimulatory influence on the collagen synthesis was followed by a decrease of the same.

As shown in Fig. 4, TGFβ significantly stimulates the collagen synthesis of fibroblasts cultured in a monolayer as well as in collagen gel. This cytokine exhibits its stimulatory effect in the monolayer approximately fourfold compared with the controls. The same influence, but not so pronounced, was observed under gel conditions.

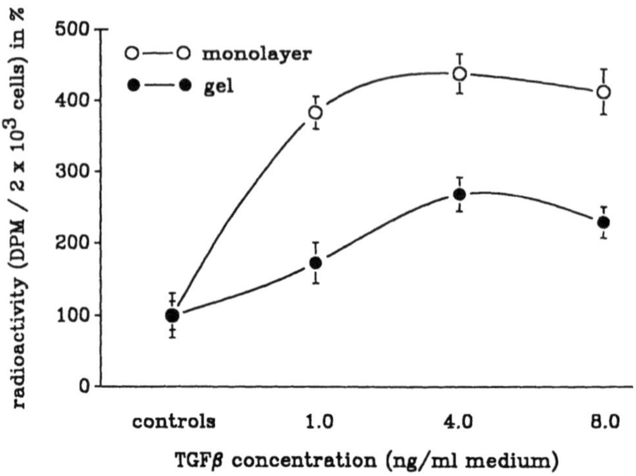

Fig. 4. Influence of transforming growth factor-β (TGFβ) on the collagen synthesis of human dermal fibroblasts cultured in a monolayer and collagen gel. The data are expressed as averages ± standard deviation of six experiments in triplicate

Discussion

Fibroblasts in vivo are usually in close contact with the surrounding extracellular matrix. This interaction is partially responsible for the modulation of several fibroblast functions.

The conventional method of studying fibroblast metabolism is the monolayer culture on plastic. In these cultures the environmental conditions are considerably different from those of the in vivo situation. Therefore, it is necessary to investigate the influence of substances which are known to interfere with the fibroblast functions in an in vivo-like situation. Elsdale and Bard [12] developed a method for culturing fibroblasts using a three-dimensional collagen gel. The cultivation of fibroblasts within such a gel resembles the in vivo situation. In contrast to the monolayer, several fibroblast functions are regulated differently in this system. For example, the protein synthesis is reduced, dramatically especially that of collagen [1, 13]. This inhibition of collagen synthesis is based not only on the transcriptional control but also on the translational regulation observed by quantification of the steady-state mRNA levels, by runoff experiments, and by the determination of the RNA half-life. This might be due to the cell-matrix interactions [1, 14]. After binding on the collagen matrix the fibroblasts develop thick actin bundles, change the organization of their cytoskeleton, and express adhesion molecules on their cell surface [15]. Additionally, fibroblasts embedded in collagen gels have been shown to reduce their proliferation markedly and to change their morphology [1, 16, 17]. In these three-dimensional culture conditions fibroblasts synthesize high amounts of proteolytic enzymes, in particular collagenase, whereas the synthesis of protease inhibitors is still reduced [1, 13]. These findings provide further evidence for the closer analogy of cultures using a three-dimensional collagen gel as the substrata to in vivo conditions.

This different regulation of fibroblasts in the monolayer versus collagen gel leads to the question of whether this difference also applies to the response to the effect of cytokines. It is also likely that cells growing in a three-dimensional network of matrix proteins expressing different adhesion molecules express different cytokine receptors, as well. This could result in a changed response to cytokines.

However, it has been suggested that the collagen synthesis of fibroblasts cultured in a monolayer or collagen gel does not lead to qualitative differences concerning their responsiveness towards cytokines. Due to the influence of the cytokines used, we discovered differences only in the amount of collagen produced. There is one exception: At a concentration of 100 U/ml, IFNγ suppresses the collagen synthesis in the monolayer, while stimulating it in the collagen gel. The reasons for such a stimulation by IFNγ are still unknown.

Our results suggest that it is important to vary the culture conditions when modulators of cell function like cytokines are tested. Only by combining different in vitro test systems can one assess the significance of the in vitro results.

In both culture systems, which themselves influence the collagen production in an opposity way, TGFβ stimulates the collagen synthesis and IFNγ suppresses it. The similar results in both test systems support the significance of the observed modulation of collagen synthesis by cytokines.

References

1. Mauch C, Hatamochi A, Scharffetter K, Krieg T (1988) Regulation of collagen synthesis in fibroblasts within a three-dimensional collagen gel. Exp Cell Res 178:493–503
2. Fessler JH, Fessler LI (1978) Biosynthesis of procollagens. Annu Rev Biochem 47:129–162
3. Mauch C, Krieg T (1990) Fibroblast-matrix interactions and their role in the pathogenesis of fibrosis. Rheum Dis Clin North Am 16:93–107
4. Nakagawa S, Pawelek P, Grinell F (1989) Long-term culture of fibroblasts in contracted gels: effects on cell growth and biosynthetic activity. J Invest Dermatol 93:792–798
5. Schor SL, Court J (1980) Different mechanisms in the attachment of cells to native and denatured collagen. J Cell Sci 41:159–175
6. Perlish JS, Fleischmajer R (1989) Immunological modulation of dermal fibroblasts in scleroderma. In: Norris DA (ed) Immune mechanisms in cutaneous disease. Dekker, New York, pp 650–624
7. Scharffetter K, Lankat-Buttgereit B, Krieg T (1988) Localization of collagen mRNA in normal and scleroderma skin by in-situ hybridization. Eur J Clin Invest 18:9–17
8. Duncan MR, Berman B (1989) Different regulation of collagen, glycosaminoglycan, fibronectin, and collagenase activity production in cultured human adult dermal fibroblasts by interleukin 1-alpha and beta and tumor necrosis factor-alpha and beta. J Invest Dermatol 92:699–706
9. Ignotz R, Massague J (1986) Transforming growth factor beta stimulates the expression of fibronectin and collagen and their incorporation into the extracellular matrix. J Biol Chem 261:4337–4345

10. Benoit J, Cormier M, Wepierre J (1988) Comparative effects of four surfactants on growth, contraction and adhesion of cultured human fibroblasts. Cell Biol Toxciol 4:111–122
11. Coustry F, Gillery P, Marquart FX, Borel JP (1990) Effect of transforming growth factor β on fibroblasts in three-dimensional lattice cultures. FEBS Lett 262:339–341
12. Elsdale T, Bard J (1972) Collagen substrata for studies on cell behavior. J Cell Biol 54:626–637
13. Nusgens B, Merrill C, Lapier C, Bell E (1984) Collagen biosynthesis by cells in a tissue-equivalent matrix in vitro. Coll Rel of Res 4:351–364
14. Hämäläinen L, Oikarinen J,Kivirikko KI (1985) Synthesis and degradation of type I procollagen mRNAs in cultured human skin fibroblasts and the effect of cortisol. J Biol Chem 260:720–725
15. Clark RAF (1990) Fibronectin matrix deposition and fibronectin receptor expression in healing and normal skin. J Invest Dermatol 94:128S–134S
16. Kono T, Tanij T, Furukawa M, Mizuno N, Kitajiama J, Ishii M, Hamada T, Yoshizato K (1990) Parallel arrangement, growth inhibition and cell cycle phase analysis of human dermal fibroblasts cultured in collagen lattice. J Dermatol 17:2–10
17. Kono T, Tanji T, Furukawa M, Mizuno N, Kitajima J, Ishii M, Hamada T, Yoshizato K (1990) Cell cycle analysis of human dermal fibroblasts cultured on or in hydrated type-I collagen lattices. Arch Dermatol Res 282:258–262

Investigation of Components of the Dermal-Epidermal Junction Zone In Vitro: Effect of Transforming Growth Factor-beta on Collagen VII Expression

A. König, and *L. Bruckner-Tuderman*

Abstract

We recently showed that the expression of collagen VII in vitro was stimulated in normal human keratinocytes and fibroblasts when both cell types were co-cultured [8]. Furthermore, in three-dimensional skin equivalents collagen VII was found primarily in the lowest epithelial cells, suggesting that these cells play a dominant role in the synthesis of collagen VII. To clarify the nature of dermal-epidermal interactions, transforming growth factor-beta (TGF-β) was investigated with regard to its effect on collagen VII expression in vitro. As shown by indirect immunofluorescence (IIF), TGF-β at concentration of 5–20 ng/ml exhibited a dose-dependent stimulation of collagen VII expression both in monocultures of keratinocytes and in co-cultures of fibroblasts and keratinocytes. The effect on fibroblasts alone was weak, and was observed only with TGF-β cncentrations of 20 ng/ml or higher. The results indicate that TGF-β may play a role in dermal-epidermal communication, and may be an important regulator of the synthesis of proteins of the dermal-epidermal junction zone.

Introduction

Collagen VII is the major structural component of the anchoring fibrils which form a three-dimensional network below the skin basement membrane (for review see [5]). So far, little is known about the regulation of collagen VII expression and about the biosynthesis of anchoring fibrils. Both fibroblasts and keratinocytes are able to synthesize collagen VII in vitro [2, 14]. However, it remains unclear which cell is responsible for its expression in vivo. We recently showed that, in vitro, the collagen VII expression in keratinocytes and fibroblasts is strongly enhanced when both cell types are co-cultured [8]. There is increasing evidence that growth factors are involved in the regulation of extracellular matrix gene expression [12]. Therefore, in this study, the effect of TGF-β on the expression of collagen VII in vitro was investigated. We show that in normal human keratinocytes, TGF- stimulates collagen VII expression in a dose-dependent manner, as assayed by IIF

A. Bernd, J. Bereiter-Hahn,
F. Hevert, H. Holzmann (Eds.)
Cell and Tissue Culture Models
in Dermatological Research

staining. A similar stimulation was observed in co-cultures of keratinocytes and fibroblasts, whereas in fibroblasts alone the effect was minute.

Materials and Methods

Primary explant cultures of human fibroblasts were initiated from normal skin in Dulbecco's modified minimal essential medum of Eagle (DMEM) with 10% fetal calf serum (FCS). Keratinocytes from trypsinized foreskin were cultured in serum-free keratinocyte growth medium (KGM) (Clonetics Corp., San Diego, CA) with 0.15 m*M* calcium. Cells from the first or second passage were used for the experiments. Co-cultures were established with fibroblasts and keratinocytes in KGM with an initial cell ratio of 1:10. For assaying collagen VII expression, the cells were cultured to subconfluency, either on coverslips or in plastic tissue culture chamber/slides (Miles Laboratories, Inc., Naperville, IL). Prior to IIF staining, ascorbic acid was added to all culture media at a concentration of 50 μg/ml for 2 days. FCS was withdrawn from cultures for the last 2 days. TGF-β2 (a generous gift from Dr. N. Cerletti, Ciba-Geigy AG Basel, Switzerland) was added to the culture media for 48 h at final concentrations of 5, 10, 20, and 30 ng/ml, respectively. For IIF staining, a primary affinity-purified polyclonal anti-human collagen VII antibody [2] and a secondary-FITC-conjugated antibody were used.

Results

The effect of TGF-β upon the expression of collagen VII was investigated in pure fibroblast or keratinocyte cultures, and in co-cultures (Table 1). In fibroblasts the basal expression of type VII collagen was very low, and was stimulated only slightly with concentrations of TGF-β of 20–30 ng/ml. The collagen VII expression in untreated keratinocytes was stronger than that in fibroblasts and was stimulated by TGF-β in a dose-dependent manner. The lowest concentration tested (5 ng/ml) showed a clear enhancement of colla-

Table 1. Effect of TGF-β on collagen VII expression in vitro

TGF-β (ng/ml)	Fibroblasts	Keratinocytes	Co-cultures
0	(+)	+	++
5	(+)	++	+++
10	(+)	++(+)	n.d.
20	+	+++	+++(+)
30	+(+)	+++	nd.

n.d., Not done

gen VII expression, and 20 ng/ml proved to be a maximal effective concentration. In contrast, untreated co-cultures of keratinocytes and fibroblasts showed a clearly enhanced collagen VII expression in both cell types compared with monocultures [8]. In co-cultures, TGF-β stimulated the collagen VII expression dose dependently, similar to the effect seen in keratinocytes but starting on a higher initial level.

Discussion

Collagen VII is the main structural protein of anchoring fibrils. These fibrils are thought to play a crucial role in the tight cohesion between the basement membrane and the underlying connective tissue, e.g., in skin. In several pathological conditions, such as hereditary and acquired epidermolysis bullosa, the anchoring fibrils are structurally or functionally defective [1, 3, 4, 7, 13, 16]. Thus, a great interest exists in understanding the regulation of collagen VII expression in vivo and the biosynthesis of anchoring fibrils. However, it is still not clear which cell type synthesizes collagen VII in vivo, and how this is regulated. In vitro, epithelial cells as well as fibroblasts from human origin have the capacity for collagen VII synthesis [2, 8, 9, 14]. In addition, human keratinocytes in vitro express components reacting with anti-anchoring fibril antibodies [6]. Finally transplantation studies in animals [11] and observations of intraepidermal retention of collagen VII in a patient with recessive dystrophic epidermolysis bullosa [13] suggest an epithelial origin of the anchoring fibrils.

To investigate the regulation of collagen VII expression, in vitro systems were employed. Recently, we showed that the very low basal expression of collagen VII in keratinocytes and fibroblasts is strongly enhanced when both cell types are co-cultured [8]. In three-dimensional skin equivalents with keratinocytes growing on a dermal substitute consisting of a lattice of collagen I and III with embedded viable fibroblasts, collagen VII was observed in the lowest parts of the epidermal equivalent [8]. These results support the hypothesis that epidermal cells are the main manufactures of collagen VII, and that collagen VII synthesis is under the control of mesenchymal-epidermal interactions, possibly via soluble mediators.

There is increasing evidence that the accumulation and maintenance of extracellular matrix components is regulated, at least to some extent, by the multifunctional cytokine TGF-β [10,12, 15].Therefore, to elucidate further the regulatory mechanisms involved in collagen VII expression, we tested the effect of TGF-β on human keratinocytes and fibroblasts. We found that in keratinocytes and in co-cultures of keratinocytes and fibroblasts the expression of collagen VII was stimulated by TGF-β in a dose-dependent manner, whereas in monocultures of fibroblasts the stimulatory effect of even high concentrations of TGF-β was only moderate. In conclusion, our results suggest that TGF-β might have a regulatory function in the expression of collagen VII in vivo.

References

1. Briggaman RA (1985) Is there any specificity to defects of anchoring fibrils in epidermolysis bullosa dystrophica, and what does this mean in terms of pathogenesis? J Invest Dematol 84:371–373
2. Bruckner-Tuderman L, Schnyder UW, Winterhalter KH, Bruckner P (1987) Tissue form of type VII collagen from human skin and dermal fibroblasts in culture. Eur J Biochem 165:607–611
3. Bruckner-Tuderman L, Mitsuhashi Y, Schnyder UW, Bruckner P (1989) Anchoring fibrils and type VII collagen are absent from skin in severe recessive dystrophic epidermolysis bullosa. J Invest Dermatol 93:3–9
4. Bruckner-Tuderman L, Niemi K, Kero M, Schnyder UW, Reunala T (1990) Type VII collagen is expressed but anchoring fibrils are defective in dystrophic epidermolysis bullosa inversa. Br J Dermatol 122:383–390
5. Burgeson RE, Lunstrum GP, Rokosova B, Rimberg CS, Rosenbaum LM, Keene DR (1990) The structure and function of type VII collagen. Ann NY Acad Sci 580:32–43
6. Goldsmith LA, McCoon P, Partridge A, Lane AT (1991) Intraepithelial anchoring fibril components. Arch Dermatol 127:53–56
7. Heagerty AHM, Kennedy AR, Leigh IM, Purkis P, Eady RAJ (1986) Identification of an epidermal basement membrane defect in recessive forms of dystrophic epidermolysis bullosa by LH7.2 monoclonal antibody: use in diagnosis. Br J Dermatolol 15:125–131
8. König A, Bruckner-Tuderman L (1991) Epithelial-mesenchymal interactions enhance expression of collagen VII in vitro. J Invest Dermatol 96:803–808
9. Lunstrum GP, Sakai LY, Keene DR, Morris NP, Burgeson RE (1986) Large complex globular domains of type VII procollagen contribute to the structure of anchoring fibrils. J Biol Chem 261:9042–9048
10. Quaglino D Jr, Nanney LB, Kennedy R, Davidson JM (1990) Transforming growth factor-beta stimulates wound healing and modulates extracellular matrix gene expression in pig skin. Lab Invest 63:307–319
11. Regauer S, Seiler GR, Barrandon Y, Easley KW, Compton CC (1990) Epithelial origin of cutaneous anchoring fibrils. J Cell Biol 111:2109–2115
12. Roberts AB, Heine UI, Flanders KC, Sporn MB (1990) Transforming growth factor-beta. Major role in regulation of extracellular matrix. Ann NY Acad Sci 580:225–232
13. Smith LT, Sybert VP (1990) Intra-epidermal retention of type VII collagen in a patient with recessive dystrophic epidermolysis bullosa. J Invest Dermatol 94:261–264
14. Stanley JR, Rubinstein N, Klaus-Kovtun V (1985) Epidermolysis bullosa acquisita antigen is synthesized by both human keratinocytes and human dermal fibroblasts. J Invest Dermatol 85:542–545
15. Wikner NE, Persichitte KA, Baskin JB, Nielsen LD, Clark RAF (1988) Transforming growth factor-beta stimulates the expression of fibronectin by human keratinocytes. J Invest Dermatol 91:207–212
16. Woodley DT, Burgeson RE, Lunstrum G, Bruckner-Tuderman L, Reese MJ, Briggaman RA (1988) The epidermolysis bullosa acquisita antigen is the globular carboxy terminus of type VIII collagen. J Clin Invest 81:683–687

*The Influence of Retinoids on Chemotaxis and Connective Tissue Synthesis of Fibroblasts**

G. Stumpenhausen, M. Kulozik, R. Hein, T. Oono, J. McLane, G. F. Bryce, C. Mauch, and *T. Krieg*

Abstract

Retinoids exert very profound effects on fibroblast functions. Often they have been shown to lead to a reduction of collagen synthesis in vitro when applied directly to fibroblast cultures. In the skin, as a more complex system, they reveal even other activities and, for example, cause an induction of cytokines in the epidermis. This induction of several cytokines could further lead to an increased collagen production in the lower dermal fibroblasts. To investigate this hypothesis, we conducted several in vitro experiments. First, we studied the influence of three different retinoids [isotretinoin (Ro 4-3780), arotinoid methyl sulfone (Ro 14-9706) and arotinoid ethyl sulfone (Ro 15-1570)] on the chemotactic activity and proliferation of fibroblasts. Further, we tested the influence of conditioned media derived from retinoid-treated keratinocytes in comparison to directly applied retinoids on collagen type I and III synthesis of fibroblasts by northern and dot blot analysis. Finally, we observed the influence of these retinoids on TGF-β1 and of Ro 14-9706 on Il-1α and β expression in human primary keratinocytes by northern blot analysis. None of the retinoids tested were chemoattractive, but all inhibited the chemotactic response to fibroblast-conditioned medium. Also, proliferation was inhibited dose dependently. While Ro 4-3780 showed a concentration-dependent reduction of the α_1(I) and α_1(III) procollagen mRNA steady-state levels, low doses of Ro 14-9706 and Ro 15-1570 first of all increased collagen mRNA levels. Only high concentrations of these retinoids decreased the collagen levels in fibroblast cultures. The inhibiting effects observed with high concentrations of retinoids were abolished by the use of conditioned media derived from keratinocyte cultures treated with the same concentration of retinoids. Furthermore, higher levels of TGF-β1 mRNA were observed in keratinocyte cultures treated with 10^{-5} *M* sulfur-containing arotinoids compared with untreated controls or cultures treated with 10^{-5} *M* Ro 4-3780. Il-1α mRNA levels were raised up to 200% compared with control when 10^{-6} *M* Ro 14-9706 was applied. Il-1β mRNA levels were also elevated, but to a lesser extent. These data demon-

---

* This work presents a major part of the M.D. thesis of G. Stumpenhausen

A. Bernd, J. Bereiter-Hahn,
F. Hevert, H. Holzmann (Eds.)
Cell and Tissue Culture Models
in Dermatological Research

strate that retinoids can act in an opposite manner on collagen synthesis when they were used in a more complex system instead of applied directly.

Introduction

The aim of this work was to investigate the different influences of retinoids on several fibroblast functions like collagen synthesis, chemotaxis and proliferation. We also wanted to look at a mediator-linked modulation of the effects on fibroblast functions caused by retinoids due to dermal-epidermal interactions.

Retinoids are a group of chemical compounds related to vitamin A. Usually, they are divided into three different generations that show special structural properties. Our studies were carried out with one first-generation retinoid and two third-generation retinoids. The retinoid of the first-generation, isotretinoin (13-*cis* retinoic acid, Ro 4-3780) is closely related to naturally occurring vitamin A. Clinically, it is already being used as a very potent remedy in several dermatoses [37]. The two retinoids of the third generation show strong structural differences from the vitamin A molecule. They have a polyaromatic structure and are therefore called arotinoids. Both these retinoids, arotinoid methyl sulfone (Ro 14-9706) and arotinoid ethyl sulfone (Ro 15-1570), have an additional sulfur-containing ligand instead of a carboxy group. These retinoids have not yet been used in therapy.

It is well known that retinoids exert effects not only on epithelial tissues [2], but also on tissues and cells of mesenchymal origin [10, 29, 42]. Long-term systemic treatment with several retinoids resulted in bone toxicity in animals and patients [12, 32, 40, 46, 47].

Ossification, calcification, and osteoporosis have been observed. It has also been shown that retionid treatment can lead to skin fragility, which is due to a reduced collagen synthesis of the dermal fibroblasts [21, 33, 35, 36, 50]. Also proliferation of fibroblasts was inhibited [19, 38]. It should be noted that these arotinoid sulfone esters did not show any bone toxicity and connective tissue damages in the animal experiments [6, 7, 24].

On the other hand, retinoids can improve dermal repair processes during wound healing [22, 27, 28, 41] and solar elastosis [8, 25, 26].

We therefore wanted to find out by in vitro experiments if the effects of retinoids on dermal fibroblasts could be modulated by epidermal mediators when applied topically. Furthermore, we tested the possibility that these arotinoids exert other effects on fibroblasts than a first-generation retinoid like isotretinoin exerts.

Results

Proliferation Assay

All the retinoids that have been tested showed a dose-dependent inhibitive effect on fibroblast growth when applied directly in the culture medium. There was no significant difference between the various retinoids. At 10^{-5} *M* concentration, proliferation of fibroblasts was reduced to 50% compared with the untreated control.

Direct Chemotaxis Assay

No chemotactic response was observed when retinoids in different concentrations were used in the lower compartment of a Boyden chamber to test their chemoattractive abilities (data no shown). Isotretinoin and the arotinoids reduced the chemotactic response of fibroblasts to fibroblast-conditioned medium when they were added to the upper and lower compartment. Again the inhibition was dose-dependent and not difference between the several compounds was observed. At 10^{-6} *M* concentration, the chemotactic response of fibroblasts was reduced to 50%–60% compared with the untreated control.

Analysis of Collagen Synthesis

Northern- and dot-blot hybridizations with probes specific for α1(I)- and α1(III)-procollagen mRNA gave different results for Ro 4-3780 and the arotinoids. According to earlier studies, a dose-dependent decrease of the steady-state mRNA levels for both collagen types was seen, when fibroblasts had been incubated with Ro 4-3780 (isotretinoin). At 5×10^{-5} *M* concentration, collagen mRNA levels were reduced to approximately 50%. In contrast, when fibroblasts had been incubated with different concentrations of Ro 14-9706 or Ro 15-1570, we observed an increase of collagen mRNA levels in a concentration range from 5×10^{-10} to 5×10^{-8} *M*. Only at 5×10^{-6} *M* concentration collagen were mRNA levels were reduced to a similar extent as with Ro 4-3780. This demonstrates a different action profile on collagen synthesis of fibroblasts by the sulfur-containing arotinoids in comparison to isotretinoin.

Two mechanisms responsible for the inhibition of collagen synthesis by retinoids are discussed. First, retinoids obviously are able to increase the activity of the aminoterminal peptidase [33]. This extracellular enzyme cleaves of the aminoterminal peptides from the procollagen molecules. The peptides exert a feedback inhibition on collagen synthesis of fibroblasts [51]. Second, it was recently shown that in fibroblasts retinoids can inhibited the lipid peroxidation induced by ascorbate [16]. Lipid peroxidation products by themselves strongly stimulate collagen synthesis, so this could explain these effects [9].

Inhibition of collagen synthesis in vitro suggested the idea of successful treatment of systemic scleroderma [31, 49] and other fibrotic diseases [11, 23] with retinoids. However, retinoids are also known to inhibit collagenase in several cell types, e.g., fibroblast-like cells [4, 5], mononuclear cells [34], and keratinocytes [3]. Also Ro 14-9706 and Ro 15-1570 decreased collagenase steady-state mRNA levels in fibroblast cultures.

We then investigated the effects of conditioned media derived from retinoid-treated keratinocytes on collagen synthesis of fibroblast cultures. The keratinocyte cultures used had been preincubated with $10^{-5} M$ of one retinoid per dish for 72 h. Then media had been renewed and cells had been incubated for further 24 h. Thereafter, media had been removed and pooled. These media still contained retinoids. They were mixed with 50% fresh medium when they were given on fibroblast cultures. A parallel experiment was done using the direct application of retinoids in the same concentration (50% of $10^{-5} M = 5 \times 10^{-6} M$). All values were calculated in comparison with an untreated control. When keratinocyte-conditioned medium with a final retinoid concentration of $5 \times 10^{-6} M$ was applied to the cells, there was an abolition of the inhibitive effect on collagen synthesis, which was observed in the direct application of retinoids. This was seen on the mRNA as well as on the protein level.

Analysis of TGF-β1 and Il-1 mRNA

These observations raised the question of whether there are several epidermal cytokines involved in the modulation of effects by retinoids on collagen synthesis of dermal fibroblasts. Therefore, we analyzed the mRNA levels for TGF-β1 and Il-1α and β in retinoid-treated keratinocytes. TGF-β [1, 18] and Il-1 [15, 20,30, 44] are known to be produced in the epidermis by keratinocytes. Both are able to stimulate collagen synthesis and other functions of fibroblasts (reviewed by [39, 43]).

When keratinocytes were incubated with $10^{-5} M$ 14-9706 or Ro 15-1570 for 48 h, there was an increase of the steady-state mRNA levels for TGF-β1. In the case of Ro 15-1570, this increase was up to ca. 200%, while Ro 14-9706 produced only a rise up to ca. 150%. In the case of Ro 4-3780 and also in lower concentrations, there was no significant increase except with Ro 15-1570, which increased TGF-β1 mRNA in $10^{-6} M$ up to ca. 130% compared with control. These data demonstrate that sulfur-containing arotinoids are able to induce TGF-β1 in human keratinocytes in high concentrations, in contrast to isotretinoin.

We also tested one arotinoid, Ro 14-9706 (arotinoid methyl sulfone) for its possible ability to induce Il-1α and β mRNA in primary keratinocytes cultured in a medium devoid of hydrocortisone. After incubation with Ro 14-9706 in concentrations from 10^{-8} to $10^{-5} M$ for 48 h, we isolated the total

RNA from these cells. There was an increase of Il-1α mRNA up to 200% when Ro 14-9706 was applied on these cells in a 10^{-6} *M* concentration. Il-1β mRNA levels were also increased, but to a lesser extent. The highest induction in this experiment was seen with 10^{-6} *M* Ro 14-9706.

These results confirm data from other authors. Retinoic acid has been shown to induce TGF-β2 mRNA and protein in mouse keratinocytes and epidermis [17]. In human epidermis there was no increase detectable on the mRNA level; nevertheless, there was increased immunoreactivity [14]. Retinoic acid was also shown to increase Il-1 and Il-3 mRNA in myeloid cell lines [48]. Later, an increase of interleukin-1 immunoreactivity in the epidermis of rats that had been treated systemically with etretinate was registered [45]. Recently, increased levels of Il-1β mRNA were found in retinoic acid-treated keratinocytes [13]. Il-1β, however, was shown to be produced by keratinocytes in an unproccesed and inactive form, in contrast to Il-1α [52].

Conclusions

Topically applied retinoids might act differently on the dermis and the dermal fibroblasts than was shown in earlier in vitro experiments. Some retinoids in low concentrations might be able to induce collagen synthesis directly. Furthermore, they are able to induce several epidermal mediators like cytokines (TGF-β and Il-1) that could induce collagen synthesis and other fibroblast functions necessary for repair processes in the dermis. Additionally, retinoids can inhibit collagen degradation by several pathways, which could also aid the formation of new connective tissue in wound healing and solar elastosis.

References

1. Akhurst RJ, Fee F, Balmain A (1988) Localized production of TGF-β mRNA in tumour promotor-stimulated mouse epidermis. Nature 331:363–365
2. Asselineau D, Bernard BA, Bailly C, Darmon M (1989) Retinoic acid improves epidermal morphogenesis. Dev Biol 133:322–335
3. Bailly C, Drèze S, Asselineau D, Nusgens B, Lapière CM, Darmon M (1990) Retinoic acid inhibits the production of collagenase by human epidermal keratinocytes. J Invest Dermatol 94:47–51
4. Bauer EA, Seltzer JL, Eisen AZ (1982) Inhibition of collagen degradative enzymes by retinoic acid in vitro. J Am Acad Dermatol 6:603–607
5. Bauer EA, Seltzer JL, Eisen AZ (1983) Retinoic acid inhibition of collagenase and gelatinase expression in human skin fibroblasts cultures. Evidence for a dual mechanisms. J Invest Dermatol 81:162–169
6. Bollag W (1984) Ro 15-1570, a new retinoid with potential clinical usefulness in dermatology and oncology. Hoffmann-LaRoche, Basel. Roche research report no 102 647
7. Bollag W, Bürgin H, Klaus M, Klecak G, Shaprio SS (1987) Ro 14-9706, arotinoid methyl sulfone, a new retinoid with potential clinical usefulness in skin aging. Investigational drug brochure. Hoffmann-La Roche, Basel. (Roche research report no B-102 977

8. Bryce GF, Bogdan NJ, Brown CC (1988) Retinoic acids promote the repair of the dermal damage and the effacement of wrinkles in the UVB-irradiated hairless mouse. J Invest Dermatol 91:175–180
9. Chojkier M, Houglum K, Solis-Herruzo J, Brenner DA (1989) Stimulation of collagen gene expression by ascorbic acid in cultured human fibroblats. J Biol Chem 264:-16957–16962
10. Chytil F (1984) Retinoic acid: biochemistry, pharmacology, toxicology, and therapeutic use. Pharmacol Rev 36:93s–100s
11. Daly TJ, Weston WL (1986) Retinoid effects on fibroblast proliferation and collagen synthesis in vitro and on fibrotic disease in vivo. J Am Acad Dermatol 15:900–902
12. Dickson J, Walls J (1985) Vitamin A and bone formation. Effect of an excess of retinol on bone collagen synthesis in vitro. Biochem J 226:789–795
13. Elder JT, Fisher GJ, Zhang QY, Eisen D, Krust A, Kastner P, Chambon P, Voorhess JJ (1991) Retinoic acid receptor gene expression in human skin.J Invest Dermatol 96:425–433
14. Fisher GJ, Elder JT, Eisen D, Higley H, Chu S, Ellingsworth L, Voorhees JJ (1990) Topical retinoic acid increases transforming growth factor-β1 immunoreactivity but not its mRNA in human epidermis in vivo. J Invest Dermatol 94:524
15. Gahring LC, Buckley A, Daynes RA (1985) Presence of epidermal-derived thymocyte-activating factor/interleukin 1 in normal human stratum corneum. J Clin Invest 76:1585–1591
16. Geesin JC, Gordon JS, Berg RA (1990) Retinoids affect collagen synthesis through inhibition of ascorbate-induced lipid peroxidation in cultured human dermal fibroblasts. Arch Biochem Biophys 278:350–355
17. Glick AB, Flanders KC, Danielpour D, Yuspa SH, Sporn MB (1989) Retinoic acid induces transforming growth factor-β2 in cultured keratinocytes and mouse epidermis. Cell Regul 1:87–97
18. Glick AB, Danielpour D, Morgan D, Sporn MB, Yuspa SH (1990) Induction and autocrine receptor binding of transforming growth factor-β2 during terminal differentiation of primary mouse keratinocytes. Mol Endocrinol 4:46–52
19. Harper RA, Burgoon T (1982) Differential effects of retinoic acid on the growth of normal fibroblast-like cells in vitro from human, swine and rabbit skin. Cell Biol Int Rep 6:163–170
20. Hauser C, Saurat JH, Schmitt A, Jaunin F, Dayer JM (1986) Interleukin 1 is present in normal human epidermis. J Immunol 136:3317–3323
21. Hein R, Mensing H, Müller PK, Braun-Falco O, Krieg T (1984) Effect of vitamin A and its derivatives on collagen production and chemotactic response of fibroblasts. Br J Dermatol 111:37–44
22. Hunt TK, Ehrlich HP, Garcia J, Dunphy J (1969) Effect of vitamin A on reversing the inhibitory effect of cortisone on healing of open wounds in animals and man. Ann Surg 170:633–638
23. Janssen de Limpens AMP (1980) The local treatment of hypertrophic scars and keloids with topical retinoic acid. Br J Dermatol 103:319–323
24. Kistler A, Sterz H, Teelmann K (1984) Ro 15-1570, a sulfur-containing retinoid devoid of bone toxicity in rats. Arch Toxicol 56:117–122
25. Kligman LH, Kligman AM (1986) The nature of photoaging: its prevention and repair. Photodermatol 3:215–227
26. Kligman LH (1989) Skin changes in photoaging: characteristics, prevention, and repair. In: Balin AK, Kligman AM (eds): Aging and the skin. Raven, New York, pp 331–346
27. Lee KH (1968) Studies on the mechanism of salicylates. III. Effect of vitamin A on the wound healing retardation action of aspirin. J Pharm Sci 57:1238–1241
28. Lee KH, Tong TG (1970) Mechanism of action of retinyl compounds on wound healing. II. Effect of active retinyl derivatives on granuloma formation. J Pharm Sci 59:1195–1197

29. Lotan R (1980) Effects of vitamin A and its analogs (retinoids) on normal and neoplastic cells. Bioch Biophys Acta 605:33–91
30. Luger TA, Stadler BM, Katz SI, Oppenheim JJ (1981) Epidermal cell (keratinocyte)-derived thymocyte-activating factor (ETAF). J Immunol 127:1493–1498
31. Maurice PDL, Bunker CB, Dowd PM (1989) Isotretinoin in the treatment of systemic sclerosis. Br J Dermatol 121:367–374
32. McGuire J, Lawson JP (1987) Skeletal changes associated with chronic isotretinoin and etretinate administration. Dermatologica 175:169–181
33. Nelson D, Balian G (1984) The effect of retinoic acid on collagen synthesis by human dermal fibroblasts. Coll Relat Res 4:119–128
34. Ohta A, Louie JC, Uitto J (1987) Retinoid modulation of collagenase production by adherent mononuclear cells in culture. Ann Rheum Dis 46:357–362
35. Oikarinen H, Oikarinen A, Tan EML, Abergel RP, Meeker CA, Chu ML, Prockop DJ, Uitto J (1985) Modulation of procollagen gene expression by retinoids. J Clin Invest 75:1545–1553
36. Oikarinen A, Vuorio E, Vuorio T (1989) Comparison of the effect of dexamethasone and 13-*cis*-retinoic acid on connective tissue biosynthesis in human skin fibroblasts. Arch Dermatol Res 281:273–278
37. Orfanos CE, Ehlert R, Gollnick H (1987) The retinoids. A review of their clinical pharmacology and therapeutic use. Drugs 34:459–503
38. Palmer HJ, Maher VM, McCormic JJ (1989) Effect of retinoids on growth factor-induced anchorage independent growth of human fibroblasts. In Vitro Cell Dev Biol 11:1009–1015
39. Pfeilschifter J (1990) Transforming growth factor-β. In: Habenicht A (ed) Growth factors, differentiation factors, and cytokines. Springer, Berlin Heidelberg New York, pp 56–64
40. Pittsley RA, Yoder FW (1983) Retinoid hyperostosis. Skeletal associated with long-term administration of 13-*cis* retinoid acid for refractory ichthyosis. N Engl J Med 308:1012–1014
41. Prutkin L (1972) Wound healing and vitamin A acid. Acta Derm Venereol (Stockh) 52:489–492
42. Roberts AB, Sporn MB (1984) Cellular biology and biochemistry of the retinoids. In: Sporn MB, Roberts AB, Goodman DS (eds) The retinoids, vol. 1. Academic, New York, pp 210–285
43. Schindler R, Dinarello CA (1990) Interleukin 1. in: Habenicht A (ed): Growth factors, differentiation factors, and cytokines. Springer, Berlin Heidelberg New York, pp 85–102
44. Schmitt A, Hauser C, Jaunin F, Dayer JM, Saurat JH (1986) Normal epidermis contains high amounts of natural tissue IL1. Biochemical analysis by HPLC identifies a MW-17 kd form with a PI 5.7 and a MW-30 kd form. Lymphokine Res 5:105–118
45. Schmitt A, Hauser C, Didierjean L, Merot Y, Dayer JM, Saurat JH (1987) Systemic administration of etretin increases epidermal interleukin 1 in the rat. Br J Dermatol 116:615–622
46. Smitt JHS, De Mari F (1984) A serious side-effect of etretinate (Tigason). Clin Exp Dermatol 9:554–556
47. Törok L, Budai S, Kása M, Kádár L (1988) Bone-X-ray in patients treated with isotretinoin: a prospective study. Br J Dermatol 118:134–135
48. Trechsel U, Evequoz V, Fleisch H (1985) Stimulation of interleukin 1 and 3 production by retinoic acid in vitro. Biochem J 230:339–344
49. Uhlmann A, Bräuninger W (1985) Successful treatment of Sharp's syndrome and progressive systemic sclerosis with aromatic retinoid. Z Hautkr 60:774–786
50. Varani J, Mitra RS, Gibbs D, Phan SH, Dixit V, Mitra R, Wang T, Siebert KJ, Nickoloff BJ, Voorhees JJ (1990) All-*trans* retinoic acid stimulates growth and extracellular matrix production in growth-inhibited cultured human skin fibroblasts. J Invest Dermatol 94:717–723

51. Wiestner M, Krieg T, Hörlein D, Glanville RW, Fietzek P, Müller PK (1979) Inhibiting effect of procollagen peptides on collagen biosynthesis in fibroblast cultures. J Biol Chem 254:7016–7023
52. Mizutani H, Black R, Kupper TS (1991) Human teratinocytes produce but do not process pro-interleukin-1 (IL-1) beta. Different strategies of IL-1 production and processing in monocytes and keratinocytes. J Clin Invest 87 (3):1066–1071

6. Models in Skin Ageing

In Vitro Studies of Fibroblast Aging: Growth Factors, Collagen and Glycosaminoglycan Synthesis – a Review

D. O. Schachtschabel

Introduction

Based on the studies of Carrel [9, 10] and Ebeling [17] with fibroblasts derived from heart explants and other tissues of chick embryos, it was assumed for about 50 years (until the late 1960s) that in vitro cultured cells proliferate indefinitely. However, if one applies rigorous culture conditions with defined culture media, the findings of Carrel and Ebeling cannot be reproduced. Actually, it was shown that chick fibroblasts have a limited life span of about 23 population doublings [25]. Immortal cell lines like L cells, HeLa, C3H, NIH3T3, and BHK 21 have abnormal properties frequently characteristic of cancer cells (for discussion see [26]). The most likely explanation for Carrel's findings appears to be the feeding technique, as was suggested by Hayflick [26]. Thus, Carrel and Ebeling used a culture medium composed of 2 volumes of chicken plasma and 1 volume of "chicken embryo juice". The latter was a supernatant prepared by sequeezing chick embryos, followed by filtration through gauze and low-speed centrifugation. The supernatant probably contained still-living cells. Thus, the "immortality" of the Carrell cultures was the result of permanent feedings of the cultures with viable, mortal fibroblasts.

Limited Life Span of Cultured Normal Cells: Relationship to In Vivo Aging

Unlike cancer cells, normal human cells with a diploid karyotype of 46 chromosomes per cell cannot be cultivated indefinitely in vitro, as shown by the careful and comprehensive study of Hayflick and Moorhead [27] with 25 cultured cell strains, derived from embryonic human skin, lung, kidney, heart, and other organs. These cultured normal fibroblasts exhibit a limited number of doublings with eventual cellular disintegration, and the authors introduced the term "senescence" as an explanation of this behavior: "that the phenomenon is attributable to intrinsic factors which are expressed as senescence at the cellular level" [27].

The growth pattern of normal fibroblasts is commonly divided into three phases [26, 27] in monolayer culture. (Normal fibroblasts cannot be culti-

A. Bernd, J. Bereiter-Hahn,
F. Hevert, H. Holzmann (Eds.)
Cell and Tissue Culture Models
in Dermatological Research

vated in suspension culture. They need attachment sites for spreading and proliferation). Phase I represents the primary culture, which can be established from cells after proteolytic disintegration (e.g., by trypsin treatment) of the tissue, or by outgrowth of cells from explanted tissue pieces. In the latter case, there is a more rapid migration rate of fibroblasts from explants (e.g., skin) of younger donors compared with older ones [70]. Further, the latent period between explantation and outgrowth of cells increases as a function of donor age [25]. The mechanism for the reduced migratory activity of cells from older organisms is not known. Phase II is applied to actively proliferating cells derived from subcultures of the confluent primary cultures. Following serial subcultivations of these phase-II cultures, they reach a period (phase III) when cell multiplication gradually declines and ultimately ends with disintegration (death) of the cells. This gradual loss of replicative capacity is characterized by both a reduced fraction of DNA-synthesizing cells and a gradually incresing cell cycle time [12]. It should be stressed that the cessation of the cell replication is not a synchronous process of all cells in a culture, but a gradual one. Thus, the proportion of phase-III cells (also termed phase-out cells, since they are at the end of the replicative life span) gradually increases with increasing population doubling level (PDL), as shown, for instance, by a decreasing percentage of ^{3}H-thymidine-labeled "DNA-synthesizing" nuclei with increasing PDL of fibroblast cultures [12]. The phase-out period of fibroblasts at the end of the replicative life span is not a rapid one; these phase-III cells "degenerate" in the course of several weeks or months. Since various metabolic changes and the decrease of the proliferative activity are especially conspicuous during the last 3–4 population doublings (PDs) before phase-out, this period was termed phase IV (terminal phase) by Macieira-Coelho and Taboury [39]. Distinct morphotypes and lineages of mitotic and postmitotic fibroblast populations have been described by Bayreuther et al. (this volume).

As mentioned above, it was suggested by Hayflick and Moorhead that this so called phase-III phenomenon represents aging at the cellular level [23, 27]. A typical example of such fibroblasts with limited life span is the WI-38 cell strain, originally isolated from the lungs of a 3-month-old female fetus [23]. The replicative life span of the WI-38 strain was reported originally as ranging from 39 to 58 passages or 50 ± 10 cumulative doublings [24]. According to Cristofalo [12], the number of PDs of the WI-38 strain amounts to 66 ± 6. These "slight" variations are due mainly to differences in the technical handling of the cultures (e.g., subcultivation protocols). However, there are hundreds of researchers who have confirmed the limited life span of Wi-38 cultures and of many other cell strains established from many different human and animal noncancerous tissues in culture. Further, this restricted proliferative activity is a typical trait not only of cultured fibroblasts, but also of vascular endothelial cells, keratinocytes, smooth muscle cells, glial cells, chondrocytes, epithelial lung cells, lymphocytes (for review [13, 38], and cells of the trabecular meshwork isolated from human and other primate eyes [61, 65, 67, 69].

The inverse relationship between human donor age and the PDs potential of cultured fibroblasts [23, 41, 70; for review 38] or human arterial smooth muscle cells [6] supports the concept of a relationship between in vivo and in vitro aging. Cloning of skin fibroblasts derived from young (20–35 years) and old (65 ± years) human donors has revealed a lower number of cells per colony originating from "old" cells, suggesting a decreased proliferative capacity of these cells [74]. While 60% of the cells derived from young subjects formed colonies with 256 or more cells, onyl 2% of fibroblasts from old donors were able to do so [53]. Further, the saturation density (cell number per unit attachment site at confluency) is significantly lower in skin cultures derived from old human donors in comparison with those from young donors [53]. This phenomenon is due to alterations in cell size and shape, and probably in locomotory activity. Thus, senescent cells usually become bigger and more polymorphic and occupy a larger surface area in comparison with "young" cells. This decreased saturation density of "senescent" human fibroblast cultures is regarded as a characteristic trait ("biomarker") of in vitro aging [12].

Further arguments in favor of a relationship between in vivo and in vitro aging are deduced from observations of a diminished replicative ability of fibroblasts derived from patients with premature aging syndromes, Hutchinson-Gilford syndrome (Progeria) or Cockayne's syndrome (for review [25, 38, 58]). There are also indications of a direct relationship between (log/log) the maximum number of PDs of fibroblasts in vitro and the maximum life span (in years) of eight species in vivo (mouse, rat, rabbit, bat, rat kangaroo, mink, horse, man) [57].

With regard to a decrease of the cellular replication potential in tissues or organs in vivo with increasing life span of the organism, several studies have been done with epidermal cells of the human skin, with different mouse epithelial cells (ear, tongue, esophagus, colon), mammary gland and erythropoietic cells of the mouse, and rat mast cells (for review [38]. In the dermis of the human abdominal skin there was a decline of 50%–60% in the number of cells (about 50% were fibroblasts) during the life span of Caucasian male or Japanese male and female subjects (for review [2]). We have reported a gradual decline of proliferative potential with final growth cessation and cell death after about 18 PDs for trabecular meshwork cells derived from a human eye [61, 69]. In this connection it is of interest that a decline of trabecular meshwork cellularity occurs in vivo after the age of 10 years in a near-linear manner (about 0.6% cell loss per year) for the remainder of 98 years of life studied [1]. A limited proliferative potential of in vivo cells was also derived from serial transplantation experiments of mammary gland cells. Thus, the ability of serially transplanted gland tissue to grow in inbred mice appears to depend on the cumulative number of PDs [16].

A typical property of aging postmitotic cells in vivo is the accumulation of lipopigments (lipofuscins). Increases of residual bodies (analogues to lipofuscin granules), autofluorescent cytoplasmic particles, lysosomes, and lysosomal enzyme activities have been found in cultured senescent fibroblasts and glial cells (for review [13]). However, the role of these "lipofuscins" in the aging process is controversial.

Growth Controls

Several growth-stimulating factors are known for fibroblasts in vitro (for review [60]). Thus for the cultivation of WI-38 fibroblasts under defined, serum-free conditions at least eight growth factors have mitogenic activity [49]: fibroblast growth factor(s) (a-FGF, b-FGF), epidermal growth factor (EGF), insulin, insulin-like growth factors I (IGF-I = somatomedin C) and II, a glucocorticoid hormone (cortisone or cortisol = hydrocortisone, or the synthetic derivative dexamethasone), platelet-derived growth factor(s) (PDGF), and thrombin. Senescent (phase-III) fibroblasts and Werner-syndrome fibroblasts exhibit a diminished responsiveness to growth factors/hormones [4, 50; for review 38]. The cellular receptors appear to respond in the same way in young cells as they do in senescent cells [14]. No significant differences in the number of receptors per μm^2 cell surface were found in the case of EGF, IGF-I, or PDGF between young and senescent fibroblasts [14, 15]. Tyrosine kinase activity is associated with the receptors for EGF, insulin, IGF-I, PDGF. A marked reduction in tyrosine phosphorylation of receptor molecules was found in detergent-treated and solubilized plasma membranes from senescent compared with young WI-38 fibroblasts [8, 15]. On the other hand, there was no difference in EGF tyrosine autophosphorylation when intact membranes of young and old cells were studies [8, 15]. The physiological significance of these findings is unknown. The authors believe that this receptor-associated protein has become more labile in senescent cells than it was in young cells, and this lability is revealed upon detergent solubilization [15]. With regard to EGF effects on the replicative life span, epitheli-al-like human keratinocytes show an increased number of PDs when they are cultured in the presence of EGF [55]. An extension of the proliferative life span of human aortic endothelial cells caused by the addition of FGF to the culture medium was also reported [30]. If these fibroblasts are cultured in the presence of glucocorticoid hormones, such as cortisone, cortisol, or the synthetic derivative dexamethasone the number of PDs increases ("delayed senescene") [11a, 36, 63, 72; for review 38]. The glucocorticoids appear to exert their cellular action, at least in part, through the production of a secondary "autocrine", low-molecular-weight, heat-stable (95°C), protease-resistant factor [18]. Glucocorticoid action might also interfere with the IGF-I receptor, as indicated by increased receptor binding of IGF-I following preincubation of cultured human skin or foreskin fibroblasts with dexamethasone or hydrocortisone [31]. Also, glucocorticoid addition to fibroblast cultures resulted in an increase of the saturation density, indicating a "relaxing" effect on the density inhibition of growth [13, 37, 38, 62, 72].

With regard to prostanoids, findings of Cristofalo et al. [14] indicate an altered regulation of arachidonic acid and prostaglandin metabolism in senescent lung (WI-38) fibroblasts. Thus, under serum-free conditions senescent cells produce substantially more prostaglandin PGE2 than young cells, and the concentrations achieved were in the range of inhibitory concentrations (ca. 0.3 ng/ml) for growth of young cells when PGE2 was added exogenously. On the other hand, Taylor et al. [76] reported a decline in bradykinin-stimulated prostacyclin synthesis in senescent human fetal lung fibroblasts and in skin fibroblasts derived from older donors.

Several studies demonstrated an involvement of calcium metabolism in cell aging in vitro. Thus, in serum-stimulated young fibroblasts the calmodulin protein level (as determined by radioimmunassay) fluctuates, while in old cells the calmodulin level remains at the basal level and shows no fluctuation following serum stimulation. Since the mRNA level fluctuates in old and young serum-stimulated fibroblasts in the same manner, some kind of failure was postulated at the post-transcriptional level [15].

Proliferation stimulatory activity of external calcium ions was demonstrated with WI-38 fibroblasts: Thus, elevation of the calcium concentration from 1 μM to 5 mM $CaCl_2$ in low-density cell cultures resulted in a 12-fold increase of cell number in a serum-free medium (MCDB104) without hormone and growth-factor supplementation [15]. Furthermore, a growth-stimulating action of lithium (Li) ions on WI-38 fibroblasts was reported by Cristofalo et al. [14]. Like dexamethasone, Li ions stimulated DNA synthesis in a synergistic manner with EGF and insulin in a serum-free medium. This dexamethasone-mimicking effect of Li ions was also observed in the presence of a specific glucocorticoid receptor antagonist (RU 486), suggesting an Li ion effect on growth by activating the glucorcorticoid-response pathway distal to the hormone binding and receptor translocation [14].

Collagen Synthesis

Collagen is a characteristic synthesis product of fibroblasts. Unique markers for collagen are the amino acids hydroxyproline and hydroxylysine. In the early 1960s it was shown that cultured human fibroblasts synthesize collagen in culture for several months after removal from the donor, and the collagen formed in vitro has the same content of proline and hydroxyproline found in collagen formed in vivo [11, 34]. Cortisol (1 μg/ml) completely suppressed collagen formation in such cultures [11]. Several factors influence this differentiated function of the fibroblast: In 3T6 cultures at saturation density, 3%–5% of the protein synthesized is collagen, while in early exponential growth phase only 0.25% of the protein made was collagen [20]. Collagen synthesis, as measured by the ratio of hydroxyproline incorporation to proline incorporation, was stimulated by addition of 20–40 μM sodium lactate (the pH of the medium was unchanged at the utilized lactate concentrations), indicating that metabolic products of cells have influences on this differentiated cell function. On the other hand, with several human foreskin fibroblast cell strains and the mouse 3T6 line, Manner [40] found that collagen synthesis was more rapid in proliferating than in stationary cultures. The percentage of collagen to total protein synthesized was higher in "old" but not in "young" human stationary fibroblast strains (the old strains had gone through approximately 50–55 PDs) or in the stationary 3T6 line, not because collagen synthesis was stimulated at high cell density but because synthesis of other proteins was more repressed than collagen synthesis [40]. Omission of serum inhibited noncollagen synthesis more strongly than collagen synthesis in "old" but not in "young" human fibroblast strains [40]. Apparently, the control mechanisms turning protein synthesis off or on affect the synthesis of collagen and of noncollagen protein to the same extent in "young" but not in "old" fibroblasts [40]. It must be mentioned that increased conversion of radioactive proline to hydroxyproline must not only represent increased collagen synthesis but can also reflect an increase of the prolyl hydroxylase activity per cell, while the rate of synthesis of collagen peptide decreases [5]. In this regard, several co-substrates and co-factors (such as ascorbate) are involved in the prolyl hydroxylase activity. A decreased induction of conversion of ^{3}H-proline to protein-bound ^{3}H-hydroxyproline by ascorbic acid was observed in senescent as compared with "young" human WI-38 fibroblasts [29]. Furthermore, the proteolytic degradation of collagen and other proteins might influence the relative proportion of protein-bound proline and hydroxyproline. Thus, the collagenolytic activity (following oxyphenylbutazone induction) of senescent WI-38 fibroblasts was lower (at PDL = 37) or not detectable (at PDL = 39–44) compared with "young" fibroblasts (PDL = 13–35). With regard to quantitative measurements of the collagen content of WI-38 fibroblasts cultures, Paz and Gallop [47] found a diminished amount of collagen (1–3.4 mg collagen/mg DNA) in late-passage cultures (PDL = 41) as compared with "younger" cultures (6.0 mg/mg DNA at PDL = 26). The measurements relate to the collagen content in the medium from 1 week at confluence. However, this age-dependent change was found only in the absence of ascorbic acid. No significant changes were found in the presence of ascorbate. Another difference between PDL = 26 and PDL = 41 was found in the absence of ascorbate in relation to cell-bound collagen. "Young" cultures contained 5%, late-passage cultures 13%–29% cell-bound collagen, the remainder released in the medium. No significant difference was found in the presence of ascorbate: The cellular collagen was between 23% (PDL = 26) and 18%–25% (PDL = 41) [47].

Age-dependent in vivo alterations of collagens have been investigated in many experimental studies for many decades (for review [33]). There are about 12 distinct genetic types of collagen (types I–XII) with characteristic tissue distribution [52]. All collagen types are composed of three peptide chains, which differ in degree of glycosylation and hydroxylation (of proline). The major compound of the connective tissue of the body is type-I collagen, which is also the predominant type in adult skin. There are changes in the proportion of the different collagen types in fetal development or, for example, in granulation tissue. Different tissues from growing mammalian organisms have different rates of collagen turnover. Collagen (primarily type I) comprises about 70% of the dry weight of adult human skin. The human skin becomes thinner and loses elasticity with age, and the absolute amount of skin collagen decreases. Further, the amount of human skin collagen solubilized by citric acid or pepsin digestion has been shown to decrease with age (for review [33]). There is an age-associated increase of the intermolecular cross-links and of nonenzymatically bound glucose to insoluble skin collagen (for review [33, 52]. Cross-links arise because of oxidation of lysine or hydroxylysine to an aldehyde, which can form, e.g., and aldol condensation with another aldehyde group of an oxidized (hydroxy-)lysine. Further, there arise stable condensation products (Pyridinolin) derived from three hydroxylysine molecules or their aldyhydes (for review [33, 52]). The oxidation process is catalyzed by an enzyme (lysyloxidase) which is secreted by the fibroblast. For the age-dependent increase of cross-links, those which arise from glycosylation are of major importance. In this regard, it is of interest that glycosylated hydroxylysine in skin collagen increases with age (for review [52]). No systematic investigations concerning age-dependent changes of cross-linking or glycosylation of collagens have been carried out with cultured fibroblasts so far.

In the whole tissue, collagen and elastic fibers are intimately associated with hyaluronic acid and proteoglycans of the ground substance. The most striking changes of dermal glycosaminoglycans (GAGs) occur in the soluble pool. While the concentration of soluble GAGs in the skin decreases from the newborn stage during infancy and stays stable from infancy through middle age, it drops again in the old-age-group, which probably causes a shrinkage of the ground substance with reduction of the water content ("turgor") (for review [48]). These GAG alterations might be a reflection of fibroblast senescence, and such changes may affect the "viscosity" of the ground substance and the passage of inorganic and organic nutrients and growth factors from the blood capillaries to the tissue cells. Further, proteoglycans probably have a function in the control of collagen fiber formation. Thus, this process might be affected by age changes of proteoglycans.

Glycosaminoglycan (GAG) Metabolism

Fibroblasts in vivo (and in vitro) synthesize cell-bound (cellular) and extracellular proteoglycans (glycosaminoglycans bound to protein) and hyaluronic acid (synonyms: hyaluronate or hyaluronan). While most glycosaminoglycan tpyes (e.g., chondroitin sulfate, dermatan sulfate, heparan sulfate) are covalently bound to core proteins ("proteoglycans"), hyaluronic acid is commonly found in vivo and in vitro as a protein-free gylcsoaminoglycan (GAG). However, interactions of proteoglycans with hyaluronate via a link protein were found, at least in cartilage. Contrary to the formation of proteoglycans, which occurs intracellularly by enzymatic polymerization of the respective UDP-bound sugar molecules, sulfation with activated sulfate, and protein synthesis in case of core protein, the synthesis of hyaluronate occurs from UPD-acetylglucosamine and UPD-glucuronic acid on the cytosolic site of the cell membrane (for review [21]).

A distinctive property of fibroblasts is their relatively high rate of GAG synthesis and excretion of hyaluronate in particular. Cultured fibroblasts exhibit a positive correlation between hyaluronate synthesis and cell proliferation (for review [43, 60]). Thus increased cell density or omission of growth factors resulted in a diminished activity of hyaluronate synthetase activity [43]. However, this is a reversible process: in the case of "young" (phase-II) fibroblasts addition of stimulating growth factors such as PDGF or dilution of cell density leads to increased hyaluronate synthesis. On the other hand, an "irreversible", progressive decrease of the synthesis rates of cell-bound and cell-released (medium) GAGs was observed in the course of in vitro aging of WI-38 fibroblasts [64, 68, 73, 78; for review 60]. The same observations were made with human skin fibroblasts (unpublished results), when phase-II cells were compared with senescent cells of the same strain. The altered GAG metabolism of senescent WI-38 fibroblasts was characterized by a change in the synthesis pattern, namely by a relative decrease in hyaluronate and an increase in heparan sulfate [73, 78].

Other researchers have made similar observations regarding decreased hyaluronate [43, 44] and increased heparan sulfate [3, 42, 44, 59] in the course of in vitro aging of human fibroblasts. Decreased synthesis rate of hyaluronate and relatively increased synthesis of heparan sulfate were also observed in the course of in vitro aging of trabecular meshwork cells of the human eye [66, 67, 69]. Cultured human skin fibroblasts show a decreased synthesis rate of hyaluronate during senescence (phase III), too (unpublished results). Further, growth factors have a stronger stimulating effect on hyaluronate synthesis of "young" (phase-II) compared with senescent (phase-III) fibroblasts (unpublished results). Age-related in vivo changes of GAGs and GAG distribution patterns are know for several tissues and organs, e.g., decreased content of hyaluronate in skin, aorta, or synovial fluid with increasing age (for review [21, 35, 48, 56]).

Heparan sulfate was suggested as an indicator of growth reduction or as an "endogenous" growth inhibitor, also because addition of structurally related heparin or heparan sulfate to the culture medium of fibroblasts had a growth-inhibiting effect [44, 68, 78].

Incubation of ^{35}S-sulfate-prelabeled WI-38 fibroblasts in the presence of exogenous heparin resulted in an increased liberation – compared with untreated controls – of ^{35}S-labeled GAGs, predominantly heparan sulfate [68]. Therefore, we suggested that exogenous heparin competes with structurally related heparan sulfate, probably located at the cell surface. This interaction with cell surface sites might interfere with membrane-associated processes in regard to cell growth and GAG synthesis, as for instance with receptor sites for growth factors and hormones, membrane-localized enzyme (of hyaluronate synthesis?), or other membrane activities. It is of interest that an increased proportion of heparan sulfate proteoglycan was found in the substratum adhesion sites of dermal fibroblasts (papillary and reticular) which had aged in vivo or in vitro [32]. Possible effects of these heparan sulfate changes on the adhesive properties of the skin were suggested [32]. With regard to the action of exogenous heparan sulfate and heparin, a reducing effect on the number of EGF receptors on smooth muscle cells should be mentioned [54]; this effect closely parallelled the antiproliferative action of these heparin(-like) molecules [54].

As mentioned above, long-term treatment of WI-38 fibroblasts with glucocorticoid hormones resulted in an increase of the number of PDs, thus counteracting the normal in vitro aging process [11a, 13, 63, 72]. Such long-term treatment with cortisol for example, revealed an elevated synthesis of cell-bound GAGs, especially pericellular hyaluronate [72]. Again, this is the opposite of the hyaluronate changes occurring during in vitro senescence of fibroblasts, which is characterized by decreased synthesis of cell-bound and pericellular hyaluronate [42, 73]. Whether these in vitro effects have a relationship to the in vivo situation cannot be ascertained at present. However, apart from these glucocorticoid effects on cell-bound hyaluronate synthesis, it should be mentioned that glucocorticoids inhibit the synthesis of extracellular GAGs (especially hyaluronate) [62] and collagen (for review [28]) of cultured fibroblasts, which might correspond to in vivo effects.

Hyaluronate (and heparan sulfate) have been implicated in cell-substratum adhesion, migratory activity, and interaction with fibronectin, collagen, or laminin of the extracellular matrix (for review [60, 63, 72, 77]). In addition, hyaluronate synthesis appears to be required for the detachment of fibroblasts from the substratum and for mitosis [7]. Also, the extracellular matrix (ECM) of fibroblasts has been reported to have growth-modulating properties (e.g., by Nakagawa et al. [46]). Although the results of in vivo studies are somewhat controversial, there are several reports about decreases in the number of skin fibroblasts (e.g., papillary dermal) and a reduction of dermal GAGs (e.g., hyaluronate) as the organism ages (for review [2, 48]).

With regard to the molecular mechanism underlying the restricted (limited) proliferative activity of normal (diploid) fibroblasts, the senescence process seems to involve repression or extinction of proliferation-promoting genes (e.g. c-*fos* proto-oncogene) with expression of antiproliferative (suppressor) genes, such as the retinoblastoma susceptibility gene (RB1), the p53 gene, or the Wilm's tumor susceptibility gene (71, 75 for review [19]). A different mechanism ("telomere abrasion") was prosposed recently, since it was observed that there was a progressive decrease in the mean telomere length (and telomeric DNA) of the chromosomes with increasing passage level of cultured human fibroblasts [22]. The mechanism for the loss of telomeric sequences with increasing number of fibroblast doublings is not known.

References

1. Alvarado J, Murphy C, Juster R (1984) Trabecular meshwork cellularity in primary open-angle glaucoma and nonglaucomatous normals. Opthalmology 91:564–579
2. Balin AK, Lin AN (1989) Skin changes as biological marker for measuring the rate of human ageing. In: Balin AK, Kligman AM (eds) Aging and the skin. Raven, New York, pp 43–75
3. Bartold PM, Boyd RP, Page RC (1986) Proteoglycans synthesized by gingival fibroblasts derived from human donors of different ages. J Cell Physiol 126:37–46
4. Bauer EA, Silverman N, Busiek DF, Kronberger A, Deuel TF (1986) Diminished response of Werner's syndrome fibroblasts to growth factors PDGF and FGF. Science 234:1240–1243
5. Berg RA, Kao WWY, Prockop DJ (1976) Prolyl hydroxylase activity in L-929 fibroblasts incubated with and without ascorbate. Biochim Biophys Acta 444:756–764
6. Bierman EL (1978) The effect of donor age on the in vitro life span of cultured human arterial smooth-muscle cells. In Vitro 14:951–955
7. Brecht M, Mayer U, Schlosser E, Prehm P (1986) Increased hyaluronate synthesis is required for fibroblast detachment and mitosis. Biochem J 239:445–450
8. Brooks KM, Phillips D, Carlin CR, Knowles BB, Cristofalo VJ (1987) EGE-dependent phosphorylation of the EGF receptor in plasma membranes isolated from young and senescent WI-38 cells. J Cell Physiol 133:523–531
9. Carrel (1912) On the permanent life of tissues outside of the organism. J Exp Med 15:516–528
10. Carrel A (1914) Present condition of a strain of connective tissue twenty-eight months old. J Exp Med 20:1–2
11. Castor CW, Muirden KD (1964) Collagen formation in monolayer cultures of human fibroblasts . The effect of hydrocortisone. Lab Invest 13:560–574

11a. Cristofalo VJ (1970) Metabolic aspects of aging in diploid human cells. In: Holeckova E, Cristofalo VJ (eds) Aging in cell and tissue culture. Plenum, New York, pp 83–119

12. Cristofalo VJ (1988) Cellular biomarkers of aging. Exp Gerontol 23:297–305
13. Cristofalo VJ, Stanulis-Praeger BM (1982) Cellular senescence in vitro. In: Maramorosch K (ed) Advances in cell culture, vol. 2. New York, Academic, pp 1–68
14. Cristofalo VJ, Phillips PD, Sorger T, Gerhard G (1989) Alterations in the responsiveness of senescent cells to growth factors. J Gerontol 44 (6):55–62
15. Cristofalo VJ, Doggeit DL, Brooks-Federich KM, Phillips PD (1989) Growth factors as probes of cell aging. Exp Gerontol 24:367–374
16. Daniel CW, Young LJT (1971) Influence of cell division on an aging process. Life span of mouse mammary epithelium during serial propagation in vivo. Exp Cell Res 65:27–32
17. Ebeling AH (1922) A ten-year-old strain of fibroblasts. J Exp Med 35:755–759
18. Finlay CA, Cristofalo VJ (1987) Autocrine stimulation of WI-38 cell proliferation in the presence of glucocorticoids. Characteristics of the stimulatory factor(s) involved in this response. Exp Cell Res 168:191–202
19. Goldstein S (1990) Replicative senescence: the human fibroblast comes of age. Science 249:1129–1133
20. Green H, Goldberg B (1964) Collagen and cell protein synthesis by an established mammalian fibroblast line. Nature 240:347–349
21. Greiling H, Haubeck HD (1991) Proteoglykane. In: Platt D(ed) Biologie des Alterns. Ein Handbuch. de Gruyter, Berlin, pp 123–133
22. Harley CB, Futcher AB, Greider CW (1990) Telomeres shortern during ageing of human fibroblasts. Nature 345:458–460
23. Hayflick L (1965) The limited in vitro lifetime of human diploid cell strains. Exp Cell Res 37:614–636
24. Hayflick L (1970) Aging under glass. Exp Gerontol 5:291–303

25. Hayflick L (1977) The cellular basis for biological aging: In: Finch CE, Hayflick L (eds) Handbook of the biology of aging. Van Nostrand and Reinhold, New York, pp 159–186
26. Hayflick L (1991) General aspects of fibroblasts cell culture. In: Platt D (ed) Biologie des Alterns. Ein Handbuch, de Gryter, Berlin, pp 54–72
27. Hayflick L, Moorhead PS (1961) The serial cultivation of human diploid strains. Exp Cell Res 25:585–621
28. Hein R, Krieg T (1987) Regulation des Stoffwechsels von Fibroblasten durch Corticoide. In: Holzmann H, Altmeyer P, Marsch W, Vogel HG (eds) Dermatologie und Rheuma. Springer, Berlin Heidelberg New York, pp 517–523
29. Houck JC, Sharma VK, Hayflick L (1971) Functional failures of cultured diploid fibroblasts after continued population doublings. Proc Soc Exp Biol Med 137:331–333
30. Johnson LK, Longenecker JP (1982) Senescence of aortic endothelial cells in vitro: influence of culture conditions and preliminary characterization of the senescent phenotype. Mech Ageing Dev 18:1–18
31. Kaplowitz PB (1987) Glucocorticoids enhance somatomedin-C binding and stimulation of amino acid uptake in human fibroblasts. J Clin Endocrinol Metab 64:563–571
32. Kent WM, Funderburg FM, Culp LA (1986) Proteoglycans in the substratum adhesion sites of human papillary or reticular dermal fibroblasts. Ageing in vivo or in vitro. Mech Ageing Dev 33:115–137
33. Kohn RR, Schneider SL (1989) Collagen changes in aging skin. In: Balin AK, Kligman AG (eds) Aging and the skin. Raven, New York, pp 121–139
34. Layman DL, McGoodwin EB, Martin GR (1971) The nature of the collagen synthesized by cultured human fibroblasts. Proc Natl Acad Sci USA 68:454–458
35. Lindner J (1976) Zur Alterung der Organe. In: Dhom G, Biologie des Alterns. Fischer, Stuttgart, pp 181–242
36. Macieira-Coelho A (1966) Action of cortisone on human fibroblasts in vitro. Experientia 22:390–391
37. Macieira-Coelho A (1983) Changes in membrane properties associated with cellular aging. Rev Cytol 83:183–220
38. Macieira-Coelho A (1988) Biology of normal proliferating cells in vitro. Relevance for in vitro aging. In: van Hahn HP (ed) Interdisciplinary topics in gerontology, vol. 23. Karger, Basel
39. Macieira-Coelho A, Taboury F (1982) A re-evaluation of the changes in proliferation in human fibroblasts during ageing in vitro. Cell Tissue Kinet 15:213–224
40. Manner G (1971) Cell division and collagen synthesis in cultured fibroblasts. Exp Cell Res 76:49–60
41. Martin GM, Sprague CA, Epstein CJ (1970) Replicative life-span of cultivated human cells. Effects of donor's age, tissue, and genotype. Lab Invest 23:86–92
42. Matuoka K, Mitsui Y (1981) Involvement of cell surface heparan sulfate in the density-dependent inhibition of cell proliferation. Cell Struct Funct 6:23–33
43. Matuoka K, Namba M, Mitsui Y (1987) Hyaluronate synthetase inhibition by normal and transformed human fibroblasts during growth reduction. J Cell Biol 104:1105–1115
44. Matuoka K, Hasegawa N, Namba M, Smith GJ, Mitusci Y (1989) A decrease in hyaluronic acid synthesis by aging human fibroblasts leading to heparan sulfate enrichment and growth reduction. Aging 1:47–54
45. McGrath CM, Soule HD (1984) Calcium regulation of normal human mammary epithelial cell growth in culture. In Vitro 20:652–662
46. Nakagawa S, Pawelek P, Grinell F (1989) Extracellular matrix organisation modulates fibroblast growth and growth factor responsiveness. Exp Cell Res 182:572–582
47. Paz MA, Gallop PM (1975) Collagen synthesized and modified by aging fibroblasts in culture. In Vitro 11:302–312
48. Perlish JS, Longas MO, Fleischmajer R (1989) The role of glycosaminoglycans in aging of the skin. In: Balin AK, Kligman AM (eds) Aging and the skin. Raven, New York, pp 153–165

49. Phillips PD, Cristofalo VJ (1988) Classification system based on the functional equivalence of mitogens that regulate WI-38 cell proliferation. Exp Cell Res 175:396–403
50. Phillips PD, Kaji K, Cristofalo VJ (1984) Progressive loss of the proliverative response of senescent WI-38 cells to platelet-derived growth factor, epidermal growth factor, insulin, transferrin, and dexamethasone. J Gerontol 39:11–47
51. Praeger FC, Cristofalo VJ (1986) The growth of WI-38 cells in a serum-free, growth factor-free, medium with elevated calcium concentrations. In Vitro 22:355–359
52. Rauterberg J (1991) Struktur, Eigenschaften und Biosynthese von Kollagen. In: Platt D (ed) Biologie des Alterns. Ein Handbuch. de Gruyter Berlin, pp 98–110
53. Reff M, Schneider EL (1981) Cell culture aging. Mol Cell Biochem 36:169–176
54. Reilly CF, Fritze LMS, Rosenberg RD (1988) Heparin-like molecules regulate the number of epidermal growth factor receptors on vascular smooth muscle cells. J Cell Physiol 136:23–32
55. Rheinwald JG, Green JH (1977) Epidermal growth factor and the multiplication of cultured human epidermal keratinocytes. Nature 265:421–424
56. Robert L, Moczar M (1982) Age-related changes of proteoglycans and glycosaminoglycans. In: Varma RS, Varma R (eds) Glycosaminoglycans and proteoglycans in physiological and pathological processes of body systems. Karger, Basel, pp 440–460
57. Röhme D (1981) Evidence for a relationship between longevity of mammalian species and life span of normal fibroblasts in vitro and erythrocytes in vivo. Proc Natl Acad Sci USA 78:5009–5013
58. Salk D (1982) Werner's syndrome: a review of recent research with an analysis of connective tissue metabolism, growth control of cultured cells and chromosomal aberrations. Hum Genet 62:1–15
59. Sames K, Lücht B (1987) Sulfate incorporation into glycosaminoglycans of human lung fibroblasts as influenced by the number of subcultures and glycosaminoglycan polysulfate. Compr Gerontol 1:80–86
60. Schachtschabel DO (1991) Fibroblastenstoffwechsel. In: Platt D (ed) Biologie des Alterns. Ein Handbuch. de Gruyter, Berlin New York, pp 73–84
61. Schachtschabel DO, Binninger E (1990) Aging of trabecular meshwork cells of the human eye in vitro. Z Gerontol 23:133–136
62. Schachtschabel DO, Sluke G (1984) Einfluß von Cortisol auf Glykosaminoglykansynthese und Wachstum von diploiden, menschlichen Fibroblasten (WI-38) in Abhängigkeit von der In-vitro-Alterung. Z Gerontol 17:141–149
63. Schachtschabel DO, Sluke G (1986) Increase of cellular hyaluronic acid synthesis following continuous exposure of cultured fibroblasts (WI-38) to hydrocortisone. Z Gerontol 19:173–178
64. Schachtschabel DO, Wever J (1978) Age-related decline in the synthesis of glycosaminoglycans by cultured human fibroblasts (WI-38). Mech Ageing Dev 8:257–264
65. Schachtschabel DO, Bigalke B, Rohen JW (1977) Production of glycosaminoglycans by cell cultures of the trabecular meshwork of the primate eye. Exp Eye Res 24:71–80
66. Schachtschabel DO, Wilke K, Wehrmann R (1982) In vitro cultures of human and monkey trabecular meshwork. In: Lütjen-Drecoll E (ed) Basic aspects of glaucoma research. Schattauer, Stuttgart, pp 211–220
67. Schachtschabel DO, Rohen JW, Wever J, Sames K (1982) Synthesis and composition of glycosaminoglycans by cultured human trabecular meshwork cells. Graefes Arch Clin Exp Opthalmol 218:113–117
68. Schachtschabel DO, Sluke G, Wever J (1986) Effects of hydrocortisone and heparin on growth and glycosaminoglycan synthesis of cultured human fibroblasts (WI-38). In: Platt, D (ed) Drugs and aging. Springer, Berlin Heidelberg New York, pp 56–69
69. Schachtschabel DO, Binninger EA, Rohen JW (1989) In vitro cultures of trabecular meshwork cells of the human eye as a model system for the study of cellular aging. Arch Gerontol Geriatr 9:251–262
70. Schneider EL, Mitsui Y (1976) The relationship between in vitro cellular aging and in vivo human age. Proc Natl Acad Sci USA 73:3584–3588

71. Seehardi T, Campisi J (1990) Repression of c-fos transcription and an altered genetic program in senescent human fibroblasts. Science 247:205–209
72. Sluke G, Schachtschabel DO (1988) Influence of long-term treatment with cortisol on proliferation, aging and glycosaminoglycan synthesis of cultured human diploid fibroblasts (WI-38). Z Gerontol 21:68–73
73. Sluke G, Schachtschabel DO, Wever J (1981) Age-related changes in the distribution pattern of glycosaminoglycans synthesized by cultured human diploid fibroblasts (WI-38). Mech Ageing Dev 16:19–27
74. Smith JR, Pereira-Smith O, Schneider EL (1978) Colony size distrubution as a measure of in vivo and in vitro ageing. Proc Natl Acad Sci USA 75:1353–1356
75. Stein GH, Beeson M, Gordon L (1990) Failure to phosphorylate the retinoblastoma gene product in senescent human fibroblasts. Science 249:666–669
76. Taylor L, Schneider EL, Smith JP, Polgar P (1981) Prostaglandin production and cellular aging. Mech Ageing Dev 16:311–317
77. Turley EA (1984) Proteoglycans and cell adhesion: their putative role during tumorigenesis. Cancer Metastasis Rev 3:325–339
78. Wever J, Schachtschabel DO, Sluke G, Wever J (1980) Effect of short- or long-term treatment with exogeneous glycosaminoglycans on growth and glycosaminoglycan synthesis of human fibroblasts (WI-38) in culture. Mech Ageing Dev 14:89–99

*Terminal Differentiation, Aging, Apoptosis, or Transformation of the WI-38 Fibroblasts in the Fibroblast Stem Cell System In Vitro**

K. Bayreuther, and *J. Gogol*

Abstract

Numerous investigations into the molecular mechanisms of the cellular aging of fibroblast populations in vitro have been undertaken since Hayflick and Moorhead reported that secondary human embryonic lung fibroblasts of the cell-lines WI-26 and WI-38 undergo a finite number of 50 ± 10 cumulative population doublings (CPD) and then die. The exhaustion of the proliferative capacity, the phase-III phenomenon, was interpreted to be aging at the cellular level. The nature of the fibroblast cell systems was not clarified. As a consequence, biological and biochemical studies were undertaken with undefined secondary fibroblast populations and many published data contradict each other. Because of this situation, the fibroblast cell systems have for some time been considered to be unsuited for studies of the molecular mechanisms of aging. Employing concepts and methods that are routinely used in investigations of the molecular biology of cells in stem cell systems, data were obtained that demonstrate that WI-38 fibroblasts develop along a unidirectional differentiation sequence of a terminal stem cell system. WI-38 transformed by SV40 (WI-38/SV40), like normal WI-38 cells, develop along a three-stage sequence of mitotic fibroblasts and a five-stage lineage of postmitotic fibroblasts in four compartments. When stem cell-system-specific concepts and methods are applied, defined fibroblast populations can be made available. Cell-biological and -biochemical studies with properly defined fibroblast populations, that are pools of cell populations with verified constant composition of fibroblast cell types and thus with identical biochemical parameters, are destined to provide reproducible results in the field of fibroblast cell biology in future.

Introduction

In 1961, Hayflick and Moorhead [8] found that cultured human embryonic lung fibroblasts undergo a finite number of serial subcultivations, or cumula-

* Supported by grants from the *Deutsche Forschungsgemeinschaft* and the *Bundesministerium für Forschung und Technologie.*

A. Bernd, J. Bereiter-Hahn,
F. Hevert, H. Holzmann (Eds.)
Cell and Tissue Culture Models
in Dermatological Research

tive population doubling (CPD), and then die. After 50 ± 10 population doublings cell death was inevitable. The culture history of these cells was divided into three phases: Phase I represented the initial culture, phase II the period of serial transfer and exponential cell increase, and phase III the death of the cells. The phase-III phenomenon was interpreted to be aging and death at the cell level. The inference from these findings was that aging is the result of events occuring at the cellular level.

As a consequence of these findings, numerous studies were undertaken with the cell lines established by Hayflick and Moorhead [8] (e.g., WI-26 and WI-38) to test the theories of the molecular mechanisms of cellular aging, the error [9] or mutation theories [6], and the differentiation [4] or program theories [12]. Distinct morphotypes and quantitative and/or qualitative disparities for a multitude of biochemical parameters are recorded for these secondary human fibroblasts in early and late passages in vitro [7]. No attempts have been undertaken to correlate the distinct fibroblast cell types with dissimilarities in biochemical parameters in secondary fibroblast populations at nonidentical passage levels. The nature of the fibroblast cell system was not clarified. Therefore, biological and biochemical studies were undertaken with undefined secondary fibroblast populations, and many biochemical results described contradict each other [19].

We have presented evidence that primary and secondary fibroblast cell systems derived from connective tissues of skin and lung of prenatal and postnatal developmental stages of Valo-chicken (unpublished work), C3H-mice (unpublished work), BN-rats [10, 13], and man [3] are stem cell systems.

In the stem cell compartment in the early stages of the developing wingbud of the chick embryo, stem cell S1 with the developmental potentials for fibroblast – chondroblast plus fibroblast – myoblast can be found. S1 proliferates and segregates stem cell S2 with the potentials fibroblast – chondroblast or fibroblast – myoblast. S2 replicates and throws off stem cell S3 with the populations fibroblast – chondroblast – fibroblast – myoblast. Two of these S3 cells are the committed fibroblast stem cells that move over to the fibroblast progenitor compartment (unpublished work). In this compartment mitotic fibroblasts (MF) of chicken, mouse, rat, and man proliferate and differentiate along the cell lineage MF I – MF II – MF III. When the proliferative capacity is exhausted, MF III proceeds to the fibroblast maturing compartment, in which the postmitotic fibroblasts (PMF) develop along the sequence PMF IV – PMF V – PMF VI. The postmitotic fibroblast PMF VI is the terminally differentiated end cell of the fibroblast stem cell system. Only the end cell expresses the full functions of the fibroblast stem cell system; all other cells in the stem cell compartment, the progenitor compartment, and the maturing compartment are restricted to renewal functions for the fibroblast end cell. After a tissue- and species-specific period of high metabolic activity, the fibroblast end cell either is removed by apoptosis as PMF VIIa in the fibroblast-degenerating compartment or transforms as PMF VIIb transformed mitotic fibroblast (TMF I) in the fibroblast-transforming compartment [1–3, 5]. Human embryonic lung fibroblast populations of the cell line WI-38 have been studied with stem cell-system-specific biological and biochemical methods. In the following, data that document terminal differentiation, aging, apoptosis, or transformation in the WI-38 fibroblast stem cell system in vitro are presented.

Results

When standardized quantitative cell culture methods are applied, WI-38 populations undergo 50–60 cumulative population doublings (CPD). When the proliferative capacity of the mitotic fibroblasts is exhausted and appropriate methods are applied, the fibroblasts do not degenerate in Hayflick's phase-III phenomenon [8] but spontaneously shift to the postmitotic fibroblasts, which can be maintained for 70–140 days in stationary culture, either to degenerate by apoptosis or to transform spontaneously thereafter.

The investigations of the WI-38 populations were started at CPD level 22. In these populations no cells of the stem cell compartment, S1, S2, or S3, were identified. In the mitotic population of the fibroblast progenitor compartment, there are three mitotic fibroblasts (MF): the spindle-shaped MF I, with a high mitotic capacity, the epitheloid MF II, with an intermediate proliferative capacity, and the larger epitheloid MF III, with a low mitotic capacity. In the postmitotic population of the fibroblast-maturing compartment three morphotypes of postmitotic fibroblast (PMF) can be found: the PMF IV, with the form of a myofibroblast, the larger epitheloid PMF V, and the largest epitheloid PMF VI, the terminally differentiated end cell in two size classes. PMF VI degenerates to PMF VII a in the fibroblast-degenerating compartment, or MF VI transforms spontaneously to PMF VII b (TMF I) in the fibroblast-transforming compartment. Mitotic fibroblast populations of WI-38 are heterogeneous populations (Table 1).

Mitotic fibroblasts of WI-38 can be cloned by cellular cloning to form pure clones of the cell types MF I or MF II or MF III or heterogenous clones of the cell types MF I + MF II or MF II + MF III (Table 2).

Fibroblasts that are not clonogenic form plated cells in a postmitotic differentiation state. In pure clones the biochemical changes related to differentiation can be analyzed. In heterogeneous clones the frequencies of the differentiation to a higher state and the direction of the unidirectional changes can be evaluated. The changes of the cell type frequencies and of the clone type frequencies as a function of the CPD level of the mass population studied are identical, establishing conclusively that the mitotic fibroblasts of WI-38 develop along the differentiation lineage MF I – MF II – MF III in the progenitor compartment (Tables 1, 2).

Microscopic studies of fixed and stained cells in low-density mass populations of WI-38 reveal that MF I, MF II, and MF III mother cells can divide either symmetrically to give rise to two identical daughter cells of MF I; MF II and MF III, or asymmetrically, to produce MF I + MF II and MF II +

Table 1. Cell type frequencies in mitotic and postmitotic WI-38 populations as a function of the CPD level of the mitotic populations and the period of time of the postmitotic populations in stationary culture

		Fibroblast cell types (%)						
		MFI	MFII	MFIII	PMFIV	PMFV	PMFVI	PMFVIIa
MF populations at								
CPD	Days							
22	0	45.6	47.0	7.2	–	–	–	–
32	15	31.0	63.9	5.1	–	–	–	–
39	43	12.0	79.5	8.5	–	–	–	–
44	48	7.2	76.5	15.7	–	0.6	–	–
46	64	8.1	73.6	17.6	0.7	–	–	–
48	68	0.5	43.8	46.7	1.6	7.5	–	–
50	77	0.4	38.8	53.9	2.4	4.1	–	–
53	112	–	20.5	61.4	1.9	13.4	2.6	–
PMF populations at								
Week	Days							
4	28	–	–	1.2	6.1	24.7	68.0	0.5
6	42	–	–	0.5	6.5	16.9	66.3	10.2
8	56	–	–	0.3	11.1	13.7	48.2	26.8
10	70	–	–	–	11.0	15.0	46.9	26.8

Table 2. Clone type frequencies of clonal populations derived from mitotic WI-38 populations at nonidentical CPD levels

	Clone types (%)					
CPD	CT I	CT I/II	CT II	CT II/III	CT III	CE (%)
20	2.2	52.4	29.3	14.0	2.0	12.3
29	1.1	60.1	27.1	9.3	2.4	12.0
38	–	47.1	44.7	4.9	3.0	8.3
46	–	36.0	56.3	5.8	1.9	4.1
50	–	13.9	36.1	40.1	9.9	0.7
52	–	–	–	67.6	32.3	0.1

MF III by differentiation division. These findings mean that there are two types of MF I and MF II mother cells, without and with a polarity that can give rise to identical or nonidentical daughter cells. The development of MF III to PMF IV must occur in the daughter cells that have arisen from symmetric divisions of MF III since no asymmetric division figures of MF III have been found. Low-density mitotic populations of WI-38 at CPD level 30 with about 80% MF II can be experimentally induced to differentiate to fibroblasts of a higher differentiation stage by a greater number of physical

(UV, electromagnetic fields [17, 18]) and chemical agents (mitomycin C, 5-fluoro-uracil, 5-bromo-2-deoxyuridin [5]). When such populations are treated with 15 J/m^2 UV or 2×10^{-7} *M* mitomycin C, MF II (80%) populations differentiate to PMF IV (85%) populations in 4 days, to PMF V (70%) populations in 8 days, and to PMF VI (85%) populations in 21 days. When spontaneously arisen or experimentally induced PMF VI are irradiated three times with 20 J/m^2 UV or exposed three times to 5×10^{-6} *M* mitomycin C with 2 days' interval between the treatments, PMF VI (80%) populations differentiate to PMF VII a in 7 days after the last treatment [5, 17].

Pure clonal populations of the mitotic fibroblasts MF I, MF II, and MF III and subpopulations with a high frequency (80%–90%) of spontaneously or experimentally induced postmitotic fibroblasts PMF IV, PMF V, PMF VI, and PMF VII a were selected, surrounded by a clone ring and labeled with [^{35}S]-methionine for 18 h, and analyzed by two-dimensional gel electrophoresis in the IEF and NEPHGE range. Cell-type-specific, [^{35}S]-methionine-labeled polypeptides were found in the nuclear and cytoplasmic fraction of the seven cell types studied. The polypeptide pattern of the spontaneously arisen and the experimentally induced postmitotic fibroblasts PMF IV, PMF V, PMF VI, and PMF VII a, are identical. These data demonstrate conclusively that the morphological and biochemical differentiation of WI-38 fibroblasts go together in the progenitor, maturing, and degenerating. Cell-type-specific marker proteins have been enriched and isolated; for some of them the amino acid sequence has been determined and monoclonal antibodies have been raised.

Populations of spontaneously transformed cells can be established as permanent cell lines. More detailed data of the cell-biological and cell-biochemical parameters of the spontaneously transformed WI-38 populations will be given elsewhere (unpublished work).

Mass populations of the permanent cell line WI-38/SV40 (VA13 2RA) at CPD level 244 are heterogeneous and composed of three mitotic fibroblasts (TMF) TMF I (0.3%), TMF II (84.6%), and TMF III (10.5%) and four transformed postmitotic fibroblasts (TPMF) TPMF IV (0.2%), TPMF V (0.2%), TPMF VI (0.1%), and TPMF VII (3.0%). The morphology of the transformed mitotic and postmitotic WI-38/SV40 cell types very much resembles that of the normal mitotic and postmitotic WI-38 cell types. Clonal populations of WI-38/SV40 fibroblasts are always heterogeneous. In the center of most clones studied TMF II can be found; the peripheral belts are made up of transformed mitotic and postmitotic fibroblasts of higher differentiation stages. In one clone mitotic and postmitotic celltypes of WI-38 can be found. These findings conclusively demonstrate that transformed WI-38/SV40, like normal WI-38 cells, develop along a three-stage sequence of transformed mitotic fibroblasts (TMF): TMF I – TMF II – TMF III, and a four stage developmental lineage of transformed postmitotic fibroblasts (TPMF): TPMF IV – TPMF V – TPMF VI – TPMF VII in four compartments. Subpopulations with a high frequency of the seven transformed mitotic and postmitotic fibroblasts were labeled with [^{35}S]-methionine, and the [^{35}S]-methionine polypeptide pattern was studies by means of 2D gel electrophoresis in the IEF and NEPHGE range. When the polypeptides of the nuclear and the cytoplasmic fractions were compared, cell-type-specific marker polypeptides for all of the seven mitotic and postmitotic WI-38/SV40 cells were found.

Discussion

When the development of WI-38 populations in vitro is studied with stem-cell-specific methods, it becomes evident that WI-38 cells differentiate along an seven-stage terminal differentiation cell lineage in four compartments. Thus WI-38 populations undergo the same differentiation process as the secondary skin and lung fibroblast population of Valo-chicken (unpublished work), C3H-mice (unpublished work), BN-rats [10], and man [1–3]. WI-38 transformed by SV40 also developed along an seven-cell-type differentiation sequence in four compartments, as has been described for transformed fibroblast populations of BN-rats [10] and C3H-mice (unpublished work). In transformed WI-38/SV40 populations the transformed mitotic fibroblast MF II was found to be the predominant cell type, with 84.6% (stem line). TMF II differentiates spontaneously to transformed fibroblasts of a higher differentiation stage only at very low frequencies. Identical data were found for TMF II in transformed fibroblast populations of BN-rats [10] and C3H-mice (unpublished work).

Now that the cell biology of the WI-38 and WI-38/SV40 cell lines is fully defined, pure clonal populations, and pure subpopulations of the spontaneously arisen and pure experimentally induced mass populations of the mitotic and postmitotic fibroblast cell types can be made available for cell-biological and -biochemical studies of fibroblasts in one differentiation stage. Furthermore, pure fibroblast populations of different mitotic and postmitotic differentiation stages can be combined, or heterogeneous mass populations with a known composition of fibroblast celltypes can be used for cell-biological and -biochemical studies of the interaction of fibroblasts in different differentiation stages. From such defined populations reproducible data about changes of biochemical parameters as a function of the differentiation state of the fibroblast have been collected [1–3, 5, 10, 11, 13–18].

As a consequence, a unifying concept of the molecular mechanisms of cellular aging and cellular neoplastic transformation has been proposed [10]. According to this concept, normal physiological cellular aging is the result of the terminal differentiation of fibroblasts in the fibroblast stem cell system, followed by either apoptosis or transformation. The course of the morphological and biochemical multistage differentiation sequence of fibroblasts can be accelerated by endogenous factors, e.g., Duchenne muscular dystrophy fibroblasts [15, 16], and by exogenous factors, like UV [17] and mitomycin C [5], and agents not degradable by lysosomes [sucrose (unpublished work)] and persistent infections with viruses [rubella virus (unpublished work)]. These treatments increase the probability for the differentiation along the genetic program of the mitotic and postmitotic fibroblast cell lineage. Normal terminal differentiation and experimentally accelerated terminal differentiation follows the same genetically programmed pathway. Spontaneous and experimentally induced differentiation follows the same cell lineage of morphotypes, and spontaneously arisen and experimentally induced cell-types have the same cell-type-specific [^{35}S]-methionine polypeptide pattern [1, 3, 5, 17, 18].

Terminally differentiated and cells PMF VI of C3H-mouse fibroblasts (unpublished work), BN-rat fibroblasts [10], and WI-38 fibroblasts show nuclear fragmentation at a frequency of 10^{-5}. Nuclear fragments surrounded by small amounts of cytoplasm bud through the cell wall and give rise to the transformed fibroblast cell lineage.

References

1. Bayreuther K, Francz PI (1990) Aging in the human dermal fibroblast stem cell system in vivo and in vitro. UCLA Symp Mol Cell Biol [New Ser] 123:205–215
2. Bayreuther K, Rodemann HP, Hommel R, Dittmann K, Albiez M, Francz PI (1988) Human skin fibroblasts in vitro differentiate along a terminal cell lineage. Proc Natl Acad Sci USA 85:5112–5116
3. Bayreuther K, Rodemann HP, Francz PI, Maier K (1988) Differentiation of fibroblast stem cells. J Cell Sci [Suppl] 10:115–130
4. Bell E, Marek LF, Levinstone DS, Merrill C, Shen S, Young IT, Eden M (1978) Loss of division potential in vitro: aging or differentiation? Science 202:1158–1163
5. Francz PI, Bayreuther K, Rodemann HP (1989) Cytoplasmic, nuclear, membrane-bound and secreted [^{35}S]-methionine-labelled polypeptide pattern in differentiating fibroblast stem cells in vitro. J Cell Sci 92:231–239
6. Hart RW, Setlow RB (1974) Correlation between deoxy-ribonucleic acid excision repair and life span in a number of mammalian species. Proc Natl Acad Sci USA 71:2169–2173
7. Hayflick L (1981) Cell death in vitro. In: Bowen ID, Lockshin G (eds) Cell death in biology and pathology. Chapman and Hall, London, pp 243–285
8. Hayflick L, Moorhead PS (1961) The serial cultivation of human diploid cell strains. Exp Cell Res 25:585–621
9. Holliday R, Tarrant GM (1972) Altered enzymes in aging human fibroblasts. Nature 238:26–30
10. Kontermann K, Bayreuther K (1979) The cellular aging of rat fibroblasts in vitro is a differentiation process. Gerontology 25:261–274
11. Limat A, Hunziker T, Boillat C, Bayreuther K, Noser F (1989) Postmitotic human dermal fibroblasts efficiently support the growth of human follicular keratinocytes. J Invest Dermatol 92:758–762
12. Martin G, Sprague CA, Norwood TH, Pendergrass WR, Bornstein P, Hoehn H, Arend WP (1975) Do hyperplastoid cell lines 'differentiate themselves to death'? Adv Exp Med Biol 53:67–90
13. Mollenhauer J, Bayreuther K (1986) Donor-age-related changes in the morphology, growth potential, and collagen biosynthesis in rat fibroblast populations in vitro. Differentiation 32:165–172
14. Niggli HJ, Bayreuther K, Rodemann HP, Röthlisberger R, Francz P (1989) Mitomycin C-induced postmitotic fibroblasts retain the capacity to repair pyrimidine photodimers formed after UV-irradiation. Mutat Res 219:231–240

15. Rodemann HP, Bayreuther K (1984) Abnormal collagen metabolism in cultured human skin fibroblasts from patients with Duchenne muscular dystrophy. Proc Natl Acad Sci USA 81:5130–5134
16. Rodemann HP, Bayreuther K (1986) Differential degradation of [^{35}S]-methionine polypeptides in Duchenne muscular skin fibroblasts in vitro. Proc Natl Acad Sci USA 83:2086–2090
17. Rodemann HP, Bayreuther K, Francz P, Dittmann K, Albiez M (1989) Selective enrichment and biochemical characterization of seven human skin fibroblast cell types in vitro. Exp Cell Res 180:84–93
18. Rodemann HP, Pfleiderer G, Bayreuther K (1989) The differentiation of normal and transformed human fibroblasts in vitro is influenced by electromagnetic fields. Exp Cell Res 182:619–621
19. Smith JR, Pereira-Smith OM (1985) Lung-derived fibroblast-like human cells in culture. In: Cristofalo V (ed) Handbook of cell biology and aging. CRC, Boca Raton, pp 375–425

Differential Gene Expression, Protein Synthesis, and Degradation in Aging Fibroblasts

H. P. Rodemann

Introduction

Since Hayflick and Moorhead [5] described the limited division potential of cultured fibroblasts, in vitro culture systems of fibroblasts have become increasingly important for investigations into the biology of aging. During the past two decades, descriptive data have been accumulated describing morphological and biochemical changes due to the cellular aging process. These data led to the description of the so-called senescent phenotype of fibroblasts. In this regard, the term "senescence" is used to characterize the loss of proliferation potential in combination with physiological changes of the "aging" cells. However, the loss of the proliferation potential does not necessarily result in the cellular impairment or imbalance of distinct physiological parameters like protein synthesis and degradation. It could also be discussed as the consequence of a terminal differentiation process, leading to a terminally differentiated postmitotic cell which has specific functions, as is the case in other terminally differentiating cell systems.

Obviously, fibroblasts have the potential to differentiate into postmitotic cells which exert cell- and tissue-specific functions [1, 3, 6]. Recent investigations by Bayreuther et al. [1, 2] (Bayreuther and Gogol, this volume) and Rodemann et al. [12, 13] and Rodemann [7] demonstrated that in vitro-cultured fibroblasts of the connective tissues of various vertebrate species (including man) differentiate along a terminal cell lineage (see also Bayreuther and Gogol, this volume). Three mitotic (cell types MFI, MFII, MFIII) and three postmitotic (cell types PMFIV, PMFV, PMFVI) fibroblast types were characterized by morphological as well as by biochemical parameters (Fig. 1) [7, 12, 13]. The terminally differentiated cell type PMFVI is the biosynthetically most active cell of the fibroblast cell system in regard to synthesis of the interstitial collagen types I, III, and V [10].

These results indicate that the cellular aging process represents a terminal differentiation process which is under the control of distinct genetic programs. In the present paper, differentiation-dependent cellular alterations of specific biochemical and molecular biological parameters will be described which can be interpreted as the result of a genetically controlled aging process of fibroblasts in culture.

A. Bernd, J. Bereiter-Hahn,
F. Hevert, H. Holzmann (Eds.)
Cell and Tissue Culture Models
in Dermatological Research

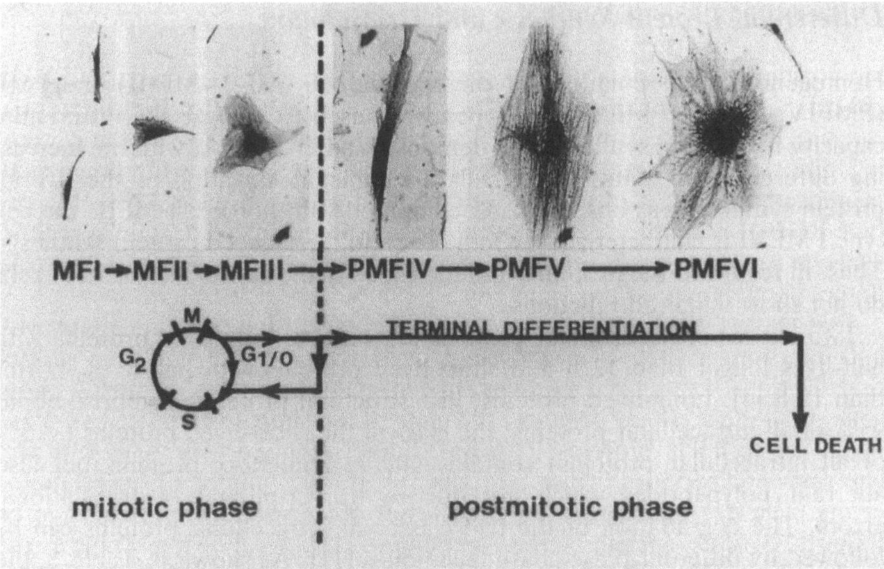

Fig. 1. Terminal cell lineage of human fibroblasts in vitro. Human fibroblasts in vitro differentiate along the terminal cell lineage MFI – MFII – MFIII – PMFIV – PMFV – PMFVI. The potentially mitotic fibroblasts MFI, MFII, and MFIII arrested in G_O can be stimulated differentially by growth factors to enter the cell cycle; the postmitotic fibroblasts PMFIV, PMFV and PMFVI cannot be stimulated. The DNA content of the PMF cells is 8C. Since all human PMF-type cells analyzed so far have one nucleus, the increased DNA content of these cells is most likely due to endopolyploidization processes and is not caused by irregular endomitosis and a block of cytokinesis

Results

Differentiation of Fibroblasts In Vitro

In the course of the so-called aging process of in vitro-cultured human fibroblasts specific cellular changes occur which are due to differentiation processes. With the increasing "age" of the cell culture the amount of mitotically active progenitor fibroblast types MFI [high mitotic activity, 25–30 cell divisions (cd)] and MFII (medium mitotic activity, 20–25 cd) decreases, while the amount of the progenitor fibroblast MFIII, the cell with lowest mitotic activity (5–8 cd), increases. The progenitor fibroblasts MFI, MFII, and MFIII are characterized by a gradual decrease in the stimulation of cell proliferation due to different concentrations of growth factors (MFI > MFII > MFIII). As a function of the terminal differentiation at higher passage numbers, the amount of postmitotic fibroblasts increases progressively. Approximately 3–6 weeks after spontaneous transition into the postmitotic phase all cells present are of the cell type PMFVI.

Differential Protein Synthesis and Degradation

Homogenous subpopulations of the specific MF (MFII, MFIII) and PMF (PMFIV, PMFV, PMFVI) cell types were used to analyze the differential capacity for protein synthesis. As demonstrated in Table 1, with the increasing differentiation state of the cells a significant elevation of the overall protein synthesis was observed. As compared with fibroblast MFII, the cell type PMFVI is characterized by four- to sixfold enhanced protein synthesis. Thus, in regard to the protein synthetic capacity the so-called senescent cells do not show impaired functions.

Two classes of intracellular proteins are known: long-lived proteins with half-lives longer than 12 h and short-lived proteins with half-lives shorter than 12 h [4]. Long-lived proteins, like structural proteins, comprise about 95% of all intracellular proteins; the class of the short-lived proteins (< 5% of all intracellular proteins) contains mainly regulatory proteins but also aberrant polypeptides which are due to transcriptional or translational errors. The degradation of the two classes of intracellular proteins can be followed by different pulse-chase techniques [11]. As shown in Table 1, the degradation of long-lived proteins does not change as a function of the differentiation state. In contrast, the degradation of short-lived proteins increases significantly from the cell type MFII to PMFVI (Table 1). Thus, the average half-life of this class of proteins decreases by approximately 2 h, from 6 to 4 h when MFII and PMFVI are compared.

Skin fibroblasts derived from patients affected with Duchenne muscular dystrophy (DMD) show a significantly enhanced intracellular degradation of short- and long-lived proteins in both MF and PMF cell types when compared with skin fibroblasts from control donors [8, 9, 11]. The pathophysiologically enhanced degradation of intracellular proteins is most likely the cause of the abnormal terminal differentiation process and impaired viability

Table 1. Protein synthesis and degradation in mitotic and postmitotic human skin fibroblasts

Cell type	Protein synthesis ($dpm/10^3$ cells)	Protein degradation (% of labeled protein degraded per hour) l.l.p.	s.l.p.
MFII	426 ± 89	0.48 ± 0.12	10.6 ± 0.4
MFIII	679 ± 123	0.57 ± 0.29	12.8 ± 0.3*
PMFIV	1567 ± 324*	0.58 ± 0.16	14.8 ± 0.4*
PMFV	1729 ± 376*	0.56 ± 0.17	15.0 ± 0.6*
PMFVI	2099 ± 448*	0.60 ± 0.21	15.1 ± 0.6*

Protein synthesis and degradation of total cellular proteins, long- (l.l.p.) and short-lived (s.l.p.) proteins, were determined in the various differentiation states by pulse-chase experiments [11].
* $p < 0.05$, $n = 8$

of DMD cells in culture. Thus, the postmitotic life span of DMD fibroblasts is significantly reduced by about 50%–70%, resulting in a premature cellular aging process and cellular degeneration.

Differential Gene Expression

In the course of the terminal differentiation, or so-called aging process, from cell type MFI to PMFVI of normal and DMD fibroblasts, 11 cell-type-specific proteins are differentially induced and sequentially expressed. In Table 2 these proteins and their physicochemical parameters are summarized. All proteins except protein PIVa are long-lived proteins. Protein PIVa, which has a half-life of approximately 1.1 h, is the first protein induced when MFIII-type cells differentiate into postmitotic PMFIV-type cells. When normal and DMD fibroblasts are compared, the half-life of protein PIVA does not differ significantly between DMD and control fibroblasts, although the half-life of all other cell-type-specific proteins is affected significantly by the enhanced degradation in DMD cells (30%–40% reduced half-life time).

The expression of protein PIVa can be induced experimentally when MFII type cells are incubated for 24 h in the presence of mitomycin C (MMC, 2×10^{-7} *M*). The MMC-treated cells differentiate within 4–5 days by 100% into PMFIV-type cells [7]. Within this time period, the expression

Table 2. Physicochemical properties and half-lives of the differentiation-type-specific proteins in hormal and DMD fibroblasts

Cell type	Protein	M_r/p_i	Protein half-live (h) Normal	DMD
MFII	PIIa	32/8.2	35 ± 1	26 ± 1
	PIIb	37/7.9	46 ± 1	33 ± 2
	PIIc	40/7.2	61 ± 1	48 ± 3
MFIII	PIIIa	38/7.0	33 ± 3	23 ± 4
PMFIV	PIVa	33/5.0	1.1 ± 0.2	0.9 ± 0.2
PMFV	PVa	31/9.2	52 ± 3	34 ± 3
	PVb	32/9.0	62 ± 2	39 ± 3
PMFVI	PVIa	33/8.0	59 ± 2	33 ± 3
	PVIb	31/7.8	38 ± 3	20 ± 3
	PVIc	30/7.3	43 ± 2	23 ± 3
	PVId	37/7.3	39 ± 1	20 ± 1

The physicochemical properties and half-lives of the cell-type-specific proteins were determined by pulse-chase experiments in combination with two-dimensional polyacrylamide M_r, Molecular mass; p_i, isoelectric point gel electrophoresis and computerized video-densitometry [7].

of PIVa correlates with the number of cells differentiating into PMFIV. These data indicate that protein PIVa could act as a regulatory protein controlling the postmitotic state of a cell. This hypothesis is further supported by the fact that poly-A-RNA isolated from PMF-type cells and microinjected into MFII-type cells induces the expression of PIVa within 2–6 h. In the PIVa-expressing MFII-type cells DNA synthesis cannot be induced by growth factors or serum stimulation. Thus, these results indicate that PMF-m-RNA contains the message for protein PIVa and the inhibition of DNA synthesis. Whether protein PIVa acts as a repressor of DNA synthesis remains to be resolved.

Differential Expression of c-FOS

Seshradi and Campisi [14] analyzed the expression and induction of c-fos in "young" and "old" human lung fibroblasts. The authors describe a significant age-dependent reduction in the induction of c-fos expression by serum stimulation. In my laboratory the expression of c-fos was analyzed in homogeneous fibroblast-populations of the differentiation states MFII, MFIII, and PMFVI (cell line HH-4) after 72 h after serum starvation of confluent cultures and various times of stimulation with culture medium containing 20& serum. As demonstrated by Northern-blot analyses, the induction of c-fos expression decreases significantly from MFII to the terminally differentiated fibroblast PMFVI. Thus, during terminal differentiation the induction of the proto-oncogene c-fos becomes differentially repressed.

Conclusion

The experiments and results described indicate that the cellular aging process of cultured fibroblasts of normal and pathological origin is a terminal multistage differentiation process. The various mitotically active (MFI, MFII, MFIII) and postmitotic (PMFIV, PMFV, PMFVI) differentiation types of fibroblasts are characterized by a differential gene expression, protein synthesis, and degradation. The fibroblast cell system differentiates along the terminal cell lineage MFI – MFII – MFIII – PMFIV – PMFV – PMFVI. Thus, in order to obtain relevant data, studies into the process of cellular senescense should not be restricted to the cells of the mitotic cell compartment (by comparing low-passage-levels cells with high-passage-level cells) but will have to include the cells of the postmitotic compartment. These cells, due to their postmitotic nature, represent the senescent or aging cells of the fibroblast cell system. This assumption is further supported by the fact that during the terminal differentiation process of fibroblasts the proto-oncogene c-fos seems to be sequentially repressed. The cell-biological and -biochemical data available so far indicate that protein PIVa, which is specific for postmitotic (senescent) fibroblasts, could act as a regulatory protein controlling the differentiation step from mitotically active to postmitotic fibroblasts. Thus, further molecular biological analyses (e.g., substractive c-DNA libraries of MF and PMF cells) are necessary to characterize the genes for protein PIVa and other factors responsible for the cellular aging process.

Acknowledgement. This work was supported by the *Deutsche Forschungsgemeinschaft,* SFB 223 "Pathomechanisms of Cellular Interactions" and Ro 527/2-2. The excellent technical assistance of Petra Wecke is gratefully acknowledged.

References

1. Bayreuther K, Rodemann HP, Hommel R, Dittmann K, Albiez M, Francz PI (1988) Human skin fibroblasts in vitro differentiate along a terminal cell lineage. Proc Natl Acad Sci USA 85:5112–5116
2. Bayreuther K, Rodemann HP, Francz PI, Maier K (1988) Differentiation of fibroblast stem cells. J Cell Sci 10:115–130
3. Bell E, Marek LF, Levinstone DS, Merrill C, Shen S, Young IT, Eden M (1978) Loss of division potential in vitro: aging or differentiation? Science 202:1158–1163
4. Goldberg AL, Dice JF (1974) Intracellular protein degradation in mammalian and bacterial cells. Annu Rev Biochem 43:835–868
5. Hayflick L, Moorhead PS (1961) The serial cultivation of human diploid cell strains. Exp Cell Res 25:585–621
6. Martin G, Sprague CA, Norwood TH, Pendergrass WR, Bornstein P, Hoehn H, Arend WP (1975) Do hyperplastoid cell lines differentiate themselves to death? Adv Exp Med Biol 53:67–90
7. Rodemann HP (1989) Differential degradation of intracellular proteins in human skin fibroblasts of mitotic and mitomycin C (MMC)-induced postmitotic differentiation states in vitro. Differentiation 42:37–43
8. Rodemann HP, Bayreuther K (1984) Abnormal collagen metabolism in cultured skin fibroblasts from patients with Duchenne muscular dystrophy. Proc Natl Acad Sci USA 81:5130–5134
9. Rodemann HP, Bayreuther K (1986) Differential degradation of ^{35}S-methionine polypeptides in Duchenne muscular dystrophy skin fibroblasts in vitro. Proc Natl Acad Sci USA 83:2086–2090
10. Rodemann HP, Müller GA (1991) Characterization of human renal fibroblasts in health and disease. II. In vitro growth, differentiation, and collagen synthesis of fibroblasts from kidneys with interstitial fibrosis. Am J Kidn Dis 17:684–686
11. Rodemann HP, Bayreuther K, Schumacher E (1987) Abnormal protein metabolism in skin fibroblasts in vitro from patients with Duchenne muscular dystrophy. Biochem Biophys Res Commun 145:1–9
12. Rodemann HP, Bayreuther K, Francz PI, Dittmann K, Albiez M (1989) Selective enrichment and biochemical characterization of seven human skin fibroblast cell types in vitro. Exp Cell Res 180:84–93
13. Rodemann HP, Bayreuther K, Pfleiderer G (1989) The differentiation of normal and transformed human fibroblasts in vitro is influenced by electromagnetic fields. Exp Cell Res 182:610–621
14. Seshradi T, Campisi J (1990) Repression of c-fos transcription and altered genetic program in senescent human fibroblasts. Science 247:205–209

Age Differences in Human Skin Collagen

J. Brinckmann, M. Brey, and *P. K. Müller*

Abstract

Aging of the skin must be divided into endogenous aging and photoaging. We investigated compositional differences in the collagens of (photo-)aged skin. The results were compared with the qualitative collagen synthesis of (photo-)aged fibroblasts. There is no difference in the relative proportion of type-III to type-I collagens in skin extracts and cell cultures from young and old sun-protected areas. In photoaged skin there is an increase of the extractable type III. This increase was not detected in cell culture. These differences may be explained by photo-induced changes during the process of cross-linking. Modified intermolecular cross-links influence the susceptibility of collagens to collagenase attack. UV-aged cells showed a higher collagenase production, so a slower cleavage of type-III collagen would result in an accumulation of this type in the tissue.

Introduction

The aging process of the skin can be divided into intrinsic, endogenous aging, and photoaging. The main feature of aging by UV light is a deep wrinkle, which does not disappear on stretching. The skin between the wrinkles is leathery. By contrast, intrinsically aged skin shows atrophy and shallow wrinkles which disappear on stretching [7]. The histological profiles are the flattened rete ridges and the hypocellularity. No changes are visible in the papillary layer; the collagen fibers in the reticular dermis appear to be more compact. Basophilic fibers are the hallmark of photoaged skin, the so-called solar elastosis. With increasing UV damage, the collagen fibers in the papillary dermis can be totally depressed, with the exception of the dermo-epidermal junction, where intact collagen bundles are visible. Fibroblasts and macrophages are numerous [5]. Collagen comprises 70%–80% of the dry weight of human skin. Elastin and proteoglycan are only minor components of the dermis. Collagens are a family of distinct genetic types, of which types I, III, IV, V, VI, and VII can be detected in the skin. The most abundant forms are types I and III. They co-exist in the same fibril and are important for the structure of the dermis.

A. Bernd, J. Bereiter-Hahn,
F. Hevert, H. Holzmann (Eds.)
Cell and Tissue Culture Models
in Dermatological Research

In this study we analyzed the proportions of collagen types III and I present in the tissue of (photo-)aged skin. These results were compared with the qualitative and quantitative collagen synthesis of (photo-)aged fibroblasts.

Material and Methods

The sun-protected skin samples were obtained from the abdomens of three patients aged 20–30 years and three [4] patients aged 70–85 years, while photoaged skin was obtained from the faces of two patients aged 70–85 years. All skin samples were surplus material of operations.

Extraction of Collagens

The skin was powdered in a hammer mill and stirred in a pepsin solution (0.1 mg/ml pepsin, 0.05% acetic acid, pH 1.8, at 4°C for 24 h). After centrifugation at 200000 g (1 h, at 4°C), supernatants were neutralized and stored at –20°C. This digestion procedure was repeated five times. All supernatants were pooled, salt-precipitated (17.5% KCl, 0.02 *M* KH_2PO_4), and dialyzed against 0.05% acetic acid. Lyophilized aliquots were redissolved in SDS sample buffer, heated to 95°C, and quenched on ice. Separation of collagen chains on polyacrylamide gels was carried out in the presence or absence of 2-mercaptoethanol. Gels were fixed and stained with Coomassie blue, and the relative amounts of the different components were measured by video densitometry.

Cell Culture

Fibroblasts were grown from skin explants and propagated as monolayers in Dulbecco's MEM supplemented with 50 μg/ml sodium ascorbate, 2 m*M* glutamine, 100 U/ml penicillin/streptomycin, and 10% fetal calf serum. Cells were maintained at 37°C in an atmosphere of 5% CO_2 and 95% air. Fibroblasts were studied at passage 5–8, subcultured by 1:2.

Quantitative Collagen Synthesis

Cells were seeded in 24-well plates (5×10^4 cells/well) and confluent monolayers were preincubated for 24 h with streptomycin-free culture medium. Labeling of cells was performed with 10 μCi/ml *L* (2–3)^{3}H-proline (45 Ci/mmol) and 150 μg/ml β-aminoproprionitril in streptomycin- and glutamine-free culture medium supplemented with 10% dialyzed fetal calf serum for 24 h. The combined cell fragments and medium were ultrafiltrated and hydrolyzed (6 *N* HCL, 110°C, 24 h). The collagen synthesis was quantified using an amino acid analyzer as ^{3}H-hydroxyproline/cell. Cell counting was done in a radioactivity-free parallel experiment.

Qualitative Collagen Synthesis

Cells were seeded in Petri dishes (1×10^5 cells) and labeled as described above. The harvested medium and cell fragments were subjected to limited pepsin digestion (0.1 mg/ml pepsin, 0.05% acetic acid, pH 1.8, 4°C, 24 h) and ultrafiltrated. The collagens were redissolved in SDS-sample buffer and seperated on polyacrylamide gels. The relative amounts of the different components were measured after fluorography by video densitometry.

Results and Discussion

Compared with aged cells, the amount of newly synthesized collagens tends to decrease in photoaged fibroblasts. These cells produced 3.6 and 11.2 Bq/1000 cells ^{3}H-hydroxyproline. Fibroblasts from unexposed aged skin produced higher amounts (8.1, 15.4, 21.3, and 28.3 Bq/1000 cells). Young, unexposed cells synthesized 8.6, 11, and 14 Bq/1000 cells ^{3}H-hydroxyproline. A decrease in collagen production in aged unexposed fibroblasts – reported in organ cultures of aged skin or using in vitro aging systems of human cells – has not been detected [3, 8]. Whether photoaged cells lose their ability to synthesize collagen or intrinsic aging is accompanied by increasing collagen production needs further investigation.

There is no difference detectable in the relative proportion of type-III to type-I collagens in young and old sun-protected skin. In both groups the concentration of type-III collagen is about 20%. In photoaged skin there is an increase of 40%, so the extractable type III accounts for 28%. This increase was not detected in cell culture. The ratio of type-III to type-I collagen in cultures of young sun-protected cells was 8%, 11%, and 12%, and in those of old sun-protected cells 8%, 10%, and 14%. Type III in cultures of photoaged fibroblasts accounted for 8% and 3%.

Comparing the results of tissue analysis and cell culture, an increase in pepsin-soluble type-III collagen in photoaged skin can be noted without a concomitant rise of the type-III synthesis in cells obtained from the analyzed tissue (Fig. 1). An explanation for these data may be changes during the process of cross-linking. The induction of covalent cross-links by irradiation of soluble, microfibrillar collagens has been reported [1]. Modified intermolecular cross-links can increase the solubility of type-III collagen or decrease the extraction of type-I collagen. Thus, the proportion of extracted collagens does not represent the real relations present in tissue. Furthermore, changes in cross-linking can influence the susceptibility of collagens to collagenase

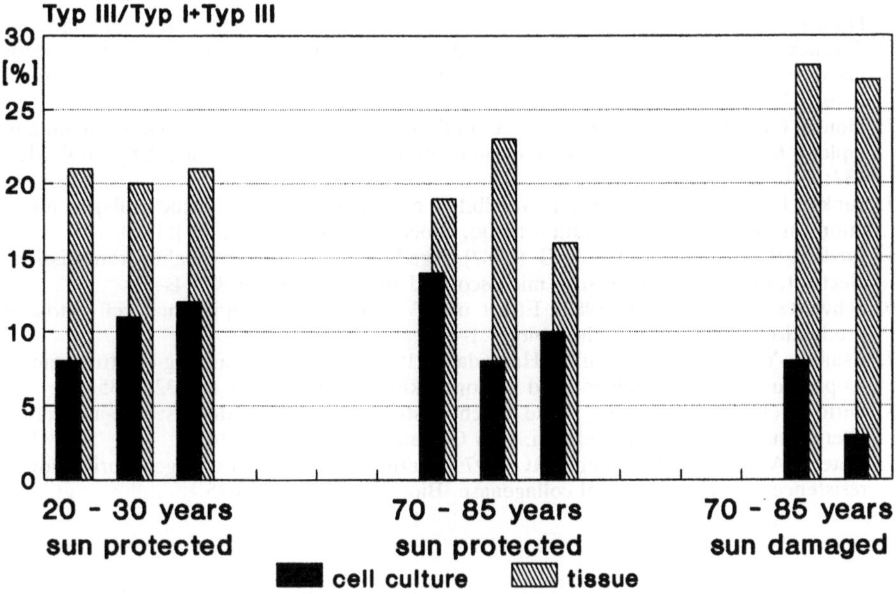

Fig. 1. The ratio of type-III to type-I collagen in tissues extracted by pepsin digestion and type III/type I synthesis in fibroblasts established from these tissues

attack. A slower cleavage of type-III collagen would result in an accumulation of this type in the tissue [9]. The modification of cross-links may be an important step which can explain changes in the extracellular matrix during photoaging. Besides this direct effect of UV light on the connective tissue, an induction of several cytokines by UV exposure in human cells (tumor necrosis factor-alpha, interleukin-1, -6) has been demonstrated [6]. TNF-alpha, IL-1, and IL-6 have been shown to decrease collagen synthesis and increase collagenase production in fibroblasts. Furthermore, some of these cytokines are able to stimulate proliferation of fibroblasts and chemotaxis of macrophages [2, 4]. This is in good agreement with histological findings. There are no changes detectable in qualitative collagen synthesis in instrinsic aged skin and fibroblast culture. In photoaged skin there is an increase of extractable type-III collagen without a concomitant rise of type-III synthesis in photoaged cells. This may be explained by photoinduced changes in the process of cross-linking and an increase of collagenase production by UV-induced cytokines.

Acknowledgement. We thank Mrs. A. Fischer for the amino acid analysis.

References

1. Dalle Carbonare M, Pathok MA (1989) Photoaging and the role of UV (290–400 nm) and reactive oxygen species in collagen cross-links. J Invest Dermatol 92:410

2. Duncan MR, Berman B (1989) Differential regulation of collagen, glycosaminoglycan, fibronectin, and collagenase activity production in cultured human adult dermal fibroblasts by interleukin 1-alpha and beta and tumor necrosis factor-alpha and beta. J Invest Dermatol 92:699–706
3. Houck CJ, Sharma VK, Hayflick L (1971) Functional failures of cultured human diploid fibroblasts after continued population doublings. Proc Soc Exp Biol Med 137:331–333
4. Junker P, Lorenzen I (1983) Reversibility of *D*-penicillamin induced collagen alterations in rat skin and granulation tissue. Biochem Pharmacol 32 (11):1753
5. Lavker RM, Zhen P, Dong G (1987) Aged skin: a study by light, transmission-electron, and scanning-electron microscopy. J Invest Dermatol 88:44s–51s
6. Schwartz T, Luger TA (1989) Effect of UV irridiation on epidermal cell cytokine production. J Photochem Photobiol 4:1–13
7. Tsuji T, Yorifuji T, Hayashi Y, Hamada T (1986) Light and scanning-electron microscopic studies on wrinkles in aged persons' skin. Br J Dermatol 114:329–335
8. Uitto J (1989) Connective tissue biochemistry of the aging dermis: age-associated alterations in collagen and elastin. Clin Geriatr Med 5 (1):127–147
9. Vater CA, Haus ED jr, Siegel RC (1979) Native cross-links in collagen fibrils induce resistence to human synovial collagenase. Biochem J 181:639–645

Influence of Cell Density on Energy Metabolism and the Actin Skeleton of Different Cell Types

A. Münnich, and *J. Bereiter-Hahn*

Introduction

Metabolic changes with age can be observed in living organisms and cells in situ, and in non-neoplastic culture cells as well. Important features of cells in aging organisms can be modelled in culture using either real "aging in culture" or the alterations taking place in single cultures exhibiting density-dependent inhibition of proliferation, which may be followed by reactivation due to further passaging of the cells. The characteristics of these two models are as follows:

Aging and Degeneration of a Culture

Subcultivation of primary cells leads to age-related changes of the cells at higher passages (see Schachtschabel, this volume), e.g., lower saturation density, production of specific proteins, incorporation of granula, a higher amount of cell debris. Finally, the whole culture degenerates; the sequence of events is irreversible [11, 12]. This behavior of a single cell type reflects processes related to the limited life span of a living organism.

Density-dependent Inhibition of Proliferation

A characteristic property of non-neoplastic cells in culture is the stagnation of proliferation at high cell densities known as "density-dependent inhibition or proliferation." This phenomenon could be related to the delayed proliferation of somatic cells, as it occurs in older rather than younger organisms [5].

With this contribution we present some investigations carried out in order to answer the question: Which factors are responsible for this density-related inhibition or proliferation?

A. Bernd, J. Bereiter-Hahn,
F. Hevert, H. Holzmann (Eds.)
Cell and Tissue Culture Models
in Dermatological Research

Material and Methods

Cells

All experiments were carried out either with 3T3 cells (an immortalized mouse fibroblast line) maintained in Dulbecco's minimal essential medium (DMEM) supplemented with 5% newborn calf serum (NCS) or with primary fibroblast cells (pMF cells) derived from NMRI mice. The pMF cells were kept in DMEM with 10% NCS. For all experiments the pMF cells were from passage 1–10 after explanation.

Methods

The following parameters were investigated in relation to cell density:

Oxygen Consumption and Aerobic Lactic Acid Production

Determination of oxygen consumption and of the simultaneous aerobic lactic acid production provide information on all energy-delivering metabolic pathways. Oxygen consumption was measured with a microperfusion respirometer [14]. Subsequently, lactic acid production was determined enzymatically from the perfusate using a commercially available test kit (Boehringer, Mannheim, FRG).

Total ATP and NAD Contents

Total ATP content and NAD/NADH ratio provide information on the energy and the redox equivalents that are available for the cell. Total ATP content was measured after extraction of the cells with hot triton X 100 using the luciferin-luciferase method. NAD content and NAD/NADH ratio were determined by scanning fluorometry of undisturbed cell cultures [6].

F-Actin

F-actin is important for cell locomotion and the maintenance of the cell shape. Both are energy-requiring processes. In addition, the actin skeleton is discussed as an association site of glycolytic enzymes [3, 4, 10, 15], and F-actin may thus be involved in compartmentation as well as in regulation of glycolysis [1]. F-actin was stained by TRITC-phalloidin and the content was subsequently determined fluorometrically [2].

Results

Both the cell types form a monolayer after reaching their saturation densities. While this applies for 3T3 cells at rather high cell densities (about 1500 cells/mm^2), the primary mouse fibroblasts already stop proliferation at around 900 cells/mm^2. Confluency, i.e., contact to other cells on all sides, is reached at 450–500 cells/mm^2 in 3T3 cell cultures and at 200 cells/mm^2 in primary mouse fibroblast cultures.

Oxygen Consumption and Aerobic Lactic Acid Production

Figure 1 represents the oxygen consumption of the two cell lines in relation to cell density. While the oxygen consumption of 3T3 cells is rather low and without any clear relation to density, the primary cells are characterized by a higher oxygen consumption and by density dependency of this parameter. Particularly in subconfluent pMF cultures with single cells or cells with partly free margins, oxygen consumption is extremely high and shows a strong decline until the border of confluency. Subsequently it remains on a fairly stable level.

The simultaneous aerobic lactic acid production resembles the course of the oxygen consumption; the only difference is a slightly higher net lactate production in sparse 3T3 cell cultures compared with those of high cell density (Fig. 2). With increasing cell densities in subconfluent pMF cultures lactate production declines dramatically and remains at a fairly constant level after reaching confluency. In most cases, lactic acid production of pMF cells is higher than that of 3T3 cells.

In summary, primary cells, in contrast to the permanent line, reveal a higher energy production which is more density dependent. Higher energy production in subconfluent cultures indicates a greater need for energy, while the constant level can be considered the necessary maintenance metabolism. Strikingly, energy metabolism changes dramatically already at the border of confluency, rather than at high cell densities, when proliferation stops.

Total ATP Content

The total ATP content of 3T3 cells does not correlate with cell density, but it remains on a fairly stable level. In primary mouse fibroblasts, on the other hand, the total ATP content is significantly correlated with cell density (Fig. 3).

Total ATP content of a cell marks the equilibrium of energy-producing and energy-requiring metabolic pathways. Therefore, a high total ATP content of pMF cells (Fig. 3) suggests that the cells produce ATP in excess, while a drop in ATP indicates an increased demand for ATP or a decreased production. The ATP content corresponds to oxygen consumption and lactic acid production, and thus indicates a decrease of ATP production with higher densities. The fairly constant total ATP content of 3T3 cells also reflects the relatively constant ATP production.

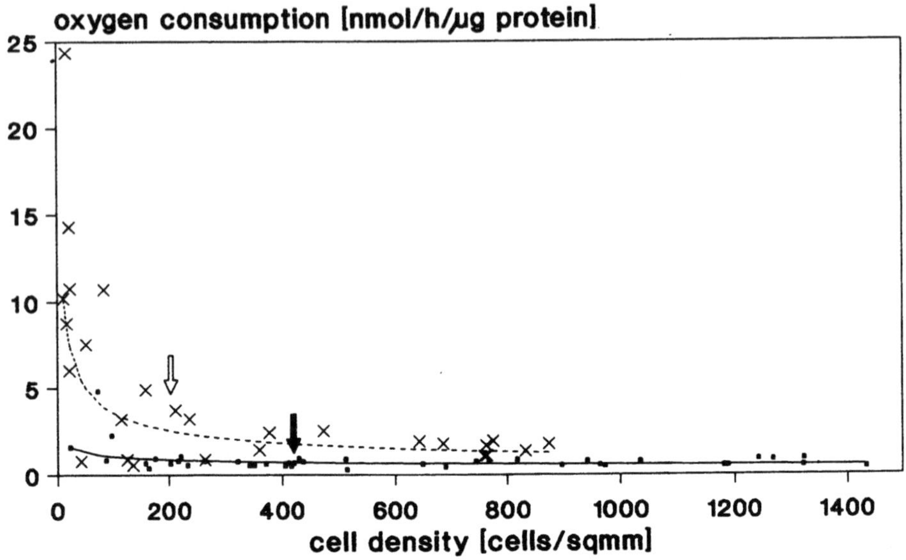

Fig. 1. Oxygen consumption of 3T3 cells (○) and pMF cells (X) in relation to cell density. *Arrows* mark the densities at which cultures become confluent (*open arrow* pMF cells). Each value represents one single experiment with the resulting 2nd-order regression curves (*broken line* pMF cells). 3T3 cells: $n = 41$, $r = -0.513$; pMF cells: $n = 28$, $r = -0.657$

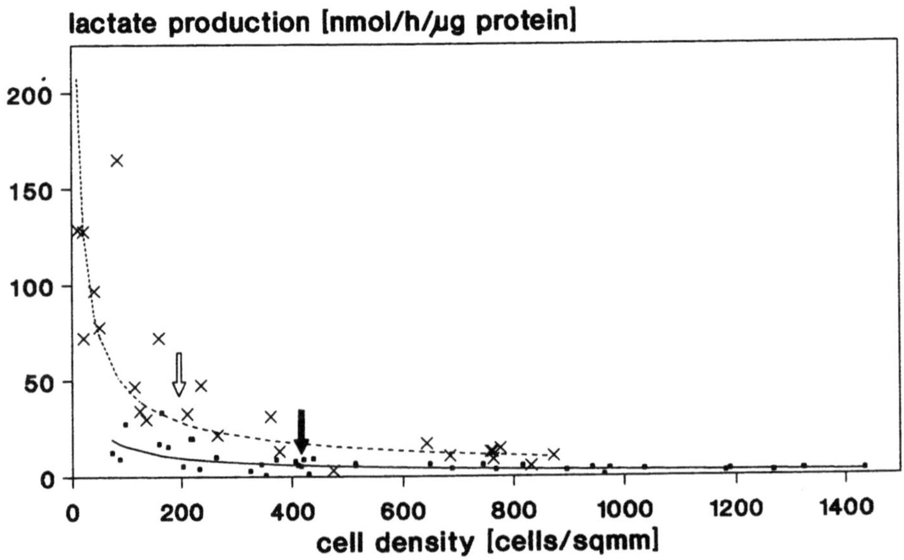

Fig. 2. Aerobic lactic acid production of 3T3 cells and pMF cells in relation to cell density. (For description of the curve see Fig. 1). 3T3 cells: $n = 39$, $r = -0.668$; pMF cells: $n = 25$, $r = -0.902$

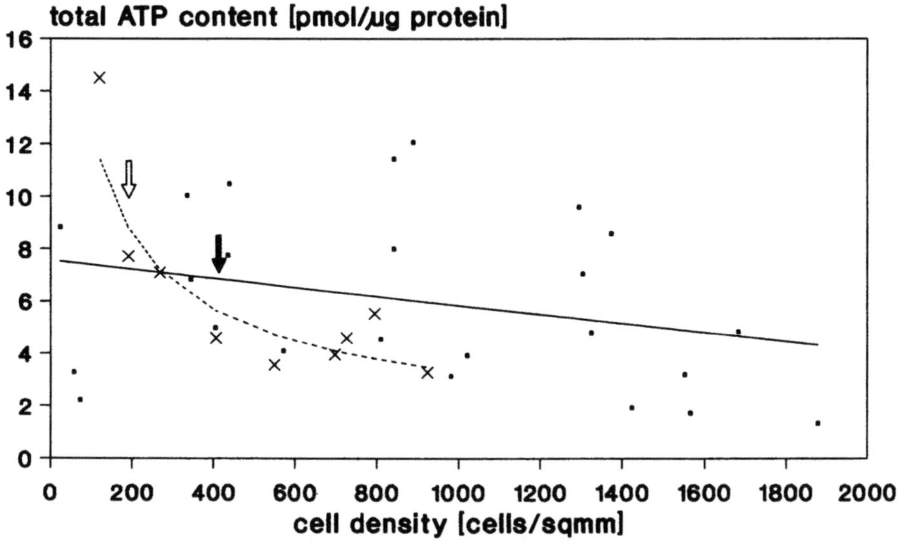

Fig. 3. Total ATP content of 3T3 cells (O) and pMF cells (X) in relation to cell density. *Arrows* mark the densities at which the cells reach confluency (*open arrow* pMF cells). The graph presents single values and resulting correlation curves (pMF cells *broken line*). 3T3 cells: $n = 24$, $r = -0.294$; pMF cells: $n = 9$, $r = -0.892$

NAD (NADH) Content and Redox State

The expression "normal NADH content" stands for the NADH content measured under steady-state conditions, while the "total NADH content" of the same cells was determined under conditions when all NAD of the cell is reduced. The percentage of steady-state NADH of the total NAD content is represented by the "redox state".

The NADH content is density dependent in both cell types. That of pMF cells is higher than that of 3T3 cells (Fig. 4). The drop in total NADH with higher cell densities indicates that fewer reduction equivalents are available for the cells, while the percent decrease of reduced NADH points to a decreased production in relation to consumption.

The high amount of reduced NAD under steady-state conditions in subconfluent cultures seems, at first view, to be contradictory to the results obtained for oxygen consumption and for lactic acid production, both NADH oxidizing metabolic pathways. However, it can be assumed that NADH is not a limiting factor at low cell densities. This interpretation is supported by the relatively high ATP content in these cultures. This general metabolic situation is in favor of the supply of the cells with intermediates for the macromolecular synthesis [8, 9], e.g., by the hexose monophosphate shunt, which is important for the synthesis of nucleic acids [13].

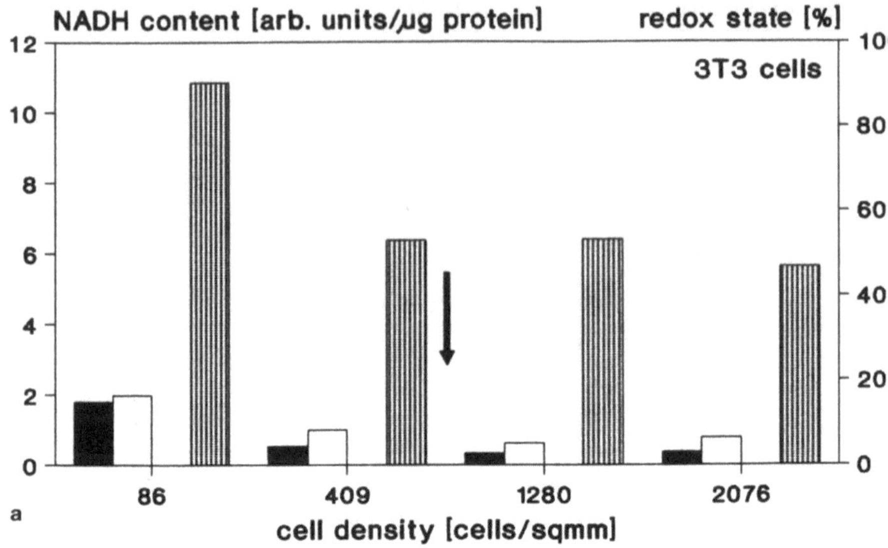

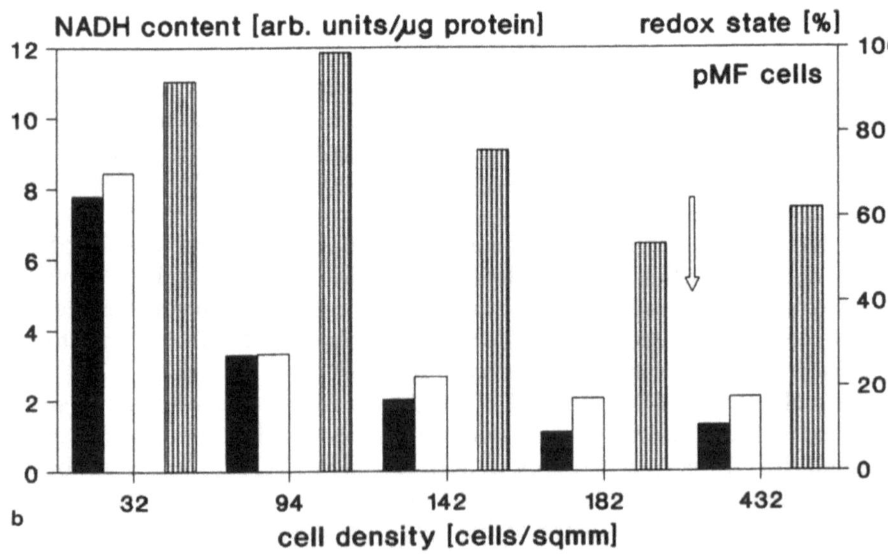

Fig. 4 a, b. NADH content and redox state in relation to cell density: **a** 3T3 cells; **b** pMF cells. ■, Normal NADH content; □, total NADH content; ▥, redox state; for explanation see text. Each *bar* represents 1–4 experiments, standard deviations are below 5%

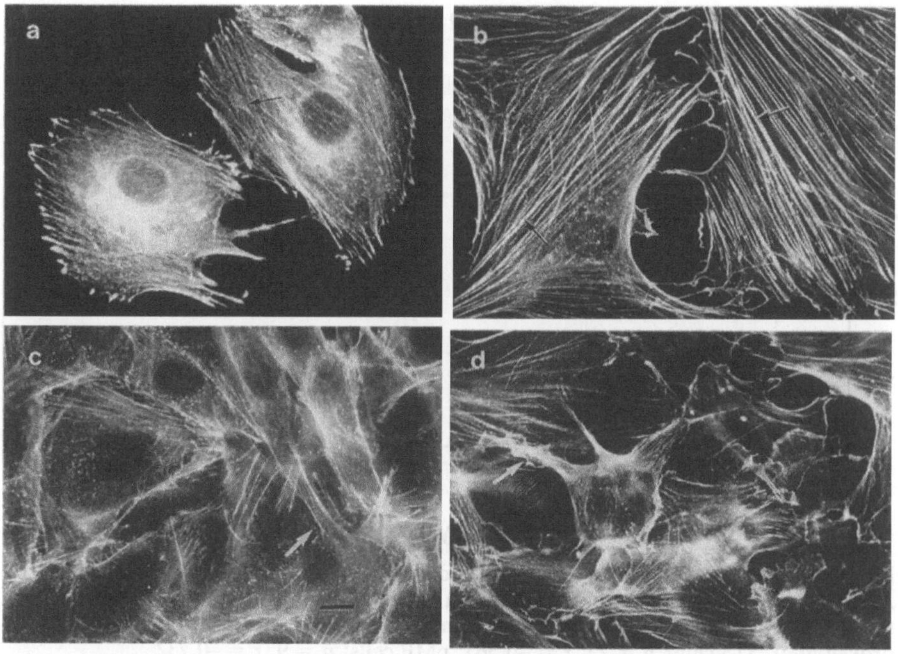

Fig. 5 a–d. F-actin structures of 3T3 and pMF cells at different cell densities. Cells were stained with TRITC-phalloidin: **a** and **c** 3T3 cells; **b** and **d** pMF cells. In subconfluent cultures (**a** and **b**) the cells are widespread and possess stress fibers *(black arrows)*. These are very prominent in the case of pMF cells. In confluent cultures (**c** and **d**) the F-actin skeleton is diminished and more diffuse; cells are more rounded up and partly overlap *(white arrows)*. *Bar:* 10 μm

F-Actin

In both 3T3 and pMF cells the actin pattern is characterized by prominent stress fibers and widely spread cells (Fig. 5a and c). In confluent, dense cultures the actin skeleton is more diffuse and diminished, the stress fibers disappear [7], and the cells begin to round up (Fig. 5 b and d). These structural changes correspond to a decrease in F-actin content with increasing density in both cell types (Fig. 6), the immortalized cells having a clearly lower F-actin content.

Conclusions

The results described above allow the following conclusions: In subconfluent cultures cells are flat and well spread and possess distinct actin fibers. These cells are characterized by a rather high energy metabolism, which delivers materials for the macromolecular synthesis as well as the energy and reduc-

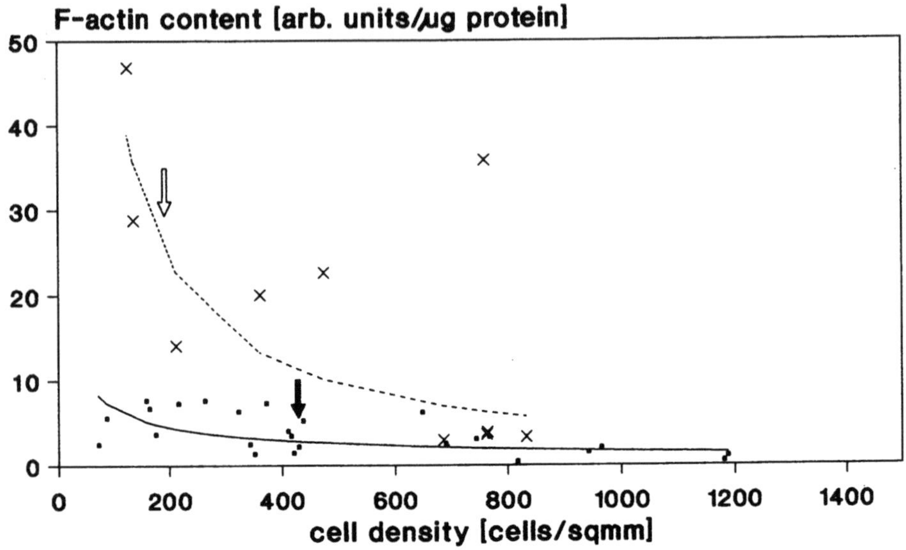

Fig. 6. F-actin content of 3T3 cells (●) and pMF cells (X) in relation to cell density. *Arrows* mark densities at which cells reach confluency (*open arrow* pMF cells). Each value represents one single experiment with the resulting correlation curve (pMF cells *broken line*). 3T3 cells: $n = 25$, $r = -0.563$; pMF cells: $n = 9$, $r = -0.732$

tion equivalents (ATP, NADH) for anabolism, catabolism, and cell movement.

With the onset of confluency, metabolism changes profoundly, persisting at high cell densities: The cells are less spread out and lie densely packed, close to each other or overlapping. The actin pattern appears diffuse, and the energy metabolism drops to a lower level.

The shape changes restrict the membrane area available for transport processes. The decrease in the activity of the energy metabolism limits the energy equivalents as well as the precursors for macromolecular synthesis, thus reducing all metabolic activities. This interpretation of the relationship between energy supply and other activities is supported by the fact that the NAD/NADH ratio is shifted to the more oxidized state when the total amount of NAD is diminished, indicating a relatively high energy demand.

The main alterations in energy metabolism are related to the glycolytic activity rather than to respiration. The intimate association between F-actin and glycolytic enzymes relates shape changes and metabolic activities to each other [1, 3, 4], in particular in the primary cells.

Reduction of metabolism, shape changes, and stagnation of proliferation are processes which occur gradually, not suddenly. Also in this respect they resemble the aging process. However, in density-inhibited cell cultures this process is reversible by trypsinization (dilution of the culture). Whether the metabolic changes described in this contribution may be assumed to reflect processes related to aging depends on whether aging is considered a differentiation process or a consequence of metabolic activities by themselves (increase of entropy).

References

1. Bereiter-Hahn J (1988) Involvement of microcompartmentation in the regulation of cell proliferation. In: Jones DP (ed) Microcompartmentation. CRC, Boca Raton, p 55
2. Bereiter-Hahn J, Kajstura J (1988) Scanning microfluorometric measurement of TRITC-phalloidin-labeled F-actin. Histochemistry 90:271–276
3. Blum S, Bereiter-Hahn J (1991) Interactions of actin filaments with LDH. Endocytobiosis Cell Res 7:231–244
4. Clarke F, Stephan P, Morton D, Weidemann J (1983) The role of actin and associated proteins in the organisation of glycolytic enzymes. In: Remedios CG, Barden JA (eds) Actin. Academic, New York, p 249
5. Hayflick L (1977) The cellular basis for biological aging. In: Finch CE, Hayflick L (eds) Handbook of the biology of aging. Van Nostrand and Reinhold, New York, pp 159–186
6. Kajstura J, Bereiter-Hahn J (1988) Scanning microfluorometric measurement of cell constituents. Principles of the method and its application to the determination of NAD content and redox state of XTH-2 cells in culture. Histochemistry 89:63–67
7. Kajstura J, Bereiter-Hahn J (1989) Loss of focal contacts accompanies the density-dependent inhibition of cell growth. Cell Biol Int Rep 13:377–383
8. Kajstura J, Korohoda W (1982) Cellular metabolism and cell contact reactions. Stud Biophys 90:199–205
9. Kajstura J, Korohoda W (1983) Significance of energy metabolism pathways for stimulation of DNA synthesis by cell migration and serum. Eur J Cell Biol 31:9–14
10. Knull HR, Laursen SE, Cottick AA (1987) Cytomatrix-glycolytic enzyme interactions. Mod Cell Biol 5:157–161
11. Kontermann K, Bayreuther K (1979) The cellular aging of rat fibroblasts in vitro is a differentiation process. Gerontology 25:261–274
12. Mollenhauer J, Bayreuther K (1986) Donor-age-related changes in the morphology, growth potential, and collagen biosynthesis in rat fibroblast populations in vitro. Differentiation 32:165–172
13. Reitzer LJ, Wice BM, Kennell D (1980) The pentose cycle. Control and essential function in HeLa cell nucleic acid synthesis. J Biol Chem 255(12):5616–5626
14. Schlage WK, Bereiter-Hahn J (1983) A microscope perfusion respirometer for continuous respiration measurement of cultured cells during microscopic observation. Microsc Acta 87(1):19–34
15. Walsh JL, Knull HR (1988) Heteromerous interactions among glycolytic enzymes and of glycolytic enzymes with F-actin. Biochim Biophys Acta 952:83–91

7. Models in Skin Pharmacology and Toxicology

(+)-Aeroplysinin-1: an Inhibitor of Intrinsic Protein Tyrosine Kinase Activity of EGF-Receptor-Kinase Complex

K. Pfeifer, M.-H. Kreuter, R. Steffen, H. C. Schröder, and *W. E. G. Müller*

Abstract

(+)-Aeroplysinin-1 was isolated from the marine sponge *Verongia aerophoba.* It was found to inhibit the phosphorylation of lipocortin-like proteins mediated by a highly purified preparation of the epidermal growth factor (EGF) receptor-tyrosine protein kinase complex from MCF-7 breast carcinoma cells. (+)-Aeroplysinin-1 blocked the EGF-dependent proliferation of both MCF-7 and ZR-75-1 human breast cancer cells and inhibited the ligand-induced endocytosis of the EGF receptor in vitro. Treatment with (+)-aeroplysinin-1 in the concentration range of 0.25–0.5 μM resulted in a time- and dose-dependent total tumor cell death in vitro. From these data we conclude that (+)-aeroplysinin-1 is a compound with strong anti-tumor effect on EGF-dependent tumor cell lines.

Introduction

(+)-Aeroplysinin-1 (Fig. 1) is a naturally occurring tyrosine metabolite from the marine sponge *Verongia aerophoba* [7, 8]. By testing a series of cell lines we ascertained human epithelial and epithelial-like tumor cells (MCF-7 breast cancer cells and WiDr colon carcinoma cells) to be most sensitive to treatment with (+)-aeroplysinin-1. A common property of these cell lines is a marked dependence on growth factors [6, 13, 14]. The epidermal growth factor (EGF) receptor, present at the surface of these cells, has been shown to be a transmembrane glycoprotein of M_r 170000 which possesses an intrinsic, ligand-dependent protein-tyrosine kinase activity (for a review see

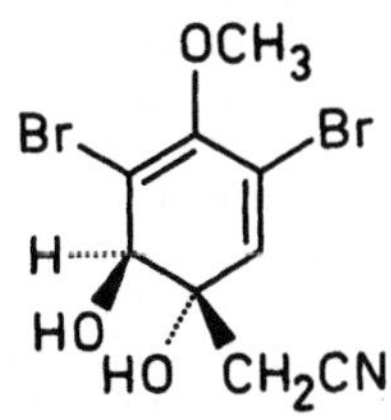

Fig. 1. Structural formula of (+)-aeroplysinin-1

A. Bernd, J. Bereiter-Hahn,
F. Hevert, H. Holzmann (Eds.)
Cell and Tissue Culture Models
in Dermatological Research

[16]). The tyrosine kinase activity of the EGF receptor is assumed to be required for EGF-induced effects on gene transcription, calcium mobilization, receptor down-regulation, and stimulation of cell growth [3].

Here we show that (+)-aeroplysinin-1, which possesses a close structural relationship to tyrosine, blocks the EGF-dependent proliferation of both MCF-7 and ZR-75-1 breast cancer cells and inhibits the ligand-induced endocytosis of the EGF receptor in ZR-75-1 cells in vitro.

Effect of (+)-Aeroplysinin-1 on EGF-induced Cell Proliferation

The data summarized in Fig. 2 show that (+)-aeroplysinin-1 selectively inhibits EGF-induced DNA synthesis in synchronized (with thymidine) and serum-derprived MCF-7 cells in the concentration range of 0.05–0.25 μ*M*. In

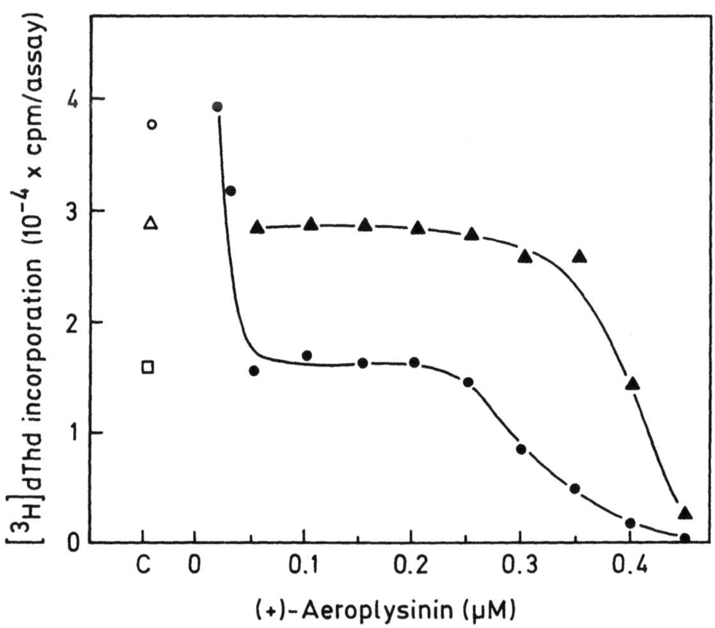

Fig. 2. Effect of (+)-aeroplysinin-1 on EGF-induced proliferation of MCF-7 human breast carcinoma cells. Cells were synchronized and grown in medium supplemented with 5% newborn calf serum (□), 10% fetal calf serum (△), or 200 ng/ml of EGF plus 5% newborn calf serum (○), or in the presence of different concentrations of (+)-aeroplysinin-1 in medium with 5% newborn calf serum plus 200 ng/ml of EGF (●). The reversibility of the inhibition was tested by incubating the cells under the same conditions in the presence of (+)-aeroplysinin-1 for 36 h; subsequently, the compound was washed out and incubation was continued for 72 h in medium containing 10% fetal calf serum without (+)-aeroplysinin-1 (▲). The incorporation of [^{3}H]dThd was determined (the DNA precursor was added 12 h prior to the end of the experiment). Under the conditions used the cell number per assay did not change. Each value represents the mean of 12 experiments; the S.D. were less than 10%. *C*, Control

an earlier report we demonstrated that (+)-aeroplysinin-1 did not display any cytostatic activity in normal human fibroblasts under otherwise identical conditions at a ten times higher concentration [12]. In this concentration range the same rates of [^{3}H]desoxy-thymidine ([^{3}H]dThd) incorporation were found as those determined for cells grown in medium supplemented with 5% newborn calf serum without the compound. The stimulatory activity of fetal calf serum represents a pool of various growth factors [1]. It was suppressed by (+)-aeroplysinin-1 to a similar extent and in the same concentration range as observed with a combination of newborn calf serum and EGF. However, in the presence of fetal calf serum and aeroplysinin the cells still displayed an at least 10% higher proliferation index than without serum (measured by [^{3}H]dThd incorporation), indicating the existence of some further processes not inhibited by (+)-aeroplysinin-1. To determine whether the inhibitory effect of (+)-aeroplysinin-1 on MCF-1 cell growth is reversible, the cells were incubated for 36 h in the presence of different concentrations of (+)-aeroplysinin-1. After treatment the compound was washed out and new medium plus 10% fetal calf serum was added, prior to further incubation with aeroplysinin for an additional 72 h. The growth-inhibitory effect of (+)-aeroplysinin-1 was found to be completely reversible (Fig. 2). Additionally, the proliferation ability of the cells was restored after incubation of the cells for 36 h in the presence of this compound at a concentration of 0.05–0.40 μ*M*. Moreover, (+)-aeroplysinin-1 dose-dependently inhibited DNA synthesis in ZR-75-1 cells. At a concentration of 0.4 μg/ml, the DNA content per culture was reduced to about 50%.

Figure 3 demonstrates the enhancement of [^{125}I]EGF binding to ZR-75-1 cells after treatment of the cells with (+)-aeroplysinin-1. The amount of bound EGF/asssay (containing 2×10^5 cells) was calculated from Scatchard plots by computer analysis. The results were based on the DNA content/assay. Addition of different concentrations of (+)-aeroplysinin-1 resulted in a decrease of both the amount of bound EGF and DNA content. However, there was a strong increase in the ratio between both values with increasing concentrations of the compound (Fig. 3). Maximal accumulation of [^{125}I]EGF was found at 1.2 μg/ml (3.53 μ*M*) of (+)-aeroplysinin-1. At higher concentrations, the amount of bound EGF per microgram DNA decreased, possibly due to the cytotoxic effect of this compound. These results indicate that (+)-aeroplysinin-1 causes an increase of either the EGF-binding affinity of the receptor or the density of the EGF binding sites.

Effect on EGF-induced Phosphorylation of Calcium-dependent Membrane-binding Proteins

Several protein substrates of EGF-dependent tyrosine protein kinases have been isolated [9]. A common property of these proteins is their ability to bind to cell membranes depending on the calcium concentration. We investigated whether the dibrominated tyrosine metabolite, (+)-aeroplysinin-1,

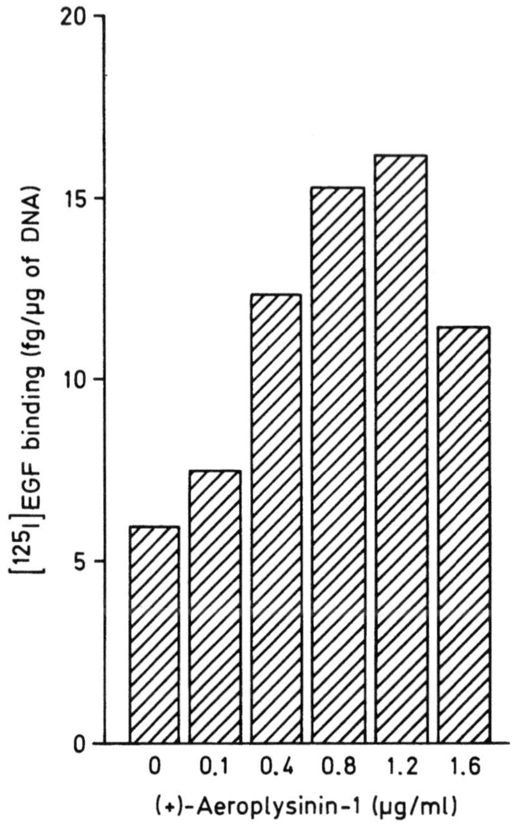

Fig. 3. Accumulation of EGF receptor by (+)-aeroplysinin-1 treatment of ZR-75-1 cells. Cells (2×10^5/assay) were incubated in the presence of different concentrations of (+)-aeroplysinin-1 for 24 h and the binding of [^{125}I]EGF (fg/µg of DNA) was determined. Mean values from eight quadruplicate determinations are given; the S.D. was less than 10%

was able to interfere with the phosphorylation of protein substrates of the EGF-receptor-associated tyrosine kinase. Calcium-dependent membrane-binding proteins from rat brain were purified by reversible binding to erythrocyte inside-out vesicles. The purified proteins were separated by SDS-gel electrophoresis and detected by Western blot using a polyclonal anti-calelectrin antiserum. Three major proteins (M_r 68000, 57000, and 34000) were found to bind to the inside-out vesicles; two of them (M_r 57000 and 34000) were recognized by the anti-calelectrin antibodies, representing calpactin I (p57) and calpactin II (p34) [2, 4].

Moreover, we determined that the solubilized shed membrane vesicle-associated protein kinase activity was markedly stimulated by EGF. The EGF-induced kinase activity of this preparation was strongly inhibited by (+)-aeroplysinin-1. At a concentration of 0.5 µM and in the presence of 20 ng/ml of EGF, this compound reduced the phosphorylation of the calcium-dependent membrane-binding protein preparation, at concentrations of the protein of 150, 80, and 50 µg/ml, by the solubilized kinase to 33%, 25%, and 60%, respectively. In the absence of EGF, the inhibition of the enzyme by (+)-aeroplysinin-1 amounted to 43%, 49%, and 26%, respectively. In control experiments it was assessed that the calcium-dependent membrane-binding proteins possess no significant endogenous protein kinase activity. Assuming a molecular weight of the EGF-dependent tyrosine kinase substrate of 68000, a K_m value of 3.47×10^6 *M* was calculated for the enzyme in the absence of inhibitor. In the presence of 0.5 μ*M* (+)-aeroplysinin-1, this value increased to 9.05×10^6 *M*, resulting in a K_i value of 0.91 μ*M* (competitive type of inhibition).

Determination of the Amino Acid Substrate for Kinase

In order to establish that the enzyme activity inhibited by (+)-aeroplysinin-1 indeed represents a tyrosine-specific protein kinase, the calcium-dependent proteins phosphorylated by the solubilized membrane preparation were immunoprecipitated by anti-calelectrin antiserum and subjected to phosphoamino acid analysis. Using solubilized protein kinase from shed membrane vesicle preparation, we found that only tyrosine was phosphorylated in the lipocortin-like proteins, recognized by anti-calelectrin antibodies. Addition of 0.5 μ*M* (+)-aeroplysinin-1 almost totally suppressed the EGF-dependent phosphorylation of tyrosine residues, while the incorporation of phosphate into threonine residues of total calcium-dependent membrane-binding protein preparation was not affected.

Discussion

Several reports demonstrated an enhancement of EGF-receptor phosphorylation in tumor cells and an involvement of the intrinsic tyrosine-specific kinase activity in this process [5]. Lipocortin-like proteins and some related proteins were shown to be major substrates for tyrosine-kinases [15]. The aim of the work presented here was to study the influence of (+)-aeroplysinin-1 on the processing of EGF receptor-associated tyrosine kinase, its substrate phosphorylation, and its signal transduction-related stimulation of cell proliferation.

Previously, we had demonstrated that (+)-aeroplysinin-1 displays a potent cytotoxic activity against tumor cells in vitro and a high antitumor activity in the L5178y mouse ascites tumor model in vivo. These effects were found to be selective over a wide concentration range, in which (+)-aeroplysinin-1 displays cytostatic activity only on nontransformed cells [12]. Now we determined that during treatment with (+)-aeroplysinin-1, MCF-7 cells lost their ability for EGF-mediated response. Inhibition of cell proliferation occurred within 36–72 h. These effects were found to be reversible after removal of the compound.

Previously, we had determined that the rate of incorporation of [^{3}H]dThd in murine spleen lymphocytes and their ability to respond to mitogens (concanavalin A for T lymphocytes and lipopolysaccharide for B lymphocytes) are not influenced at a 20 times higher dose than that which effectively inhibits EGF response and strongly suppresses the [^{3}H]dThd incorporation rate of MCF-7 cells [12]. From these observations we got the idea that (+)-aeroplysinin-1 might affect growth-factor receptor-associated enzymes and, in consequence, gene expression in transformed cells. One parameter of activation of EGF-receptor-associated tyrosine-specific kinase is the ligand-induced endocytosis of the receptor. We found a dramatic increase of EGF binding to (+)-aeroplysinin-1-treated cells compared with the controls, despite a suppressed DNA synthesis, indicating an accumulation of the EGF receptor. Because substrate phosphorylation is required for the EGF signal transduction process, initiated by autophosphorylation and subsequent C-kinase phosphorylation of the EGF receptor [10], we conclude that (+)-aeroplysinin-1 inhibits one step in this processing. Determinations of protein kinase activity of a solubilized membrane preparation of MCF-7 cells using lipocortin-like proteins as substrates revealed a nearly complete inhibition of tyrosine-specific phosphorylation in the presence of (+)-aeroplysinin-1 at concentrations as low as 0.5 μM. The activity of the solubilized kinase was found to respond to treatment with EGF. When we used an immunoprecipitation procedure, lipocortin-like proteins, which are inhibitors of phospholipase A2 activity [11], were detected as the main phosphorylated substrates. The phosphorylation of these proteins was markedly enhanced after treatment with EGF; it was abolished in the presence of 0.5 μM (+)-aeroplysinin-1. Phosphoamino acid analysis revealed phosphotyrosine to be the only phosphorylated amino acid in immunoprecipitates of lipocortin-like proteins, obtained by using an anti-calelectrin antiserum which also recognized calpactin I and calpactin II.

From its ability to inhibit the ligand-dependent intrinsic protein tyrosine kinase activity of EGF receptor, the sponge tyrosine metabolite (+)-aeroplysinin-1 appears to be a promising compound in a rational chemotherapy. Future studies must show if (+)-aeroplysinin-1 also inhibits other protein tyrosine kinase activities found to be present in many oncogene products.

Acknowledgements. This work was supported by a grant from the *Bundesministerium für Forschung und Technologie* (no. 0319207A8, under the coordination of the *Kernforschungsanlage* Jülich and of the *Internationales Büro GKSS,* Geesthacht).

References

1. Bozzola M, Schimpff RM (1985) Neonatal serum growth-promoting activity measured in lymphocytes: comparison of cord capillary blood. Horm Metab Res 17:418–420
2. Brugge JS (1986) The p35/p36 substrates of protein-tyrosine kinases as inhibitors of phospholipase A2. Cell 46:149–150
3. Chen WS, Lazar CS, Poenie M, Tsien RY, Gill GN, Rosenfeld MG (1987) Requirement for intrinsic protein tyrosine kinase in the immediate and late actions of the EGF receptor. Nature 328:820–823

4. Crompton MR, Owens RJ, Totty NF, Moss SE, Waterfield MD, Crumpton MJ (1988) Primary structure of the human, membrane-associated Ca^{2+}-binding protein p68: a novel member of a protein family. EMBO J 7:21–27
5. Earp HS, Hepler JR, Petch LA, Miller A, Berry AR, Harris J, Raymond VW, McCune BK, Lee LW, Grisham JW, Harden TK (1988) Epidermal growth factor (EGF) and hormones stimulate phosphoinositide hydrolysis and increase EGF receptor protein synthesis and mRNA levels in rat liver epithelial cells. J Biol Chem 263:13868–13874
6. Engel LW, Young NA (1978) Human breast carcinoma cells in continuous culture: a review. Cancer Res 38:4327–4339
7. Fattorusso E, Minale L, Sodato S (1970) Aeroplysinin, a new bromo-compound from *Aplysina aerophoba.* Chem Commun 7C:751–752
8. Fattorusso E, Minale L, Sodato S (1971) Aeroplysinin-1, an antibacterial bromo-compound from the sponge *Verongia aerophoba.* J Chem Soc [Perkin I]:16–17
9. Glenney JR (1986) Two related but distinct forms of the M_r 36000 tyrosine kinase substrate (calpactin) that interact with phospholipid and actin in a Ca^{++}-dependent manner. Proc Natl Acad Sci USA 83:4258–4262
10. Glenney JR, Chen WS, Lazar CS, Walton GM, Zokas LM, Rosenfeld MG, Gill GN (1988) Ligand-induced endocytosis of the EGF receptor is blocked by mutational inactivation and by microinjection of antiphosphotyrosine antibodies. Cell 52:675–684
11. Gramzow M, Schröder HC, Fritsche U, Kurelec B, Robitzki A, Zimmermann H, Friese K, Kreuter MH, Müller WEG (1989) Role of phospholipase A2 in the stimulation of sponge cell proliferation by the homologous lectin. Cell 59:939–948
12. Kreuter MH, Bernd A, Holzmann H, Müller-Klieser W, Maidhof A, Weißmann N. Kljajić Z, Batel R, Schröder HC, Müller WEG (1989) Cytostatic activity of aeroplysinin-1 against lymphoma and epithelioma cells. Z Naturforsch 44c:680–688
13. Osborne C, Hamilton B, Nover M (1982) Receptor binding of epidermal growth factor by human breast cancer cells. J Clin Endocrinol Metab 55:86–93
14. Osborne CK, Hobbs K, Trent JM (1987) Biological differences among MCF-7 human breast cancer cell lines from different laboratories. Breast Cancer Res Treat 9:111–121
15. Veale D, Kerr N, Gibson GJ, Harris AL (1989) Characterization of epidermal growth factor receptor in primary human non-small cell lung cancer. Cancer Res 49:1313–1317
16. Yarden Y, Ullrich A (1988) Growth factor receptor tyrosine kinases. Annu Rev Biochem 57:443–478

In Vitro Cultivation of Human Dermal Microvascular Endothelial Cells as a Model for the Dermatopharmacology

Zb. Ruszczak, and *R. Stadler*

Introduction

The cultivation of endothelial cells began with the experiments of Lewis [35], who described the first successful outgrowth of vascular epithelial cells from explant cultures in 1922. In the early 1970s, Maruyama [38] described a technique for isolation and cultivation of endothelial cells from the large vessels. He found that the large-vessel epithelial cells could be enzymatically released from the umbilical vein by exposure to trypsin and that these cells could be cultured on coverslips coated with collagen. In the same year, 1973, Jaffe et al. [31] described a similar technique for separation and in vitro cultivation of human umbilical vein endotheliocytes. Their protocol served then as a basis for the development of further modifications of isolation and in vitro cultivation of endothelial cells from different animal and human organs.

In 1978 the first description of the isolation and growth of cells obtained from the dermal microvasculature of an experimental animal was reported by Davison and Karasek [10]. This technique, based on mechanical prepreparation of superficial epidermal-papillary dermis splits and a combined enzymatic-mechanical separation of vascular fragments, served as a foundation for the subsequent development of human microvascular cell research.

Human dermal microvascular endothelial cells (HDMEC) were subsequently isolated from the foreskin by Folkman et al. [24] and from foreskin and adult skin by Davison et al. [11, 12]. Enzymatically and mechanically separated cells were cultivated in vitro using high concentrations of human serum and conditioned media from tumors to achieve optimal endothelial cell growth. In all these experiments, however, a relatively high contamination with other cell types, in particular with fibroblasts, occurred, and several technically difficult manipulations were necessary to obtain a satisfactory purified endothelial cell population.

Procedures to isolate and propagate endothelial cells from various organs in culture have been described [12–14, 51]. Although similar in basic morphology, detailed comparative studies have demonstrated substantial genetic, immunologic, and metabolic diversities between endothelial cells from various tissues [2, 29, 32, 56]. The heterogeneity of endothelial cell populations is related to the organ of origin (in the case of skin, also to the part of

A. Bernd, J. Bereiter-Hahn,
F. Hevert, H. Holzmann (Eds.)
Cell and Tissue Culture Models
in Dermatological Research

the body), vessel origin, vessel caliber, and whether fetal or postembryonic tissue is used. In this report we describe the procedure we have recently developed to isolate, identify, and cultivate a highly purified population of human dermal microvascular endothelial cells of children and adult skin.

New information regarding the function of endothelial cells has been obtained from recent studies on the effect of cytokines on endothelial cells in culture. These studies performed predominantly on *macro*vascular endothelial cells derived from large vessels revealed that both human and animal endothelial can be modulated by interleukins, interferons, and tumor necrosis factor [7, 15, 23, 33, 41, 43, 49]; in addition, they produce some of these cytokines, i.e. as a response to viral infection [22]. Long-term cultivation of human dermal *micro*vascular endothelial cells has been used only sporadically for cytokine screening [15, 30, 42, 46].

In this report we briefly discuss the influence of some proinflammatory cytokines such as interleukin 1-alpha (IL 1-alpha), interleukin 1-beta (IL 1-beta), granulocyte/macrophage-colony stimulating factor (GM-CSF), and tumor necrosis factor alpha (TNF-alpha) and present some details on the influence of interferon alpha (rIFN-alpha), interferon beta (rIFN-beta), and interferon gamma (rIFN-gamma) on the morphology, growth kinetics, and expression of class-II antigens and ICAM-1 on HDMEC cultured in vitro in low-serum concentration medium. The cytokines are of great importance not only as potent cell-cell interaction modulators in vivo and in vitro, but also because of their effects in the treatment of different inflammatory and neoplastic skin diseases.

Isolation of Endothelial Cells and Culture Conditions

HDMEC were isolated from surgically removed normal foreskin obtained from newborns and children up to 9 years old and from adult skin biopsies, as we have described recently [46]. Briefly, after being washed in sterile Ca^{2+}- and Mg^{2+}-free phosphate-buffered saline (PBS; Biochrom KG, Berlin, FRG), skin fragments were cut into small pieces in PBS containing 400 U/ml penicillin, 400 U/ml streptomycin, 0.5 μg/ml amphotericin B, washed again in PBS, and then trypsinized overnight (0.3% trypsin in PBS; 4°C) to remove the epidermis. After further washing microvascular fragments were released into the medium by gentle scraping and were filtrated through a 100-μm sterile nylon mesh (Reichelt Chemie-Technik, Heidelberg, FRG) to separate tissue fragments and debris. The cell suspension obtained was washed with fresh Iscove's medium (Biochrom KG) and the endothelial cell population was separated from other cell types by continuous Percoll (Sigma, St. Louis, CA, USA) gradient centrifugation. The Percoll gradient was prepared by centrifugation of a 35% Percoll/PBS solution at 30000 g for 15 min. The cell suspension containing both endothelial and nonendothelial cell populations was placed on the top pf the gradient and was centrifuged at 400 g for 15 min using Percoll density marker beads (Sigma) for standardization. The endothelial cells were removed from the density gradient between 1.033 and 1.047 and were seeded in plastic tissue culture dishes (Falcon, Becton Dickinson, Heidelberg, FRG) covered with human fibronectin (10 μg/ml; Biochrom) with a seeding density of 2 × 10^4 cell/cem^2.

Endothelial cell basal medium (EBM; Clonetics Co., San Diego, CA, USA), supplemented with 10 ng/ml epidermal growth factor (EGF), 0.5 μg/ml hydrocortisone, 100 U/ml penicillin, 100 U/ml streptomycin, 0.25 μg/ml amphotericin B (all purchased from Clonetics), and 2% fetal calf serum (FCS; Biochrom) was used to support cell growth. This medium was tested using other cell populations such as keratinocytes and human dermal fibroblasts isolated from the same skin samples and was found to almost selectively support the growth of microvascular endothelial cells (unpublished data from our laboratory). The medium was replaced every 2 days. At a preconfluent stage the cells were trypsinized (0.3% trypsin in PBS), washed in supplemented EBM, frozen in culture medium supplemented with 10% FCS and 10% dimethyl sulfoxide (DMSO; Biochrom), and stored in liquid nitrogen until use. For the experiments, second-passage HDMEC were cultivated at 37°C in an atmosphere of 95% air and 5% CO_2 with 100% humidity in supplemented EBM.

Endothelial Cell Identification

HDMEC were identified by morphological and immunocytochemical criteria using light and electron microscopy according to the methods described previously [26, 46]. The growing cells formed monolayers of partially polygonal, but mostly slightly elongated, transparent cells in both primary and in secondary cultures. For the study of immunohistochemical phenotype the cultures were stained with a monoclonal antibody against Factor VIII-related antigen (Factor VIIIr-Ag, DAKO, Hamburg, FRG), with the lectin *Ulex europaeus* agglutinin I (UEA-I, DAKO), and with monoclonal antibodies against specific endothelial cell markers such as Pal-E (Dunn, Asbach, FRG), EN-4 (Dunn) and BMA-120 (Behring, Hamburg, FRG).

Standards for Drug Screening

For proliferation assays, APAAP immunocytochemistry, and light and electron microscopic examination HDMEC were usually treated with 0.1 IU/ml–10^4 IU/ml of the test substance(s) for 2–10 days as described previously by us for interferons, interleukins, and TNF [16, 17, 46]. To control the effect of a drug, discontinuation of the treatment and/or combination with neutralizing antibodies was performed.

Proliferation Assay

HDMEC were seeded in uncoated 96-multiwell plastic tissue culture plates (Falcon) with a density of 2.5×10^3 cells/cm^2 and were cultured with supplemented EBM. Following 12–14 h of cell attachment, medium containing different concentrations of the test substance(s) was added. After 2, 4, 6, 8, and 10 days of culture the cells were incubated for 1 h with 4-methylumbelliferyl heptanoate (MUH, Serva, Heidelberg, FRG) at 37°C, and measurements of the absolute fluorescent units (AFU) were then made using a Fluoroscan II device (Flow Lab., Meckenheim, FRG) as described previously by us [46, 53]. MUH is hydrolyzed by cell estereases, and the amount of measured fluorescence is proportional to the number of viable cells. In a number of experiments using different MUH concentrations and incubation periods, a 1-h incubation with 100 μg/ml MUH was found to give the most reliable correlation with HDMEC cell numbers and with DNA synthesis, measured by ^{3}H-thymidine incorporation, similar to the results in epidermal keratinocytes [53].

Immunocytochemistry

After treatment with 0.3% trypsin, HDMEC were washed twice with supplemented EBM (2% FCS), attached to uncoated glass slides using a Cytospin II unit (Shandon, Runcorn, Great Britain), air dried, fixed for 10 min in acetone and incubated for 30 min at room temperature with monoclonal antibodies (MoAb) against HLA-DR (L 243, Becton Dickinson, Mountain View, CA, USA), HLA-DQ (SK 10, Becton Dickinson), and ICAM-1 (84H10, Immunotech, Marseille, France). The cultured cells were further stained according to the alkaline phosphatase anti-alkaline phosphatase (APAAP) technique [8].

Results and Discussion

Studies on the expression of class-II antigens and cell adhesion molecules are important pathways to a better understanding of the interactions between endothelial cells, lymphocytes, and macrophages and of the function of endothelial cells as an antigen-presenting, immunologically active population. Recent studies have shown that endothelial cells are able to express primary transplantation antigens in vivo, and that microvascular endothelial participate more actively in the immune response process than any other vascular and stromal cell type, representing the major potentially immunogenic organ [4, 27, 44, 57].

The class-II alloantigens HLA-DR and HLA-DQ are expressed on endothelial cells in inflammatory disorders, whereas in normal papillary and subpapillary vessels these antigens are only minimally expressed by endothelial cells [1, 3, 21, 28]. Intercellular adhesion molecule-1 (ICAM-1; CD 54) is only slightly expressed by dermal microvascular endothelial cells under normal conditions; however, ICAM-1 was found on endotheliocytes in inflammatory skin diseases and after stimulation of normal skin with interferon gamma [3, 19–21, 28].

We have established a long-term culture of HDMEC using selective density gradient centrifugation and a cell culture medium supporting endothelial cell growth. The endothelial origin of the cultured cells was confirmed by the presence of Factor VIII-r Ag and by labeling with UEA-I; all cells were positively stained for both markers. The cells used for the experiments wee additionally stained with MoAb to specific endothelial cell markers: Pal-E, EN-4, and BMA-120 showed positive reactions. Electron microscopy revealed the presence of Weibel-Palade bodies.

HDMEC showed the typical morphology of human endothelial cells cultured in low-serum concentration medium in the absence of potent growth modulators such as cholera toxin, 3-isobutylmethyl-xanthine, adenosine 2′:3′-cyclic monophosphate (cAMP), insulin and human or human prepartum maternal serum (Fig. 1 A).

Recently we studied the influence of recombinant human interferon alpha-2a (rIFN-alpha; specific activity 2 × 10^8 IU/mg), recombinant human interferon beta-1 (rIFN-beta; specific activity 2 × 10^8 IU/mg), and recombinant human interferon gamma (rIFN-gamma; specific activity 2.3 × 10^7 IU/mg) on HDMEC in culture [46]. All interferons tested induced alter-

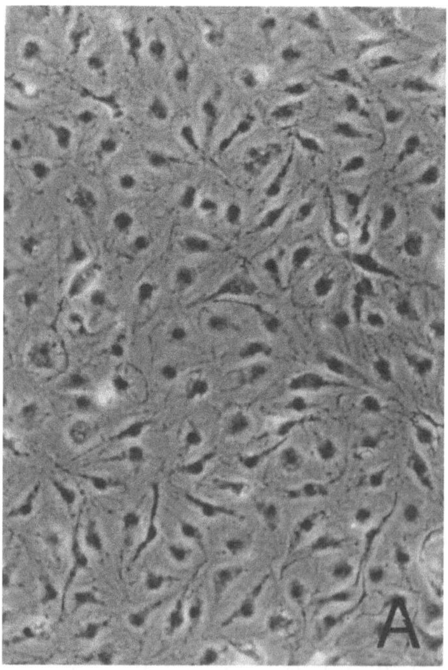

Fig. 1 A–C. Examples of drug influence on the morphology of HDMEC in vitro. Phase-contrast microscopy of HDMEC after 8 days of culture in supplemented low-serum concentration EBM (Clonetics Co.): **A** without additional drugs; **B** treated with interferons *(1)* 10^2 IU/ml rIFN-alpha, *(2)* 10^2 IU/ml rIFN-beta, *(3)* 10^2 IU/ml rIFN-gamma; and **C** treated with 10 IU/ml TNF-alpha

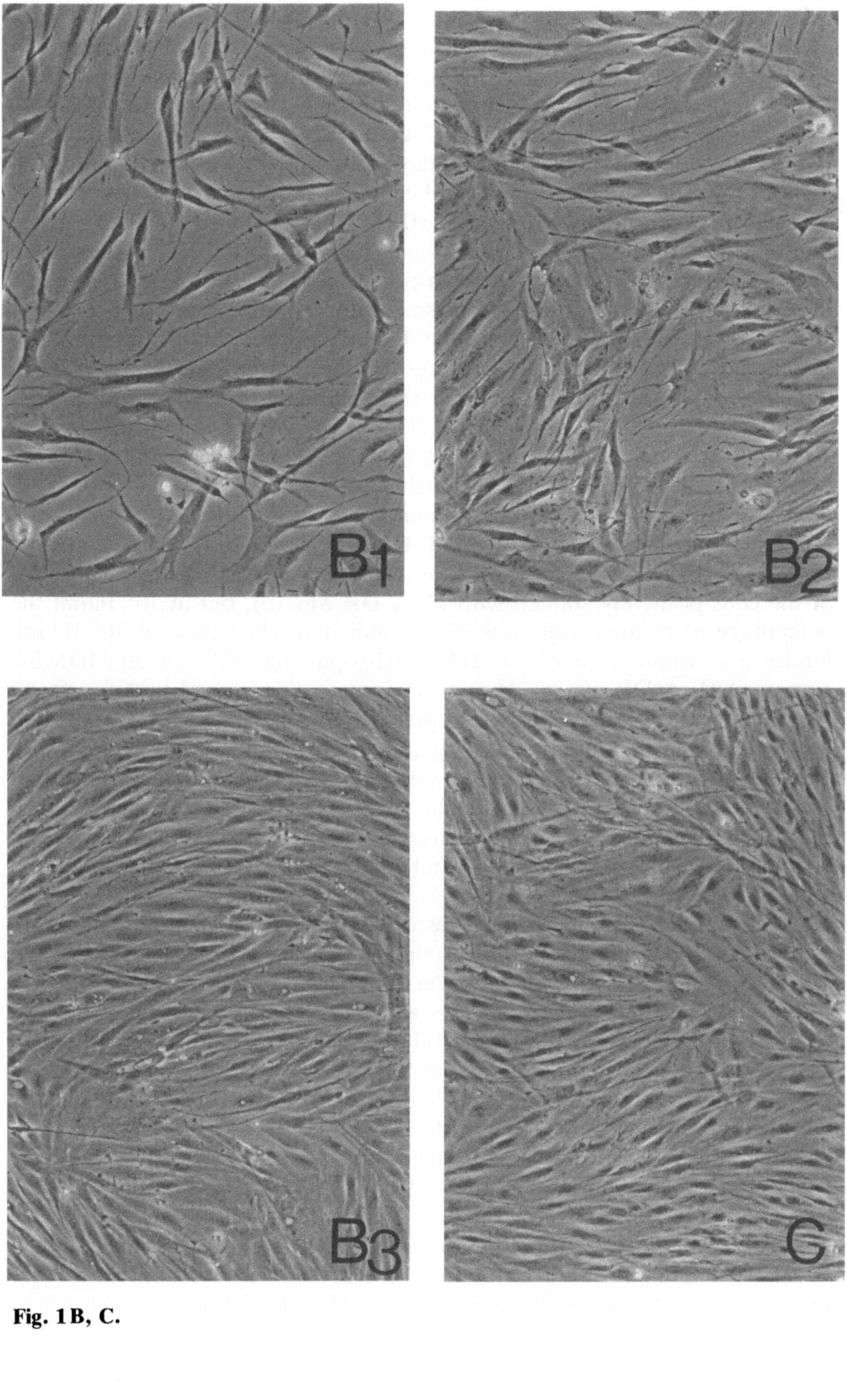

Fig. 1 B, C.

ations of HDMEC morphology in a dose- and time-dependent manner. The cells became spindle-shaped without altering their positivity for Factor VIIIr-Ag and their labeling with UEA-I (Fig. 1 B, 1–3). This effect was observed as early as 24 h after administration of rIFN and was blocked by antibodies against rIFN or by elimination of rIFN from the growth medium. Also the ultrastructure of HDMEC cultured in low-serum medium was typical for endothelial cells. After 6 days of treatment with 10^3 IU/ml rIFN-alpha, reduction of membrane-bound cellular transport activity and marked cytoplasmic alterations were found. rIFN-gamma induced such features as multiplication of Golgi complexes and dilatation of RER cisterns and decreased number of membrane transport vesicles, indicating up regulation of protein synthesis and intracytoplasmic retention of cell products.

All rIFNs exhibited dose- and time-dependent inhibition of the proliferation of HDMEC in vitro, as measured by the fluorometric assay. Growth inhibition was most prominent under rIFN-alpha and was observed as early as after 2 days of treatment with 10^4 IU/ml. The growth-inhibitory effect of rINF-beta first appeared after 4 days with 10^3 IU/ml; IFN-gamma induced only a slight growth inhibition at a high dose level.

Only rIFN-gamma exerted immunomodulatory effects on cultured HDMEC [46]. Treatment of HDMEC for 4 days with 10 IU/ml IFN-gamma resulted in a slight induction of intracytoplasmic HLA-DR expression (8% of the cells positively stained with HLA-DR MoAb), but at 10^2 IU/ml the percentage of positive cells was 77.5% and it reached 85% at 10^4 IU/ml. Under a constant dose of 10^2 IU/ml rIFN-gamma, 73% of all HDMEC expressed HLA-DR after 3 days; this percentage increased to 93% after 6 days and to 100% after 10 days. Expression of HLA-DQ on HDMEC was only minimally stimulated by rIFN-gamma [46]. Similar to the induction of class-II antigens, only rIFN-gamma was found to modulate the expression of ICAM-1 on the surface of HDMEC in vitro. After 4 days' treatment with 10 IU/ml rIFN-gamma, 12% of the cells were positively stained with ICAM-1 MoAb, increasing to 78% at 10^2 IU/ml and to 93% after 4 days of treatment with 10^3 IU/ml [46].

Our group has also recently investigated the effect of proinflammatory cytokines interleukin 1-alpha (IL 1-alpha), interleukin 1-beta (IL 1-beta), interleukin 6 (IL 6), granulocyte/macrophage colony-stimulating factor (GM-CSF) and tumor necrosis factor-alpha (TNF) on cell proliferation and expression of ICAM-1 on HDMEC [17]. Interleukin 1, formerly described as a lymphocyte-activating factor, has been reported to affect the proliferation of human umbilical vein endotheliocytes (HUVEC), and IL 1 inhibits the proliferation of growth-arrested HUVEC and HUVEC stimulated by endothelial cell growth factors. IL 1 has also been described to induce the expression of ICAM-1 on HUVEC [9, 48, 52], IL 6, a multifunctional cytokine, has been reported to be produced by HUVEC in vitro [45] as well as by HDMEC (manuscript in preparation). GM-CSF was recently shown to affect the migration and proliferation of HUVEC in vitro and to increase the adherence of neutrophils to EC cultures [5, 18]. TNF has been shown to inhibit the proliferation of various cell lines including human keratinocytes and HUVEC [49], while it inhibited the growth of HDMEC in vitro [16]. In addition, TNF may induce the expression of ICAM-1 on a variety of cell types such as keratinocytes, melanocytes, and HDMEC [16]. Recently the effects of anti-inflammatory agents such as acitretin, dexamethasone, and indomethacin on the TNF-related ICAM-1 induction on HDMEC were also reported by our group [17].

Microvascular endothelial cells of various origin express organ-specific antigens, and the specificity at the level of capillary endothelium is a significant element in selectie cell recognition, immunological reactivity, and response to mediators [2, 29, 32, 56, 60]. In the studies presented here we found that HDMEC are targets for the action of interferons and other cytokines in vitro [15–17, 46], and that rIFN-alpha and rIFN-beta induced a dose- and time-dependent growth inhibition (IFN-alpha > INF-beta); rIFN-gamma, however, influenced cell growth only at high dose levels and after long exposure [46]. Other authors reported that rIFN-gamma markedly inhibits the growth of human umbilical vein endotheliocytes in a dose-dependent manner [25, 55, 59]. Our results further suggest that rIFN-alpha may have a cytotoxic effect on HDMEC [46]. This may better explain the influence of rIFN-alpha on endothelial cell proliferation in HIV-associated Kaposi's sarcoma, as reported previously by our group [40]. This observation was recently supported by us, using an in vitro treatment of HIV-associated Kaposi's sarcoma-derived cells with rIFN-alpha [47]. Interestingly, rIFN-beta also acts as an antiproliferative agent on endothelial cells in vitro. The expression of class-II antigens can be induced or modulated on various endothelial cell populations after incubation with mitogens and different cytokines [6, 28, 33, 34, 36, 37, 39, 44, 47, 58]. In our experiments rIFN-gamma was the only interferon which modified the immunophenotype of HDMEC in vitro. In contrast, HLA-DQ expression was always markedly lower than HLA-DR under the same culture conditions. It seems that rIFN-gamma markedly stimulates ICAM-1 expression [3, 42, 44, 54]. In our experiments, a minimum of 10 IU/ml rIFN-gamma was able to stimulate ICAM-1 expression on HDMEC. These findings confirm previous studies reporting that normal endothelial cells of various origin in culture usually do not express ICAM-1, but that this molecule appears after incubation with IFN-gamma [19, 28].

It was recently shown by our group that another cytokine, IL 1-alpha, induced dose- and time-dependent stimulation of HDMEC proliferation [17]. The minimal effective concentration was 10 IU/ml, maximum stimulation of HDMEC proliferation was observed after 10^3 IU/ml IL 1-alpha. Higher concentrations of IL 1-alpha did not lead to increased cell growth. Also, IL 1-beta and GM-CSF were found to be potent dose- and time-dependent stimulators of HDMEC proliferation with minimal effective concentrations of 1–10 IU/ml [17]. In contrast to the results obtained with HDMEC, HUVEC proliferation was not stimulated by IL 1-alpha or IL 1-beta; moreover, in high concentrations a slight inhibitory effect of IL 1-beta on HUVEC DNA synthesis was found [17]. In contrast to these results, HDMEC did not show any alteration of growth under treatment with IL 6 and TNF [16, 17].

It was also recently demonstrated that IL 1-alpha markedly increased the expression of ICAM-1 in HDMEC with a minimal concentration of 1 IU/ml [17]. IL 1-beta was more potent than IL 1-alpha; increased labeling with MoAb against ICAM-1 was observed in concentrations as low as 0.1 IU/ml, and maximal stimulation was observed after 10^2 IU/ml. Treatment with GM-CSF as well as with IL 6 did not influence ICAM-1 expression, in contrast to TNF, which has been shown to induce ICAM-1 in HDMEC in concentrations as low as 1 IU/ml [16, 17]. IL 1-alpha, IL 1-beta, and TNF led to alterations of cell morphology with induction of a spindle-shaped appearance already after 1 day of treatment (Fig. 1C). In contrast, IL 6 and GM-CSF did not lead to major alterations.

The data reported here indicate that dermal microvascular endothelial cells are a major target for the action of proinflammatory cytokines, with pronounced effects on cell proliferation and ICAM-1 expression. Remarkably, HDMEC responded to each of the tested cytokines in a specific way, indicating that inflammatory cytokines may exert type-specific effects.

Our results correspond well to the response of human microvascular endothelial cells in vivo after topical IFN-gamma administration [28, 54]. It seems that together with other cytokines, such as tumor necrosis factor-alpha IL 1, IFN-gamma plays an important role in the immune activation of microvascular endothelium and in the mediation of lymphocyte and macrophage adhesion. We have found that IL 1 is a major, multifunctional HDMEC-targeted cytokine, while TNF seems to be involved predominantly in the regulation of HDMEC antigen expression [15, 17]. GM-CSF mainly promotes angioproliferation in vitro, whereas dermal endothelial cells seem not to be a major target for IL 6 [17]. Our findings concerning HDMEC growth inhibition by rIFN-gamma and growth stimulation by IL 1 differe considerably from those in other studies in which endothelial cells from large vessels were used [50, 59]. As these cell populations are not only functionally and metabolically, but also immunologically different from endothelial cells of the terminal vascular bed, HDMEC may represent a more appropriate model for further research on the role of dermal endothelium in the mediation of immunological phenomena and dermal/epidermal interactions.

Conclusions

The microvascular endothelium, because of its strategic location, plays a central role in the normal physiology and biochemistry of the skin. When dysfunctions occur they are associated with a number of pathologic changes such as inflammation, immune disorders, and hyperproliferation.

In the past, umbilical vein endotheliocytes have been used to study functions that are more closely associated with the microvasculature and usually not observed in the large blood vessels. This practice has continued mainly because of the easy separation technique, the availability of umbilical veins, the general purity of the cell population obtained, and the assumption that HUVEC growth in vitro accurately simulates that associated with the small blood vessels. Recent comparative studies, including results presented here, reveal that this perspective may be incorrect. Using the current procedure, the ability to isolate and cultivate cells from skin microvasculature provides a more appropriate and physiologically related system to evaluate microvessel function. Because of organ specificity the isolation and long-term in vitro cultivation of HDMEC seems to be a good and adequately tested model not only for in vitro–in vivo correlations but – of great importance – also for clinic-related studies of the pathology of several dermatological diseases as well as new methods to control them pharmacologically.

Acknowledgements. The authors thank Dr. M. Detmar, Department of Dermatology, University Medical Center, Steglitz, Berlin, for his long-time cooperation, and Mrs. M. Hoffmann and Mrs. M. Geisler, Berlin, as well as Mr. W. Schaefer, Minden for their excellent technical assistance.

References

1. Allen MJ, Barker JNWN, McDonald DM (1989) Immunophenotypic alterations induced by immune interferon in human skin. J Invest Dermatol 93:538
2. Auerbach R, Alby L, Morrissey LW, Tu M, Joseph J (1985) Expression of organ-specific antigens on capillary endothelial cells. Microvasc Res 29:401–411
3. Barker JNWN, Allen MH, McDonald DM (1989) The effect of in vivo interferon-gamma on the distribution of LFA-1 and ICAM-1 in normal human skin. J Invest Dermatol 93:439–442
4. Burger DR, Ford D, Vetto RM, Hamblin A, Goldstein A, Hubbard M, Dumonde DC (1981) Endothelial cell presentation of antigen to human T cells. Hum Immunol 3:209–230
5. Bussalino F, Wang JM, Defilippi P, Turrini F, Sanario F (1989) Granulocyte and granulocyte colony-stimulating factors induce human endothelial cells to migrate and proliferate. Nature 337:471–473
6. Capobianchi MR, Ameglio F, Tosi R, Dolei A (1985) Differences in the expression and release of DR, BR, and DQ molecules in human cells treated with recombinant interferon gamma; comparison to other interferons. Hum Immunl 13:1–11
7. Cavender DE, Edelbaum D, Ziff M (1989) Endothelial cell activation induced by tumor necrosis factor and lymphotoxin. Am J Pathol 134:551–560
8. Cordell JL, Falini B, Erber WN, Ghosh A, Zainalabideen A, MacDonald S, Pulford KAF, Stein H, Masony DY (1984) Immunoenzymatic labelling of monoclonal antibodies using immune complexes of alcaline phosphatase and monoclonal anti-alkaline phosphatase (APAAP-complexes). J Histochem Cytochem 32:219–229
9. Cozzalino F, Torcia M, Aldinucci D, Zicke M, Almerigogna F, Bani D, Stern DM (1990) Interleukin 1 is an autocrine regulator of human endothelial cell growth. Proc Natl Acad Sci USA 87:6487–6491
10. Davison P, Karasek M (1978) Factors affecting the growth and morphology of rabbit skin endothelial cells in vitro. Clin Res 26:208–209

11. Davison P, Bensch K, Karasek M (1980) Isolation and growth of endothelial cells from the microvessels of the newborn human foreskin in cell culture. J Invest Dermatol 75:316–321
12. Davison P, Bensch K, Karasek M (1983) Isolation and long-term serial cultivation of endothelial cells from the microvessels of the adult human dermis. In Vitro 19:937–945
13. De Bault LE, Kah LE, Frommes SP, Cancilla PA (1979) Cerebral microvessels and derived cells in tissue culture: isolation and primary characterisation. In Vitro 15:473–487
14. deLecuw AM, Barelds RJ, de Zanger R, Knock DL (1982) Primary cultures of endothelial cells of the rat liver. Cell Tissue Res 223:201–215
15. Detmar M, Ruszczak Zb, Imcke E, Orfanos CE (1989) Effects of tumor necrosis factor alpha (TNF) on cultured microvascular endothelial cells derived from human dermis. J Invest Dermatol 93:546
16. Detmar M, Imcke E, Ruszczak Zb, Orfanos CE (1990) Effects of tumor necrosis factor-alpha (TNF) on cultured microvascular endothelial cells derived from human dermis. J Invest Dermatol 95:219s—222s
17. Detmar M, Tenorio S, Hettmannsperger U, Ruszczak Zb, Orfanos CE (1992) Cytokine regulation of proliferation and ICAM-1 expression of human dermal microvascular endothelial cells in vitro. J Invest Dermatol 98:147–153
18. Devereux S, Bull HA, Campos D, Saib R, Linch DC (1989) Granulocyte macrophage colony-stimulating factor induced changes in cellular adhesion molecule expression and adhesion to endothelium: in vitro and in vivo studies in man. Br J Haematol 71:323–330
19. Dustin ML, Springer TA (1988) Lymphocyte function-associated antigen-1 (LFA-1) interaction with intercellular adhesion molecule-1 (ICAM-1) is one of at least three mechanisms for lymphocyte adhesion to cultured endothelial cells. J Cell Biol 107:321–331
20. Dustin ML, Singer KH, Tuck DT, Springer TA (1988) Adhesion of T lymphoblasts to epidermal keratinocytes is regulated by interferon gamma and mediated by intercellular adhesion molecule (ICAM-1). J Exp Med 167:1323–1340
21. Editorial (1989) Mechanisms of cutaneous inflammation. Interactions between epidermal cytokines, adhesion molecules and leukocytes. Arch Dermatol 125:1406–1412
22. Einhorn S, Eldor A, Vlodavsky I, Fuks Z, Panet A (1985) Production and characterization of interferon from endothelial cells. J Cell Physiol 122:200–204
23. Ferry B (1987) Interferon gamma stimulates the class-II antigen expression in cultured endothelial cells. Transplantation 44:499–503
24. Folkman J, Haudenschild CC, Zetter Br (1976) Long-term culture of capillary endothelial cells. Proc Natl Acad Sci USA 76:5217–5221
25. Friesel R, Komoriya A, Maciag T (1987) Inhibition of endothelial cell proliferation by gamma-interferon. J Cell Biol 104:689–696
26. Gelderblom H, Kocks C, L'âge-Stehr J, Reupke H (1985) Comparative immunoelectron microscopy with monoclonal antibodies on yellow fever virus-infected cells: preembedding labelling versus immunocryoultramicrotomy. J Virol Methods 10:225
27. Geppert TD, Lipsky PE (1985) Antigen presentation by interferon gamma-treated endothelial cells and fibroblasts: differential ability to function as antigen-presenting cells despite comparable la expression. J Immunol 135:3750–3762
28. Griffiths CEM, Voorhees JJ, Nickoloff BJ (1989) Characterisation of intercellular adhesion molecule-1 and HLA-DR expression in normal and inflamed skin: modulation by recombinant gamma-interferon and tumor necrosis factor. J Am Acad Dermatol 20:617–629
29. Gumkowski F, Kaminska G, Maminski L, Morrissey LW, Auerbach R (1987) Heterogeneity of mouse vascular endothelium. Blood Vessels 24:11–23
30. Hughes CCW, Male DK, Lantos PL (1988) Adhesion of lymphocytes to cerebral microvascular cells: effects of interferon gamma, tumor necrosis factor and interleukin-1. Immunology 64:677–681

31. Jaffe EA, Nachmann RL, Becker CG, Minick GR (1973) Culture of human endothelial cells derived from umbilical vein. J Clin Invest 52:2745–2756
32. Kumar AU, West DC, Ager A (1987) Heterogeneity in endothelial cells from large vessels and microvessels. Differentiation 36:57–70
33. Lapierre LA, Fiers W, Pober JS (1988) Three distinct classes of regulatory cytokines control endothelial cell MHC antigen expression. J Exp Med 167:794–804
34. Leeuwenberg JF, van Damme J, Meager T, Jeunhomme TM, Buurman WA (1988) Effects of tumor necrosis factor on the interferon gamma-induced major histocompatibility complex class-II antigen expression by human endothelial cells. Eur J Immunol 18:1469–1472
35. Lewis WH (1922) Endothelium in tissue culture. Am J Anat 30:39–59
36. Manyak CL, Tse H, Fischer P, Coker L, Sigal NH, Koo GC (1988) Regulation of class-II MHC molecules on human endothelial cells. Effect of IFN and dexamethasone. J Immunol 140:3817–3821
37. Markus BH, Colson YL, Fung JJ, Zeevi A, Duquesnoy RJ (1988) HLA antigen expression on cultured human arterial endothelial cells. Tissue Antigens 32:241–253
38. Maruyama Y (1973) The human endothelial cell in tissue culture. Z Zellforsch 60:69–79
39. Masuyama J, Minato N, Kano S (1986) Mechanisms of lymphocyte adhesion to human vascular endothelial cells in culture. T lymphocyte adhesion to endothelial cells through endothelial HLA-DR antigens induced by gamma-interferon. J Clin Invest 77:1596–1605
40. Mayer-da-Silva A, Stadler R, Imcke E, Bratzke B, Orfanos CE (1987) Disseminated Kaposi's sarcoma in AIDS: histogenesis-related cell populations and influence of long-term treatment with rIFN alpha. J Invest Dermatol 89:618–624
41. Montesano R, Orci L, Vassalli P (1985) Human endothelial cell cultures: phenotypic modulation by leukocyte interleukins. J Cell Physiol 122:424–435
42. Nickoloff BJ, Reusch MK, Bensch K, Karasek MA (1988) Preferential binding of monocytes and Leu 2 + T lymphocytes to interferon gamma-treated cultured skin endothelial cells and keratinocytes. Arch Dermatol Res 280:235–245
43. Norioka K, Nakagawa K, Onozuka S, Hara M, Kitani A, Hirose T, Hirose W, Harigai M, Suzuki K (1988) The effect of lymphokines on growth and phenotype of human vascular endothelial cells. J Natl Def Med Coll 13:71–79
44. Pober JS (1988) Cytokine-mediated activation of vascular endothelium. Physiology and pathology. Am J Pathol 133:426–433
45. Podor TJ, Jirik FR, Laskutoff DJ, Carson DA, Lotz M (1989) Human endothelial cells produce IL-6. Lack of response to exogenous IL-6. Ann NY Acad Sci 557:374–385
46. Ruszczak Zb, Detmar M, Imcke E, Orfanos CE (1990) Effects of rIFN-alpha, -beta and -gamma on the morphology, proliferation and cell surface antigen expression of human dermal microvascular endothelial cell in vitro. J Invest Dermatol 95:693–699
47. Ruszczak Zb, Detmar M, Stadler R, Bratzke B, Orfanos CE (1991) rIFN-alpha inhibits the proliferation and rIFN-gamma upregulates cell surface antigen expression of HIV-associated Kaposi's sarcoma-derived cells in vitro. Arch Dermatol Res 283:49–50
48. Saegusa Y, Ziff M, Welkovich L, Cavender D (1990) Effect of inflammatory cytokines on human endothelial cell proliferation. J Cell Physiol 142:488–495
49. Sato N, Goto T, Haranaka K, Satomi N, Nariuchi H, Mano-Hirano Y, Sawasaki Y (1986) Action of tumor necrosis factor on cultured vascular endothelial cells: morphologic modulation, growth inhibition and cytotoxicity. J Natl Cancer Inst 76:1113–1121
50. Sato N, Goto T, Tsuruoka N, Nishihara T, Beitz JG, Calabresi P, Frackelton AR Jr (1989) Actions of TNF and IFN on cultured vascular endothelium. J Invest Dermatol 93:575
51. Simionescu M, Simionescu N (1978) Isolation and characterisation of endothelial cells from the heart microvasculature. Microvasc Res 16:426–452

52. Smith CW, Rothlein R, Huges BJ, Mariscalo MM, Rudolff HE, Schmalstieg FC, Anderson DC (1988) Recognition of an endothelial determinant for CD18-dependent human neutrophil adherence and transendothelial migration. J Clin Invest 82:1746–1756
53. Stadler R, Detmar M, Stephanek K, Bangemann C, Orfanos CE (1989) A rapid fluorometric assay for the determination of keratinocyte proliferation in vitro. J Invest Dermatol 93:532–534
54. Thornhill MH, Williams DM, Speight PM (1989) Enhanced adhesion of autologous lymphocytes to gamma interferon-treated human endothelial cells in vitro. Br J Exp Pathol 70:59–64
55. Tsuruoka N, Sugiyama M, Tawaragi Y, Tsujimoto M, Nishihara T, Goto T, Sato N (1988) Inhibition of in vitro angiogenesis by lymphotoxin and interferon-gamma. Biochem Biophys Res Commun 155:429–435
56. Turner RR, Beckstead JH, Warnke RA, Wood GS (1987) Endothelial cell phenotypic diversity. In situ demonstration of immunologic and enzymatic heterogeneity that correlates with specific morphologic subtypes. Am J Clin Pathol 87:569–575
57. Wagner CR, Vetto RM, Burger DR (1985) Subcultured endothelial cells can function independently as fully competent antigen-presenting cells. Hum Immunol 13:33–47
58. Wedgwood JF, Hatam L, Bonagura VR (1988) Effect of interferon gamma and tumor necrosis factor on the expression of class-I and class-II major histocompatibility molecules by cultured human umbilical vein endothelial cells. Cell Immunol 111:1–9
59. Weimann BJ, Takacs F (1989) Different effects of recombinant human interferons on cultured human vascular cells. J Invest Dermatol 93:584
60. Yamamoto M, Shimokata K, Nagura H (1988) An immunohistochemical study on phenotypic heterogeneity of human pulmonary vascular endothelial cells. Virchows Arch [A] 412:479–486

Effect of Nicotine on the Differentiation of Human Keratinocytes In Vitro

C. Theilig, A. Bernd, F. E. Görmar, A. Ramirez-Bosca, J. Bereiter-Hahn, B. Keller-Stanislawski, N. Rietbrock, and *H. Holzmann*

Abstract

The influence of nicotine on cell differentiation was investigated. Cultures of human keratinocytes (HaCaT) were stained with an antibody against pan-keratin. They showed a more ordered keratin pattern with a nicotine concentration starting at 0.1 μg/ml in contrast to the control cultures. The expression of the keratins 1 and 10/11, which are regarded as differentiation markers [13, 15], increased after treatment with low nicotine concentrations (0.1–10 mg/ml medium). The number of "cornified envelopes" increased at a concentration of 0.1–100 μg nicotine/ml medium. In two-dimensional SDS-gel electrophoresis, cytokeratin 1 appeared at a nicotine concentration of 1 μg/ml medium. This was not found in parallel controls. This demonstrates that low concentrations of nicotine have a supporting influence on the keratinization and cornification of HaCaT keratinocytes.

Introduction

With increasing tobacco consumption, typical changes in the area of the hard palate are registered more frequently. These spots are covered with different-sized, sometimes more rectangular or sometimes more polygonal, opal-white, flat, eminent papules. They have a red dimple in their center, the excretory duct of a salivary gland. This phenotype is called leukokeratosis nicotinica palati [17] (Fig. 1). The elimination of tobacco consumption slowly leads to a degeneration of the changes in the area of the hard palate [8]. In order to determine if these changes were due to nicotine, to other substances in cigarette smoke, or to heat, the effect of nicotine on the differentiation of human keratinocytes in culture was studied.

Differences in the distinctiveness of the cytokeratin patterns are noticeable during keratinocyte differentiation in the epidermis. In the basal layer primarily cytokeratins 5 and 14 are expressed. They are also found with a decreasing tendency in stratum spinosum and in stratum granulosum [1, 16].

The cytokeratins 1,2 and 10/11 first appear in the suprabasal keratinocytes of the stratum spinosum and the stratum granulosum and act as markers for a terminal differentiation [5, 11, 12, 19]. A further differentiation step

A. Bernd, J. Bereiter-Hahn,
F. Hevert, H. Holzmann (Eds.)
Cell and Tissue Culture Models
in Dermatological Research

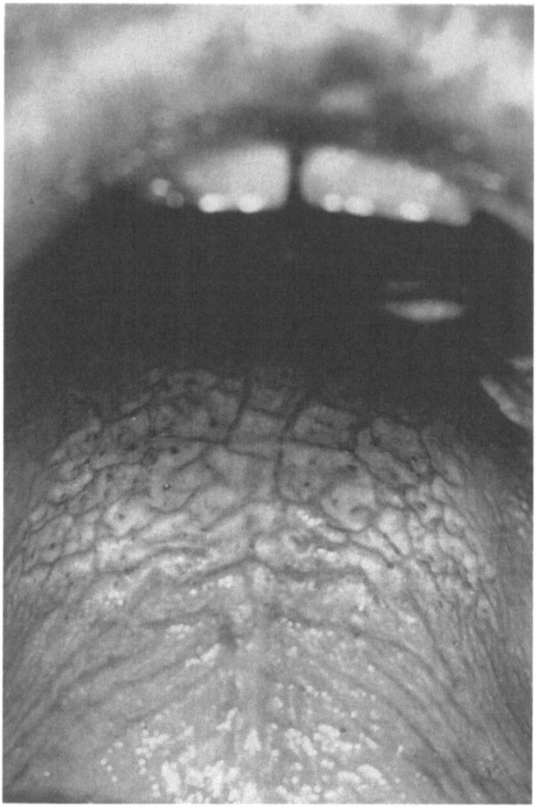

Fig. 1. Typical picture of leukokeratosis nicotinica palati

during the terminal differentiation of epidermal keratinocytes is the forming of cornified envelopes [18]. They consist of a structure which is synthesized by the plasma membrane-bound enzyme transglutaminase [14]. The determination of keratins 1 and 10/11 and the determination of the number of cornified envelopes are therefore good markers for cell differentiation of keratinocytes.

Material and Methods

Cell Culture

In our experiments, we used human keratinocytes which were spontaneously immortalized at a low calcium content and an increased temperature [6]. These cells are called HaCaT cells (human adult low calcium high temperature) and were kindly given to us by Prof. Fusenig (German Center for Cancer Research, Heidelberg, FRG). They are cultivated in plastic culture bottles (Nunc, Wiesbaden, FRG) using Hank's medium with 5% fetal calf serum, 1% of an antibiotic mixture (penicillicin/streptomycin; 10000 U/10000 μg/ml), and carbonate buffer. The cultures are kept at 37°C in a 10% CO_2 atmosphere.

Fluorescent Staining of the Keratin Filaments

The keratinocytes are sown in lab-tek at a density of 4×10^4 cells/cm^2. After 24 h nicotine is added in different concentrations and the cultures are incubated for 72 h. The cells are fixed with a methylacetone mixture 1:1 for 10 min at 4°C, rinsed with PBS, and incubated overnight at 4°C, with 500 μl keratin antibody solution (Sigma; Deisenhofen, FRG) 1:100. After three rinses the cells were incubated over night with 50 μl TRITC-marked second antibodies, washed, and embedded in Mowiol.

Cell Differentiation

In order to determine the degree of cell differentiation, keratins 1 and 10/11 were stained as described above, counted, and related to the total number of the cells. The classification of the total number is achieved by specifically staining the DNA with a fluorescent color (bisbenzimide H 33342, Hoechst, Frankfurt, FRG).

Preparation of the Keratins

The cells were sown in culture vessels with a growing area of 175 cm^2 and were incubated for 3 weeks at different nicotine concentrations. Afterwards, the keratinocytes were mopped off and centrifuged (150 g, 10 min, room temperature). The sediments were resuspended in ice-cooled lysis buffer (10 m*M* Tris/HCl pH 7.5, 1% triton-X-100, 5 m*M* EDTA, 140 m*M* NaCl, 5 m*M* EGTA) and lysed for 20 min at 0° C. Then the probes were centrifuged for 20 min at 4° C and 27000 g. The pellet was resuspended by sonification (3 × 3 s × 100 W, 0° C) in "high salt buffer" (10 m*M* Tris/HCl pH 7.5, 1.5 *M* KCl, 5 m*M* EDTA, 0.5% triton-X-100, 140 m*M* NaCl, 2 m*M* EGTA, 0.5 m*M* DTE). After a further centrifugation (27000 g, 20 min, 4° C) the pellets were resuspended in PBS and the protein content was measured using a method from Lowry et al. [10]. The cytoskeleton proteins were finally dissolved in a concentration of 20 μg/μl in electrophoresis buffer (50 m*M* Tris/HCl pH 6.8, 1% SDS, 5% mercaptoethanol) and boiled for 5 min at 97°C.

Two-dimensional Gel Electrophoresis

For the 2D gel electrophoresis (modified method according to Bowden et al. [4]) the cytoskeleton probes were adjusted to 9.5% urea, 5% CHAPS, 5% mercaptoethanol, and 5% ampholinemix (1 part LKB 3.5–5.1, 1 part LKB 3.5–10.2, 2 parts LKB 5–8). Afterwards, the probes were applied to polyacrylamide tube gels (4% polyacrylamide, 9 *M* urea, 5% ampholinemix, and 2% CHAPS). After 25 000 Vh the gels were equilibrated for 10 min in 2% SDS buffer (0.05 *M* Tris/HCl pH 6.8, 2% SDS, 1% mercaptoethanol, 1 m*M* EDTA) and then 10 min in 0.2% SDS buffer (0.05 *M* Tris/HCl pH 6.8, 0.2% SDS, 1 m*M* EDTA). Then the proteins were separated in the second dimension by a modified SDS gel electrophoresis according to Laemmli [9]. The slap gels had a linear polyacrylamide gradient between 7.5% and 17.5%. Separation was performed at 20 mA/gel. The proteins were dyed with 2% Coomassie blue G350 in 20% acetic acid. Bovine serum albumin (68 kD, pI 7.2), phosphoglycerate kinase (43 kD, pI 7.4), and actin (42 kD, pI 5.4) served as standard proteins.

Cornified Envelopes

In order to determine the number of cornified envelopes, the cells were sown out in multiwell dishes with a density of $4 \times 10^4/cm^2$ and incubated with nicotine (0.1–400 µg/ml medium). The medium was changed twice a week and the used medium was collected and saved. After 3 weeks the cells were trypsinized, added to the saved media, and centrifuged at 2000 g for 10 min. The pellets were dissolved in 2% SDS and 10 m*M* dithiothreitol for 10 min at 100° C. The cornified envelopes remain as complete structures after this treatment. In order to prevent aggregations of the envelopes, a DNAse treatment was carried out after cooling off. Finally the number of cornified envelopes was determined in a Neubauer counting chamber.

Results

Immunofluorescent staining of the keratins was done in order to determine morphological changes caused by the influence of nicotine. The immunofluorescent investigations showed a higher degree of order of the keratin filaments as the effect of 1–10 µg nicotine/ml medium and more cell-cell contacts, which indicates a higher degree of differentiation. At a nicotine concentration of 100 µg/ml medium, a disorderliness and later reduction of the keratin filaments (Fig. 2a, b, c) became visible. This observation suggests that a closer investigation of the influence of nicotine on the differentiation behavior could be of interest.

The expression of keratins 1 and 10/11 was determined as the parameter for the assessment of the cell differentiation. For this purpose keratin fila-

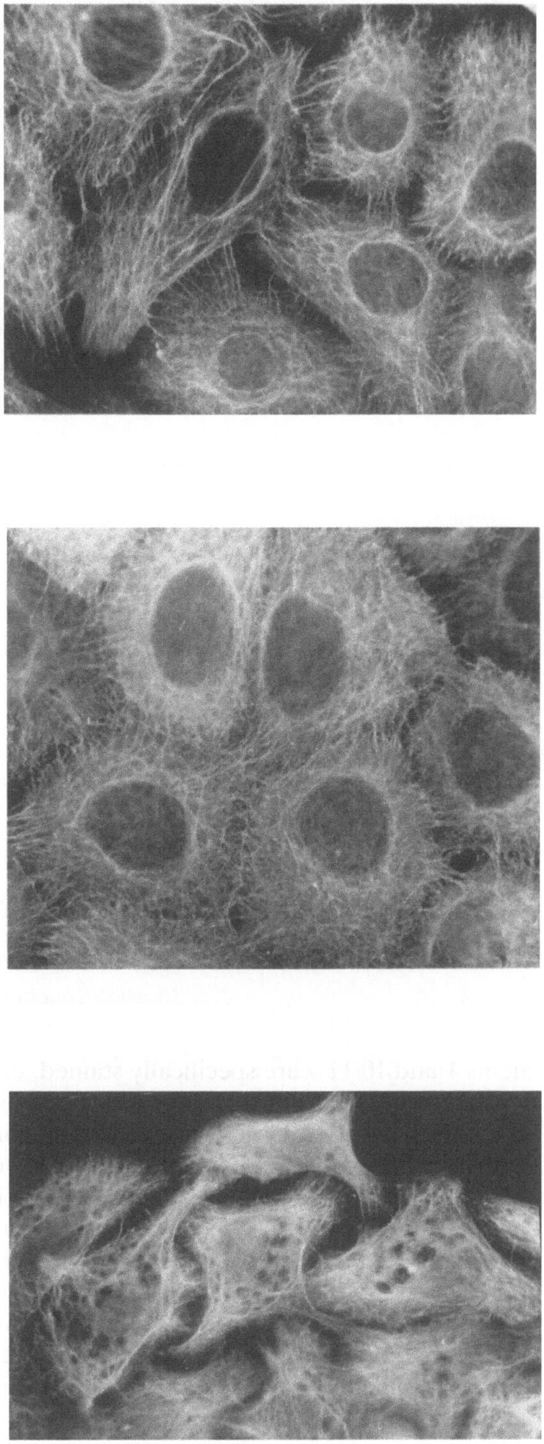

a

b

c

Fig. 2 a–c. Immunofluorescent staining of the keratin filament of HaCaT cells: **a** control; **b** after a 72-h treatment with 0.1 μg/ml nicotine at 37°C; **c** after a 72-h treatment with 1000 μg/ml nicotine at 37°C

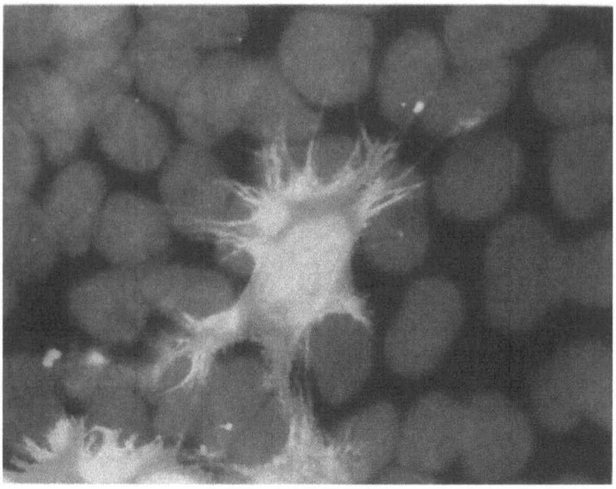

Fig. 3a, b. a The effect of nicotine on the expression of keratins 1 and 10/11. The HaCaT cells were treated with nicotine and afterwards stained with antibodies against keratin 1 and 10/11. The nuclei were stained with bisbenzimide H 33342. Some cells expressed keratin 1 and 10/11 *(red fluorescence)*. **b** The number of cells which expressed keratins 1 and 10/11 is given as a percentage of the total number of the cells. Every point of the curve is a mean of four parallel tests

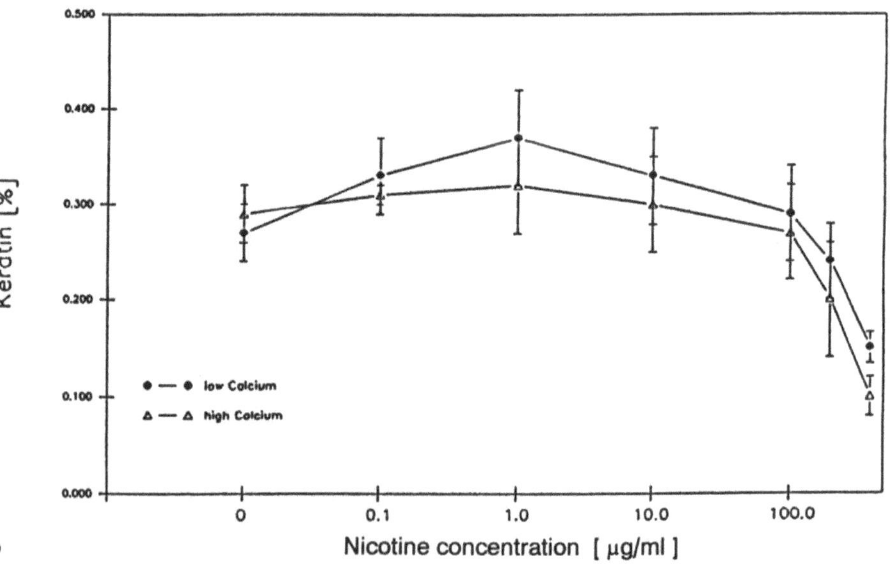

b

ments 1 and 10/11 were specifically stained, counted, and applied to the total amount of keratin (Fig. 3a). The cells demonstrated an increased expression of keratins 1 and 10/11 at a lower nicotine concentration compared with the control cultures. The cells exposed to a lower calcium concentration were influenced more than the cells that were kept in a higher calcium content. At concentrations of 100 μg nicotine/ml medium and more only a few keratin filaments were still visible. The isolated keratins of the growing nicotine-influenced keratinocytes were separated by the 2D gel electrophoresis. The control cultures showed no spots that indicated the existence of keratin 1. The electrophoresis of the keratins, however, influenced by nicotine in a concentration of 1 μg/ml medium, showed a clear spot in the position of keratin 1 (Fig. 4a, b).

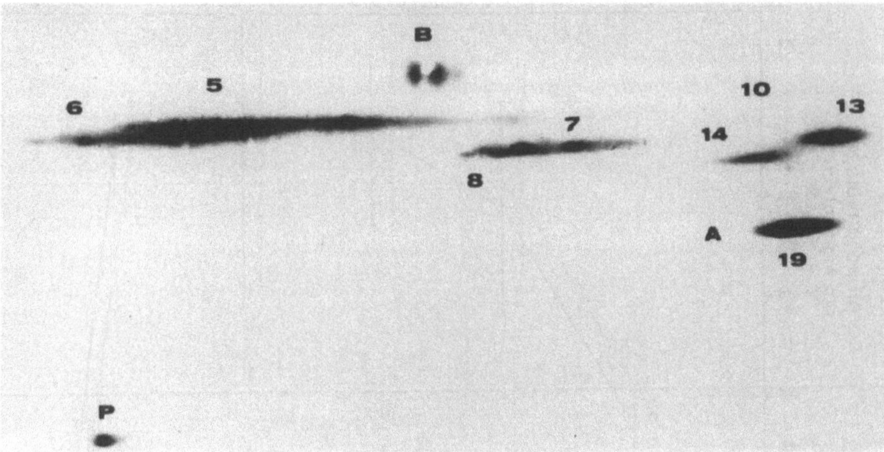

a

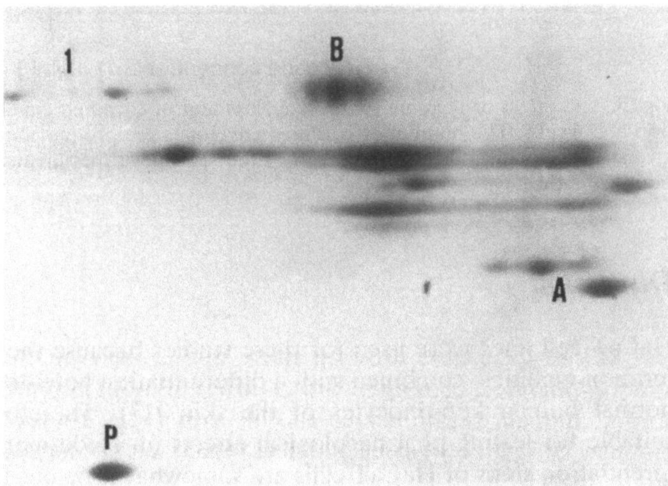

b

Fig. 4a, b. a Gel electrophoresis of an uninfluenced control culture. The keratins are numbered according ot the catalogue proposed by Moll [11]. **b** After 3 weeks' incubation with 1 μg nicotine/ml medium; the spot of keratin 1 is very distinct. *A*, Actin; *B*, bovine albumin; *P*, phosphoglycerate kinase

The electrophoresis of keratins influenced by 1000 μg nicotine/ml medium showed a reduction of many keratins. The cornified envelopes, which were used as markers for terminal differentiation of keratinocytes, were counted and applied to the total protein content of each respective culture. The number of cornified envelopes of cells influenced by 0.1–100 μg nicotine/ml medium was more than twice as high as that in the control cultures (Fig. 5). At a higher nicotine concentration a drastic decline in cornified envelopes was observed as reaction to a toxic influence.

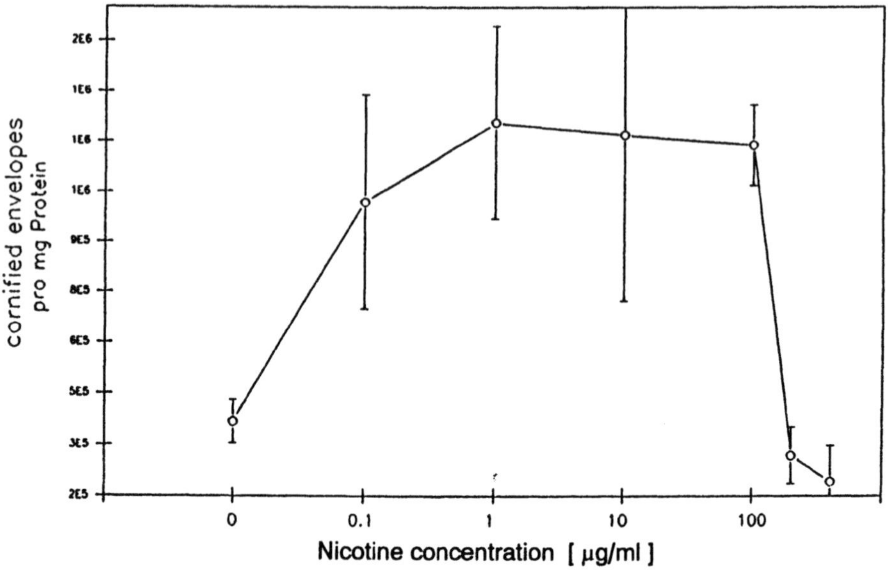

Fig. 5. The effect of nicotine on the development of cornified envelopes after an incubation of 3 weeks. The number of cornified envelopes per protein content of each culture is given. Every point indicates the mean of four parallel experiments. The standard deviations are indicate

Discussion

HaCaT cell lines were used for these studies because they have good proliferation qualities, combined with a differentiation potential similar to that of normal human keratinocytes of the skin [17]. Therefore, they are very suitable for testing pharmacological effects on epidermal cells [2]. The differentiation steps of HaCaT cells are somewhat different from those in vivo. In the human epidermis differentiation takes place in stratified cell layers. In culture, the various types of cells differentiate simultaneously and side by side. Furthermore, keratins were synthesized by HaCaT cells, which do not occur in the normal epidermis [3]. In vitro it is therefore necessary to determine several differentiating parameters in order to observe the changes in differentiation caused by nicotine.

The habitual intraoral use of snuff leads to plaques with a higher degree of keratinization [7]. The assumption that this is a nicotine-specific effect is supported by these studies on keratinocytes in culture. The keratin filaments seemed to be more ordered when influenced by low nicotine concentrations, and keratins 1 and 10/11 were increasingly expressed. Furthermore, the number of cornified envelopes occurring in the horny keratinocytes was higher. It is always difficult to apply conclusions based on in vitro systems to in vivo systems because of the interactions between cells and tissue in the living organism. The fact that a stronger keratinization and cornification of the

HaCaT cells take place under the influence of nicotine justifies the conclusion that nicotine is probably responsible for leucokeratosis nicotinica palati.

References

1. Abernathy JL, Hill RL, Goldsmith LA (1977) Epsilon-(gamma-glutamyl) lysine cross-links in human stratum corneum. J Biol Chem 252:1837–1839
2. Bernd A, Breuer M, Dold K, Holzmann H (1990) Uptake-kinetics and the effect of minoxidil on human keratinocyte-cultures. Arzneimittelforschung 40:413–416
3. Boukamp P, Petrussevska RT, Breitkreuz D, Hornung J, Markham A, Fusenig NE (1988) Normal keratinization in a spontaneously immortalized aneuploid human keratinocyte cell line. J Cell Biol 106:761–771
4. Bowden P, Quinlan R, Breitkreutz D, Fusenig N (1984) Proteolytic modification of acidic and basic keratins during terminal differentiation of mouse and human epidermis. Eur J Biochem 142:29–36
5. Dale B, Holbrock KA, Kimbal JR, Hoff M, Sun TT (1985) Expression of epidermal keratins and filaggrin during human fetal skin development. J Cell Biol 101:1257–1269
6. Franke WW, Schiller DL, Moll R, Winter S, Schmid E, Engelbrecht I (1981) Diversity of cytokeratins. Differentiation-specific expression of cytokeratin polypeptides in epithelial cells and tissues. J Mol Biol 153:933–959
7. Görmar FE, Bernd A, Bereiter-Hahn J, Holzmann H (1990) A new model of epidermal differentiation: induction by mechanical stimulation. Arch Dermatol Res 282:22–32
8. Holzmann H (1978) Leukokeratosis nicotinica palati – eine Präkanzerose? Med Welt 29:544
9. Laemmli UK (1970) Cleavage of structural proteins during the assembly of the head of bacteriophage T4. Nature 227:680–685
10. Lowry OH, Rosenbrough NS, Farr AL, Randall RJ (1951) Protein measurement with the folin phenol reagent. J Biol Chem 193:265–275
11. Moll R, Franke W, Schiller D (1982) The catalog of human cytokeratins: patterns of expression in human epithelia, tumors and cultured cells. Cell 31:11–24
12. Moll R, Moll I, Franke WW (1986) Intermediärfilamente als Kriterium bei der Diagnostik von Hauttumoren. Pathologie 7:164–174
13. Pindberg JJ, Renstrub G (1963) Studies in oral leukoplakias. II. Effect of snuff on oral epithelium. Acta Derm Venereol (Stockh) 43:271
14. Ryle CM, Breitkreutz D, Stark HJ, Steinert PM, Roop D, Fusenig NE (1989) Density-dependent modulation of synthesis of keratins 1 and 10 in the human keratinocyte line HaCaT and in ras-transfected tumorigenic clones. Differentiation 40:42–54
15. Schmidt R, Reicher U, Michel S et al (1985) Plasma membrane transglutaminase and cornified envelope competence in cultured human keratinocytes. FEBS Lett 186:201–204
16. Staquet MJ, Haftek M, Cordier G, Thivolet J (1987) Keratin filament composition of human epidermal spinous and granular cell fractions selected by flow cytometric sorting. Arch Dermatol Res 279:273–275
17. Tappheimer S (1941) Über tabakbedingte Leukokeratosen des Gaumens. Arch Dermatol Syphilol 181:173
18. Tseng SGC, Jarvinen MJ, Nelson WG, Huang JW, Woodcock-Mitchel J, Sun TT (1982) Correlation of specific keratins with different types of epithelial differentiation: monoclonal antibody studies. Cell 30:361–372
19. Woodcock-Mitchel J, Eichner R, Nelson WG, Sun TT (1982) Immunolocalization of keratin polypeptides in human epidermis using monoclonal antibodies. J Cell Biol 95:580–588

A New In Vitro Test System and Phototoxicological Studies on Cell Cultures Based on the Melamine Substrate Foil

C. Westphal Frösch, and *A. Rück*

Cells change their behavior and their rates of proliferation and differentation when isolated from their natural environment and forced to survive in an artificial one. Nevertheless, if weighed against control data recorded in vivo, experiments using cells which grow in culture certainly remain an invaluable source of knowledge for cellular and molecular biology.

A number of cell functions, e.g., cell adhesion and cell-cell contact, develop only if cells adhere to a substrate. Common substrates in cell culture are polystyrene glass and glass coated with gelatine, polystyrene, or extracellular matrix proteins such as collagen, fibronectin, laminin, etc. Experimental work on cell cultures has proven that not all cell types adhere and differentiate sufficiently on purchaseable substrates.

The new melamine substrate foil has made it possible for cells to adhere quicker and to differentiate earlier [3, 4]. The reason for this may be that the structure of the melamine molecule (Fig. 1) is similiar to that of proteins. Here are some examples of how cells develop on glass and on melamine-coated glass. The adhesion of F-9 teratocarcinoma cells is shown in different stages (Fig. 2). After 30 min cells already adhere on the melamine-coated site, the right of the glass, whereas the glass substrate shows only cell clusters 3 h later. Until recently, the cell line F9 was cultured on gelatine-coated polystyrene.

The melamine substrate is even better, because now cell functions can be studied microscopically. Heart muscle cells of newborn rats differentiated and started to beat much earlier on the melamine side of the glass slide (Fig. 4). An additional reaction of mucous cells of a cow is apparent in Fig. 5. There is absolutely no adhesion on the glass slide in complete contrast to the melamine, where the cells proliferate extensively.

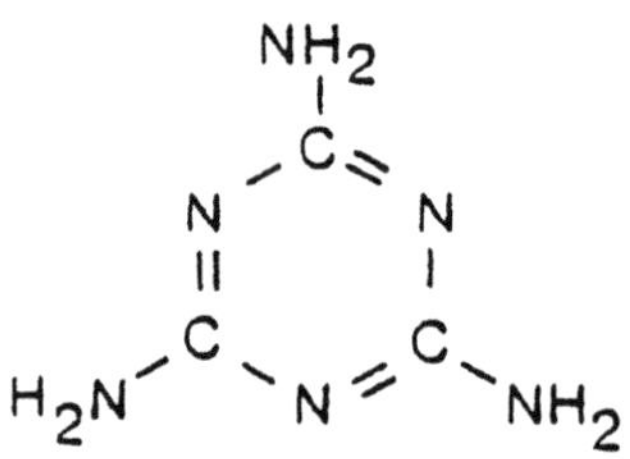

Fig. 1. Structure of the aminoheterocyclic compound triaminotriazine, or melamine

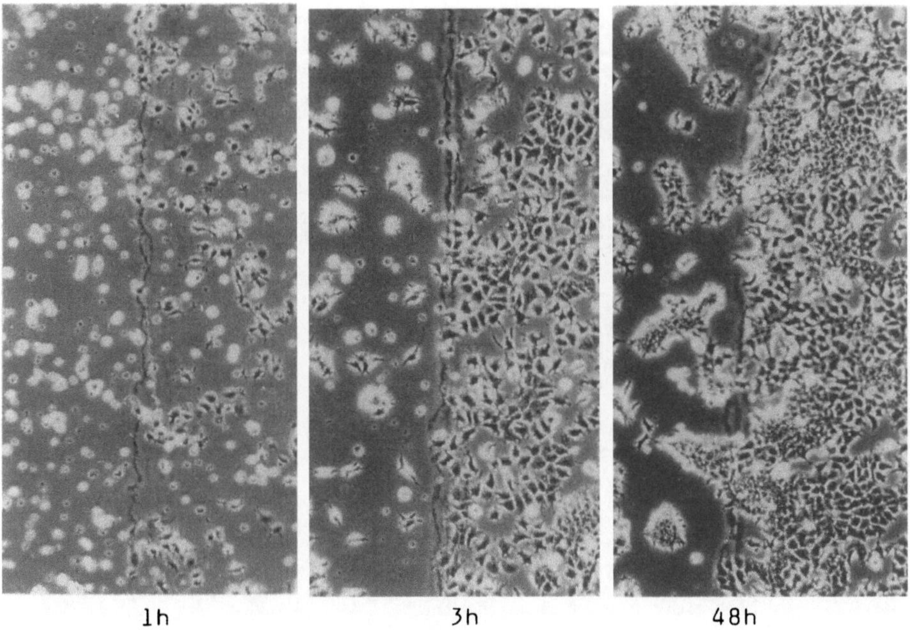

Fig. 2. The adhesion of F9 teratocarcinoma cells is shown after *1, 3,* and *48h.* The cells adhere faster on the melamine-coated side of the glass

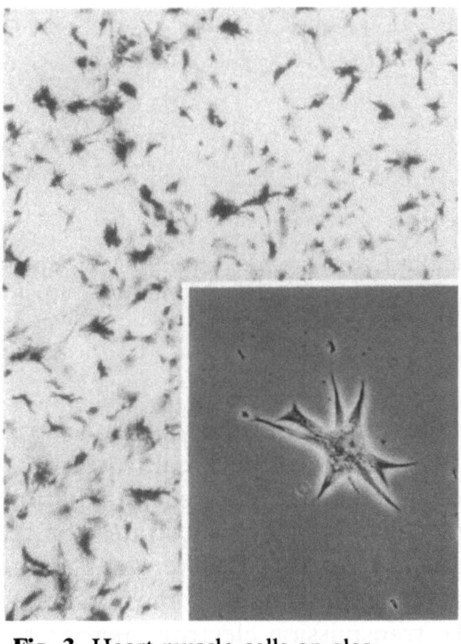

Fig. 3. Heart muscle cells on glas

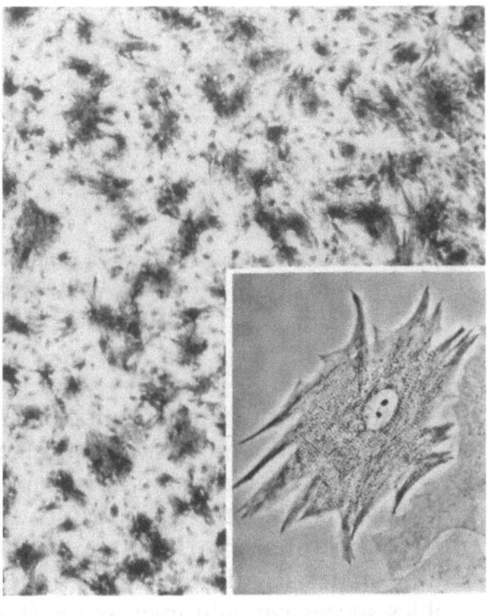

Fig. 4. Heart muscle cells on melamine-coated side of the glass. These cells are differentiated. The *inset* already shows myofibril

Fig. 5. Scanning-electron microscopy of mucous cells of a cow. The upper part shows the bare glass side; on the melamine side these cells proliferate extensively

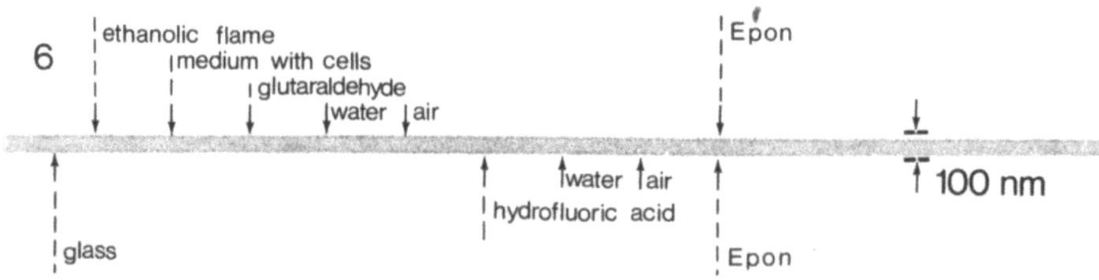

Fig. 6. Cross-sectioned melamine foil, embedded in Epon. From *left* to *right* indications are given concerning the materials and reagents interacting successively with this particular foil during the entire whole-cell precedure

Only recently, human leukemic blast cells were made adherent for several days on melamine, the first time this cell type had ever been cultivated on a substrate [5]. This opens up a new approach in cancer research. In addition to its advantages in the field of cultivating cells, melamine substrates can be processed with ease for light and electron microscopy. Unlike other glass-coating materials, the melamine foil can be removed from the glass and enables successive light microscopy (LM), transmission-electron microscopy (TEM), and scanning-electron microscopy (SEM) studies. A cross-sectioned 80-nm melamine foil embedded in Epon is shown in Fig. 6.

For cell cultures glass slides were dipped in the melamine solution and immediately flamed for hardening and sterilization [3, 4]. Cells grown on the foil are either used for LM studies or fixed with glutaraldehyde and consequently air dried for whole-mount TEM preparation or dehydrated for embedding and SEM procedures.

An example of how this method can be applied even for individual cells is given in Fig. 7. Two cells in contact are selected first by LM (inset), then an EM grid is layed upon them and after the foil is removed from the glass, these cells can be imaged by TEM and SEM in succession.

A further application of this technique is the ultrastructural localization of cell-cell contact molecules by immuno-electron-microscopy. The first example is the localization of VLA-3, a member of the integrin family. This surface protein is involved in cell-to-cell contact. Figure 8 shows a TEM picture of a whole cell preparation of human osteosarcoma cells in contact with each other. These cells were incubated in vivo with monoclonal antibodies against VLA-3 [1]. Immunogold-stained antibody-binding sites can be observed as black spots in TEM. Clearly, and with hardly any background staining, the clustered VLA-3 reactivity is restricted to the ends and crossing points of filopodia which contact a neighboring cell.

Another example is the immunogold labeling of human leukemic blast cells with antibodies against ICAM-1 molecules. If these are cocultivated with mouse fibroblast cells, labeling of the ICAM-1 is visible only on the leukemic cell (Fig. 9).

The ease with which delicate cells can be cultured on the melamine substrate foil, and the suitability of this material for various kinds of microscopic studies, suggest that the technique reported here is an interesting alternative to other cell culture methods. Photodynamic therapy (PDT) is a new approach to cancer treatment, combining the application of light and photosensitizers. For most cancer treatments chemotherapy is still applied although it interferes with the multiplication system of cells and can cause severe side effects. Ideally, cancer therapy should be based on high selectivity of the therapeutic agent to the transformed cells and high cytotoxic efficiency to the target tumor.

Administration of the photosensitizer to a cancer patient followed by illumination induces a selective photodynamic reaction. The treated area undergoes necrosis, with minimal changes to the surounding tissue. The only side effect is skin photosensitivity.

The photoactivation of the localized photoactive substance induces damage to cellular organelles, membranes, specific proteins, and the nucleus. Singlet oxygen or hydrocyl radicals are the active species involved in this process.

The main problems unsolved so far are how to increase the selectivity of the photosensitizers and how to diminish the photosensitivity of the skin. Basic research in this field tries to find new photosensitizers which are nontoxic in the dark, penetrate tissues with ease, and allow a maximum absorption of red light and a minimum of skin sensitivity.

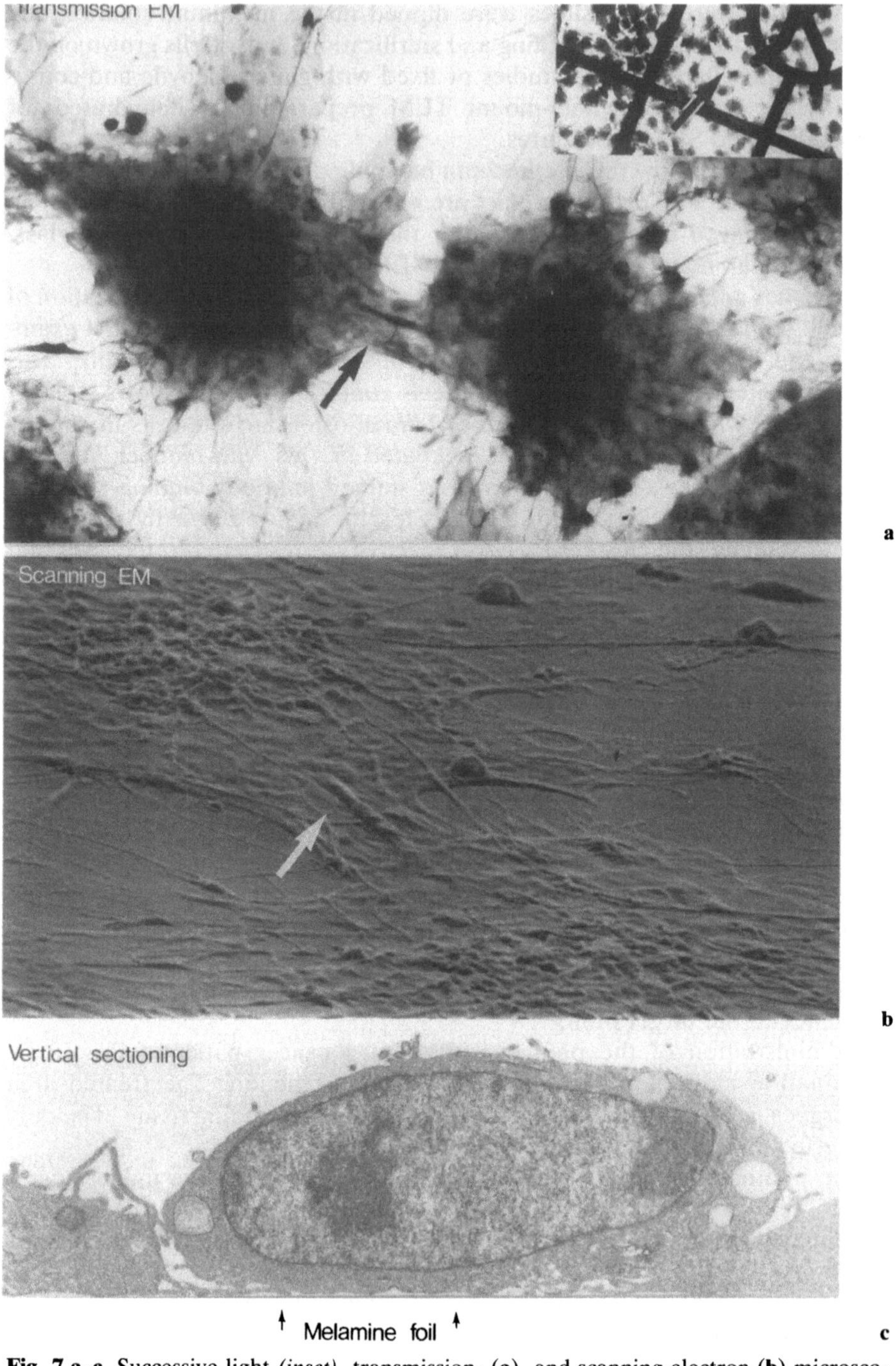

Fig. 7 a–c. Successive light *(inset)*, transmission- (**a**), and scanning-electron (**b**) microscopy (EM) studies of the same cells *(arrows)*. A section of two embedded cells on the melamine foil is shown *below* (**c**)

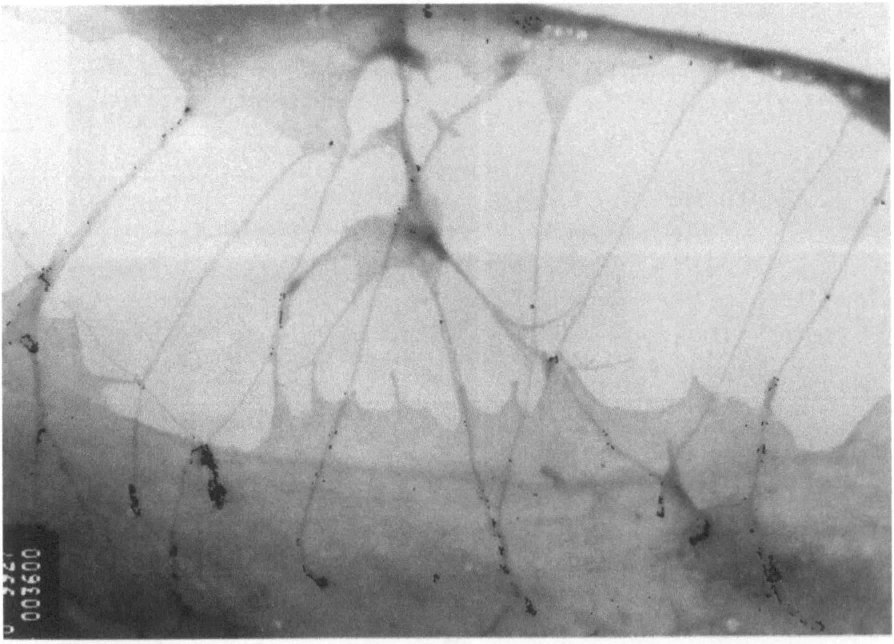

Fig. 8. A melamine-whole-mount electron microscopic picture of two osteosarcoma cells in contact with each other. A cell-to-cell contact protein VLA-3, a member of the integrin family, is immunogold labeled

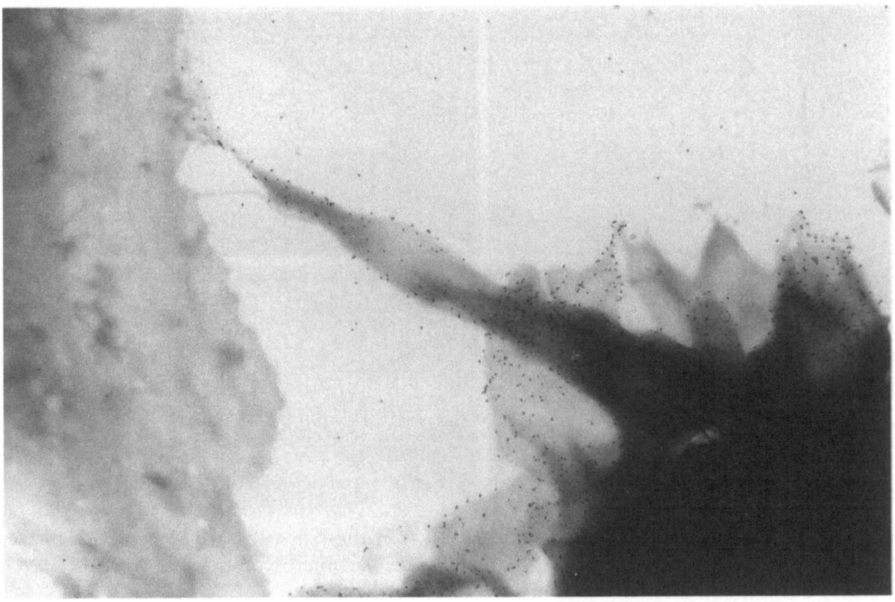

Fig. 9. A melamine-whole-mount electron microscopic picture of a human leukemic blast cell incubated with antibodies against the cell adhesion molecule ICAM-1

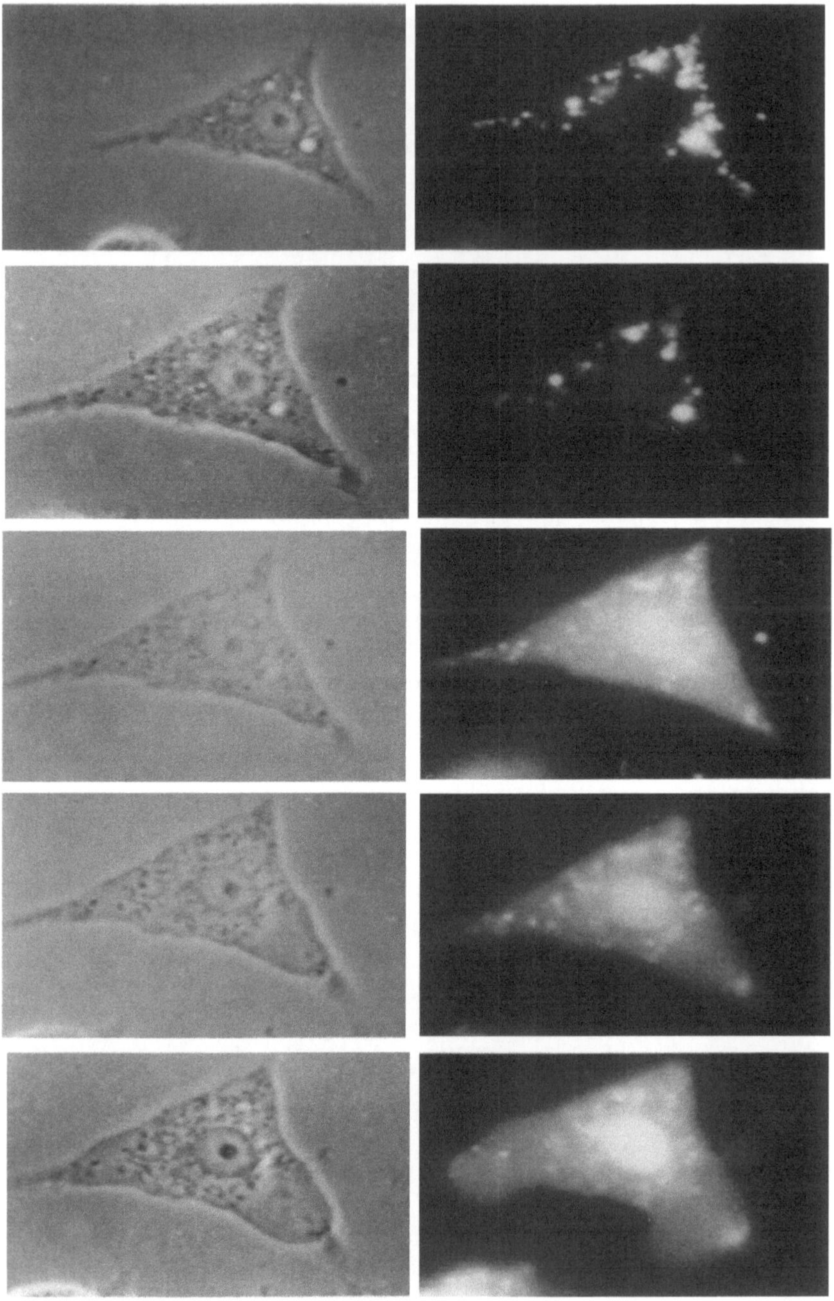

Fig. 10. Epithelial cells incubated with $TPPS_4$. On the *left,* cells are shown with phase-contrast microscopy, on the *right,* with fluorescence microscopy. Cells were observed before, during, and after irradiation with red laser light

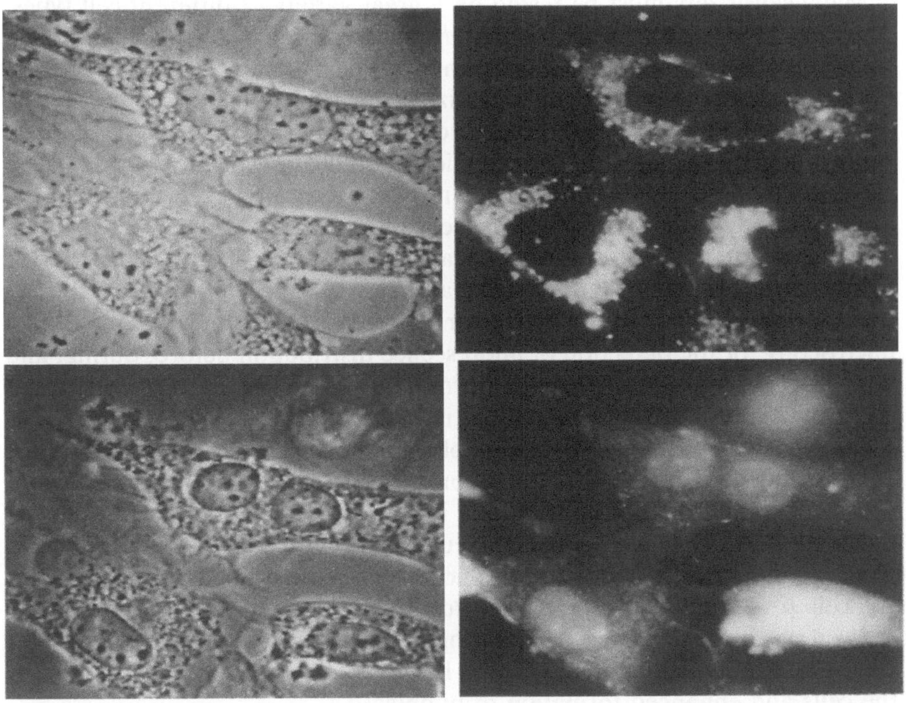

Fig. 11. Glioblastoma cells incubated with $TPPS_4$. On the *left*, phase-contrast microscopy before and after irradiation. On the *right*, fluorescence microscopy before and after irradiation

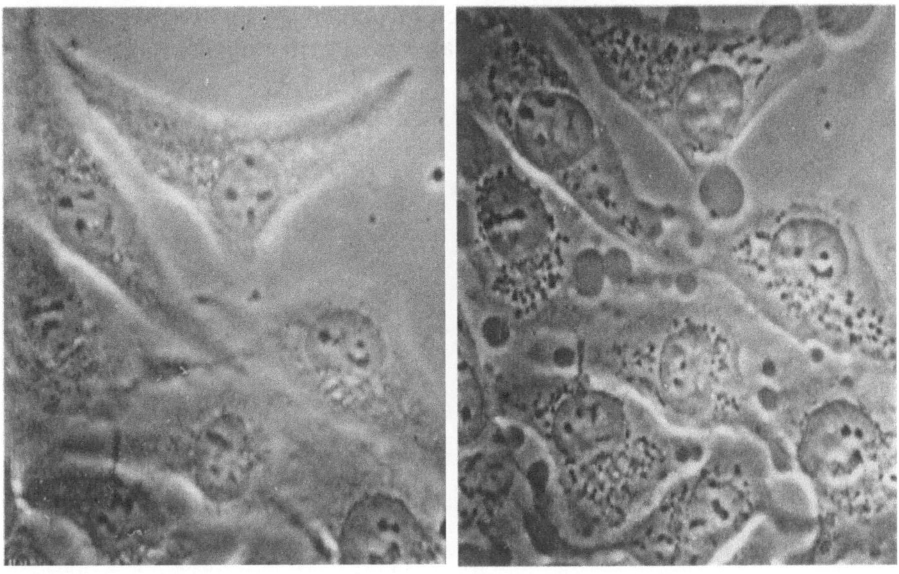

Fig. 12. Keratinocytes incubated with $TPPS_4$. Phase-contrast microscopy before *(left)* and after *(right)* irradiation

Many sensitizers must be tested on a great variety of different cell types. The melamine foil can help to speed up the screening of possible substances because cells proliferate and differentiate much better. We observed the fluorescence variations of the sensitizers during irradiation to follow the light-induced photodynamic process during PDT. The change in fluorescence intensity depends on the chemical modification of the sensitizer and the change of the chemical environment of the dye. However, the site of the initial photodynamic reaction of the sensitizer is hardly known.

We evaluated the intracellular localization and the fluorescence during light exposure of meso-tetra-4-sulfonatophenylporphyrin ($TPPS_4$) [2]. The fluorescence intensity of this sensitizer was measured using video-intensified microscopy for the detection of fluorescence distribution before, during, and after light exposure in order to perform PDT. The photodynamic activity of $TPPS_4$, which is a cationic sensitizer, is well known.

In epithelial cells (Fig. 10) and gliobalstoma cells (Fig. 11) incubated for 24 h with 10 μg/ml $TPPS_4$ the fluorescent spots, before light exposure showed a similar distribution. During irradiation with 630 nm light a drastic increase of fluorescence in the nucleus, particulartly in the nucleoli, took place. Parallel to the formation of fluorescence in the nucleus, morphological cell deformations were visualized. In contrast to this, keratinocytes showed no fluorescence changes during and after irradiation with laser light (Fig. 12). Phase-contrast microscopy, however, shows large deformations of the cells and enhanced formation of bubbles.

With the help of the melamine foil it is possible for the first time to document the results of PDT on monolayer of leukemic cells. The phase-contrast study (Fig. 13) demonstrates the morphological effect on human leukemic cells during irradiation, when incubated with $TPPS_4$. During irradiation the primary granules grow and fuse with one another, and this results in a final rupture of the cell.

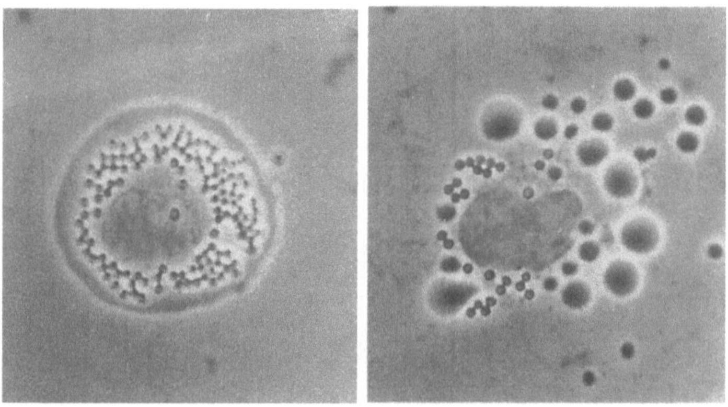

Fig. 13. Human leukemic cells incubated with $TPPS_4$. Phase-contrast microscopy before *(left)* and after *(right)* irradiation

References

1. Kaufmann R, Klein CE, Frösch D, Westphal C, Weber L (1989) Integrin VLA-3: ultrastructural localization and regulation of synthesis in human cell cultures. J Cell Biol 109:1807–1815
2. Rück A, Köllner T, Dietrich A, Strauß W, Schneckenburger H (1992) Fluorescence formation during PDT in the nucleoli of cells incubated with cationic and anionic water-soluble sensitizers. J Photochem Photobiol B (in press)
3. Westphal C, Hörler H, Pentz S, Frösch D (1988) A new method for cell culture on an electron-transparent melamine foil suitable for successive LM, TEM and SEM studies of whole cells. J Microsc 150:225–231
4. Westphal C, Hörler H, Pentz S, Frösch D (1989) A prolific, EM- compatible melamine substrate foil for cell culture. In Vitro 25:393–394
5. Westphal C, Krauter J, Heil G (1991) Leukemic cells on melamine substrate foils (letter). In Vitro Cell Dev Biol 27A:597–598

In Vitro Assays for Evaluation of the Effects of 8-Methoxypsoralen and UV Irradiation on Human Lymphocytes and Granulocytes

W. Gast, M. Rytter, T. Walther, and *U.-F. Haustein*

Abstract

The psoralen derivative 8-methoxypsoralen (8-MOP) is clinically used in the photochemotherapy (PUVA) of various dermatoses. With and without additional photoactivation, 8-MOP exerted impairment of several functions of (stimulated) human lymphocytes and granulocytes in vitro. These findings suggest and explain the efficacy of PUVA treatment by different action mechanisms on plasma membranes and intracellular structures.

Introduction

8-Methoxypsoralen (8-MOP) is widely used in the photochemotherapy (PUVA) of several dermatoses, including psoriasis and mycosis fungoides. Lymphocytes and granulocytes are now accepted as one of the important targets in PUVA; however, the mode of action of this effective treatment is still under discussion. Thus, we studied both the functions and the viability of lymphocytes and granulocytes.

We tried to clarify aspects of the cellular effects of PUVA therapy. This paper presents a relatively simple in vitro model for the evaluation of (photosensitizing) 8-MOP reactions under conditions similar to those prevailing during photochemotherapy (Fig. 1).

Material and Methods

The experimental procedure has been described previously in detail [16]. The following parameters of white blood cells from healthy donors were investigated:

Lymphocytes

1. Viability, using a 1% aqueous solution of trypan blue (trypan blue exclusion method)

A. Bernd, J. Bereiter-Hahn,
F. Hevert, H. Holzmann (Eds.)
Cell and Tissue Culture Models
in Dermatological Research

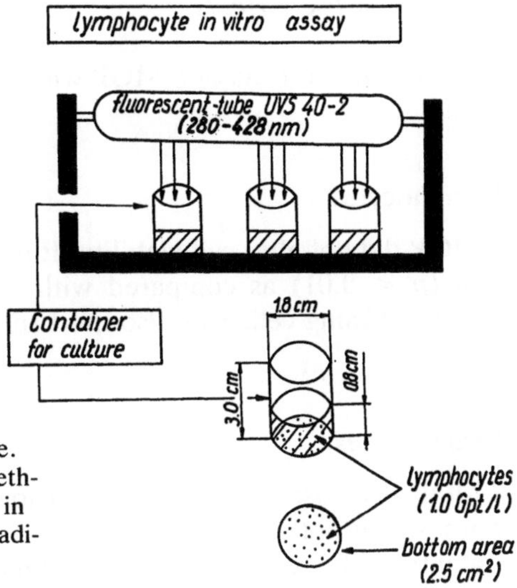

Fig. 1. Setup of the experimental procedure. Isolated peripheral lymphoid cells and 8-methoxypsoralen *(8-MOP)* solution were mixed in the container and then subjected to UV irradiation

2. The formation rate of lymphoblasts after stimulation by phytohemagglutinin (PHA; 0.25 µl/ml)
3. The ability of T and B lymphocytes to form rosettes with sheep (SE) and mouse erythrocytes (ME), respectively

Granulocytes

4. Viability, with the trypan blue exclusion method
5. The ability to phagocytose *Candida albicans*
6. The capacity to reduce nitro blue tetrazolium (NBT) as an indicator of the activity of intracellular redox systems
7. Spontaneous migration and chemotactic activity against casein

8-MOP was dissolved in aqueous 0.11 mol/l ethanol in concentrations of 0.92, 2.3, 4.6, 9.2, and 23.0 µmol/l. The purity of the psoralen sample was checked by means of mass spectroscopy, UV spectrophotometry, and thin-layer chromatography. After 10 min of 8-MOP incubation, the UV irradiation (dose 10.17 kJ/m^2) was performed by means of a UVA fluorescent tube 40-2 (emission spectrum: 280–428 nm; 7% UVB). The statistical evaluation was carried out using the Mann-Whitney U-test.

Results

In several control checks we ruled out any modifying influence of the 8-MOP solvent.

8-MOP Dark Effects

The following effects of 8-MOP without any additional UV irradiation were found:

Lymphocytes

8-MOP did not alter cell viability. It inhibited PHA-induced cell transformation ($p < 0.01$) as compared with the controls. The number of SE/ME rosette forming cells was also markedly reduced by 8-MOP ($p < 0.01$; Fig. 2).

Granulocytes

8-MOP decreased the phagocytic capacity and the chemotaxis of granulocytes significantly ($p < 0.01$) in a dose-dependent manner, whereas cell viability, NBT reduction, and spontaneous migration were unaffected (Fig. 3).

8-MOP/UV effects (PUVA)

The combined effect of psoralen and UV radiation caused lethal damage to cultivated human lymphocytes. There was a significant ($p < 0.01$) decrease

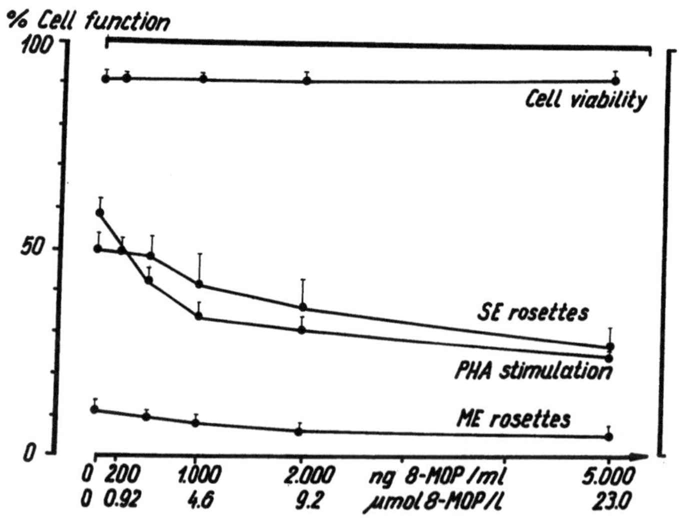

Fig. 2. The dose-dependent effect of 8-methooxypsoralen *(8-MOP)* on several cell functions of human lymphocytes. The cells were treated for 10 min with the indicated concentrations of 8-MOP ($n = 10$). *SE,* sheep erythrocytes; *ME,* mouse erythrocytes; *PHA,* phytohemagglutinin

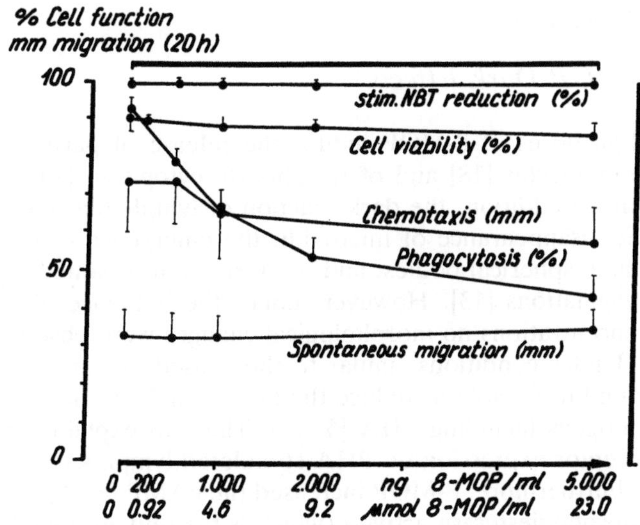

Fig. 3. Several cell functions of human granulocytes after 10 min of 8-metoxypsoralen *(8-MOP)* incubation in various concentrations (n = 10). *NBT,* nitro blue tetrazolium

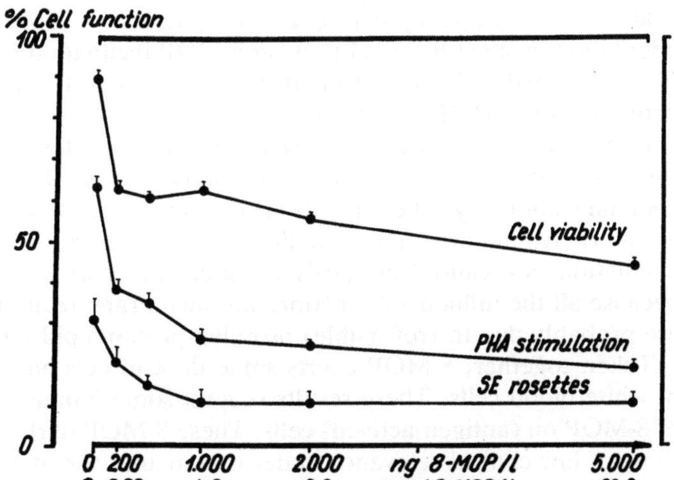

Fig. 4. Influence of 8-MOP plus UV irradiation (10.17 kJ/m^2) on several cell functions of human lymphocytes; 8-MOP incubation time 10 min (n 10). *PHA,* phytohemagglutinin; *SE,* sheep erythrocytes

of PHA-stimulated lymphocyte proliferation in cultures treated by 8-MOP plus UV irradiation in comparison to those exposed to 8-MOP only. Moreover, we found the number of SE rosette-forming cells more reduced after photosensitization than after 8-MOP incubation alone ($p < 0.01$; Fig. 4).

Discussion

8-MOP Dark Effects

High doses of 8-MOP induce the release of parts of the glycocalyx from granulocytes [18] and of lymphocyte cytoplasm [14] into the culture medium. In addition, the dark reaction of lymphoblasts with 8-MOP resulted in the disappearance of microvilli, the emergence of plasma membrane-associated spherical bodies, and the formation of lamellar fungiform membrane evaginations [13]. However, under the influence of 8-MOP in therapeutic concentrations no morphological changes were observed [11, 19].

Under conditions similar to those used in our experiments, 8-MOP was found to be able to reduce the blast transformation rate induced by various mitogens including PHA [5, 12]. This was explained by a decrease in IL-2 receptor expression on PHA-stimulated lymphocytes [5].

Furthermore, 8-MOP increased the cAMP level [1] and decreased cAMP-phosphodiesterase activity (which is the only enzyme known to catalyze the hydrolysis of cAMP) in mononuclear leukocytes [15]. Moreover, 8-MOP was able to completely reduce the PHA-triggered cytokine release (macrophage-slowing factor) from lymphocytes [17].

Investigations using lymphocyte surface markers (monoclonal antibodies against the CD2 receptor) revealed a significant decrease of the T-cell population in short-term cultures on 8-MOP incubation lasting up to 3 h. In PHA-activated cells the premature expression of the DR antigen was depressed by 8-MOP as well [11].

Concerning the reduced chemotaxis of granulocytes due to 8-MOP exposure used in our experiments, the results presented are concordant with communications by others [3]. It seems most likely that the observed alterations of lymphocyte and granulocyte functions by 8-MOP (without UV irradiation) are caused primarily by interactions with membrane structures because all the influenced functions are membrane related [8]. These effects are probably due to (reversible) psoralen-protein/lipid interactions [13].

Taken together, 8-MOP exerts some dark effects on surface membranes of white blood cells. These results suggest some immunosuppressive effects of 8-MOP on (antigen-actived) cells. These 8-MOP dark effects are assumed to be of low clinical relevance under therapeutic conditions because of their mostly small and transient effects.

8-MOP/UV Effects (PUVA)

In nonstimulated, as well as in PHA-activated, lymphocyte cultures we found formerly a significant increase of the number of nonviable cells between 2 and 6 h after UV irradiation [9]. In stimulated cultures this peak was followed by a second maximum 36 h after the beginning of PHA stimulation, i.e., in the S-phase of the cell cycle. The first effects can be observed very early and are supposed to be due to membrane alterations (lipid/protein photo-oxidation [2], modification of cell membrane DNA [7]), while the 36-h maximum of increase in dead cells is supposed to be caused by nuclear damage, such as mono- and bifunctional photoadducts with pyrimidine bases of DNA [19]. According to our studies of the 8-MOP action spectrum of phototoxicity [9] and those on inhibition of DNA synthesis of lymphocytes [6], the most effective wavelength was about 330 nm.

The mechanisms of PUVA treatment are still unclear. However, there is some evidence that radical-released oxygen-limited processes play an important role. The well-known antioxidant propyl gallate was found to be able to reduce the 8-MOP/UV-caused killing of lymphocytes. The protective efficacy does not seem to be an absorption effect. Propyl gallate may act as a radical killer against oxygen-dependent PUVA mechanisms such as lipid peroxidation and protein photo-oxidation [10].

UV irradiation enhanced the 8-MOP-caused decrease in the percentage of SE rosette-forming cells and PHA-induced blast transformation. A maximum inhibition of lectin-stimulated proliferation was found seen when 8-MOP was added to the cultures in S-phase, 36 h after the beginning of PHA incubation. We believe that these irradiation effects are caused by nuclear alterations, predominantly by initiating DNA cross-links [9].

Although the fundamental molecular biological action of the photosensitizer 8-MOP on the cells is not fuly understood, it is certain that different mechanisms are involved in the complexity of PUVA efficiency, including targets on membranes, cytoplasm, and nuclear components (Fig. 5).

Similar effects may be considered for in vivo photochemotherapy. This should affect primarily white blood cells localized in the skin, rather than

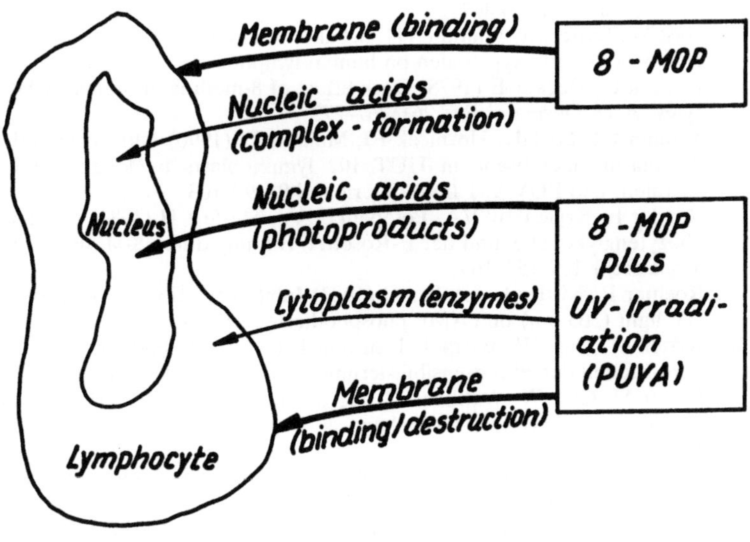

Fig. 5. Cellular reaction mechanisms caused by (photosensitized) 8-methoxypsoralen and *PUVA (8-MOP)* treatment

those circulating in the blood stream and lymph vessels. The latter constantly evade the UV exposure so that they are affected by only 1%–5% of the UV radiation dose [4].

References

1. Albrightson CR, Fertel RE, Brown BV, Stephens R (1985) Psoralens increase the concentration of cyclic AMP in human cells in vitro. J Invest Dermatol 85:264–268
2. Boehm F, Meffert H, Bauer F (1986) PUVA therapy damages psoriatic and normal lymphoid cells within milliseconds. Arch Dermatol Res 279:16–19
3. Bredberg A, Forsgren A (1984) Effects of in vitro PUVA on human leucocyte function. Br J Dermatol 111:159–168
4. Cohen LF, Kraemer KH, Waters HL, Kohn KW, Glaubiger DL (1981) DNA cross-linking and cell survival in human lymphoid cells treated with 8-methoxypsoralen and long wavelength ultraviolet radiation. Mutat Res 80:347–356
5. Cox GW, Orosz CG, Fertel RH (1987) 8-Methoxysporalen inhibits lymphocyte proliferation in vitro in the absence of ultraviolet radiation. Int J Immunopharmacol 9:475–481
6. Gasparro FP, Berger CL, Edelson RL (1984) Effect of monochromatic UVA light and 8-methoxypsoralen on human lymphocyte response of mitogen. Photodermatology 1:10–17
7. Gasparro FP, Dall'Amico R, O'Malley M, Heald PW, Edelson RL (1990) Cell membrane DNA: a new target for psoralen photoadduct formation. Photochem Photobiol 52:315–321
8. Gast W, Haustein UF (1991) 8-methoxypsoralen acts on lymphocyte membranes in the dark. J Invest Dermatol 96:293
9. Gast W, Barth J, Rytter M (1983) PUVA-Wirkungen auf Lymphozyten. Dermatol Monatsschr 169:217–226
10. Gast W, Biella U, Fickweiler E, Loeschburg M, Haustein UF (1987) Effects of the antioxidant propyl gallate on PUVA reactions in vitro. In: Braun-Falco O, Stuettgen G, Orfanos CE (ed) Proc. 17th World Congress of Dermatology Berlin, May 24–29. Braun, Karlsruhe, p 407
11. Gast W, Rytter M, Walther TH, Schaller J, Hofmann CH, Haustein UF (1988) Dark effects of 8-methoxypsoralen on human lymphocytes. Stud Biophys 124:173–178
12. Lischka G, Decker E (1978) Dark effect of 8-methoxypsoralen (8-MOP) on lymphocytes. Arch Dermatol Res 261:267–272
13. Malinin GI, Lo HK, Hornicek FJ, Malinin TI (1990) Ultrastructural modification of the plasma membrane in HUT 102 lymphoblasts by long-wave ultraviolet light, psoralen, and PUVA. J Invest Dermatol 95:97–103
14. Roeder B, Krylenkov VA, Potapenko AY, Meffert H (1983) Modifikation der Lymphozytenglykokalyx und der E-Rosetten-Bildung durch 8-MOP und UVA. Dermatol Monatsschr 169:157–165
15. Roemer R, Kittler L, Loeber G (1983) Inhibitory effects of furocoumarins plus near UV light (365 nm) on cAMP phosphodiesterase activity. Stud Biophys 94:33–36
16. Rytter M, Gast W, Barth J, Haustein L (1984) Methodische Aspekte zu einem in-vitro-Modell der Photosensibilisierung. Z Med Lab Diagn 25:392–396
17. Rytter M, Gast W, Walther TH, Herold W, Haustein UF (1989) 8-Methoxypsoralen inhibits the release of macrophage-slowing factor from lymphocytes. Stud Biophys 133:109–112
18. Schmidt H, Weigel C (1988) Zur Wirkung von Methoxalen auf menschliche Lymphozyten. Dermatol Monatsschr 174:20–27
19. Schmoeckel CH, Scherer R, Kern B, Braun-Falco O (1978) The cytolytic effect of PUVA treatment on PUVA-stimulated human peripheral lymphocytes. Acta Derm Venereol (Stockh) 58:203–211

Comparative Cytotoxicity Test with Human Keratinocytes, HaCaT Cells, and Skin Fibroblasts to Investigate Skin-Irritating Substances

A. Hoh, and *K. Maier*

Abstract

Irritation of the skin involves reactions of the cells that form the human dermis and epidermis. Particularly the keratinocytes, which are affected first by toxic substances, are now known to be involved in a variety of inflammatory and immune responses of the skin. For this reason, keratinocytes should be the most appropriate target cells for assessing sensitivity of the skin to toxicants. In this study normal human keratinocytes, the spontaneously transformed keratinocyte cell line HaCaT, and 3T3 mouse fibroblasts were compared to assess their potential as models for skin irritancy testing. Twelve chemicals with a wide range of skin-irritating potential were tested using the neutral red uptake and the kenacid blue staining. In most cases the normal human keratinocytes showed less sensitivity than the HaCaT cell line or the 3T3 cells. Nevertheless, all three cell types correlate in their ranking with in vivo Draize data. These results indicate that the choice of cell type is not crucial for assessment of the primary effect of toxic substances.

Introduction

Present methods using live animals for testing irritancy provide only limited information about the toxicological properties of a chemical [17]. Additionally, the application of chemicals to the skin or to the eye of animals (Draize test) is perceived as potentially painful. Therefore, a number of in vitro cytotoxicity assays have been developed and many efforts are being made to improve test systems for specific investigations. During the past several years, particularly cell lines have been used extensively to detect toxicity at the cellular level [2–6, 8, 10, 13–16].

However, the cells that are first affected by toxic compounds are – in the case of skin irritation – human keratinocytes. It is possible to culture human normal keratinocytes such that they resemble in vivo epidermis to a certain extent. We used a confluent multilayered keratinocyte culture which consists of proliferating basal cells and differentiated cell layers, thus corresponding to the epidermis in vivo, where chemicals have to pass a few cell layers while

A. Bernd, J. Bereiter-Hahn,
F. Hevert, H. Holzmann (Eds.)
Cell and Tissue Culture Models
in Dermatological Research

penetrating into the skin. Furthermore, keratinocytes exhibit the metabolism of the skin more precisely than other cell types do. The 12 selected chemicals give a broad range of irritating potential. Human normal keratinocytes were tested in comparison with the spontaneously transformed keratinocyte cell line HaCaT [7] and with 3T3 mouse fibroblasts using the neutral red uptake and the kenacid blue test.

HaCaT cells were chosen because they possess several characteristics of keratinocytes; however, they stratify less, and thus they represent target cells which are interesting to be compared with keratinocytes.

Materials and Methods

Keratinocyte Cultures

Human keratinocytes were isolated from skin specimens, obtained by surgery. After the skin was rinsed in PBS (Gibco) containing 50 μg/ml gentamicin, 100 μg/ml streptomycin (Flow) and 100 U/ml penicillin, the epidermis was minced with scissors, resulting in pieces of 16–26 mm^2. They were then transferred to a Petri dish containing 0.3% trypsin in PBS and incubated at 4° C over night. The pieces were transferred to DMEM/FCS and the epidermis was removed from the dermis; the basal cells were scraped from the dermis. After suspension of the pieces by pipetting the solution and centrifuging, the cells were counted. Cells were seeded with a density of 2.8×10^4/cm in tissue culture flasks (Nunc) onto a feeder monolayer of lethally irradiated (60 Gy) BALB/c mouse 3T3 fibroblasts (FLOW) (2.8×10^4/cm). They were grown in DMEM/HAM F12 (Gibco) supplemented with 10% fetal calf serum, hydrocortisone (0.4 μg/ml), insulin (5 μg/ml), triiodothyronine (2×10^{-9} *M*), transferrin (5 μg/ml), adenine (8×10^{-4} *M*), (all obtained from Sigma), forskolin (5×10^{-6} *M*, Hoechst) gentamicin (50 μg/ml, Serva), glutamine (Gibco), penicillin (100 U/ml, Flow), streptomycin (100 μg/ml, Flow), and epidermal growth factor (10 ng/ml, Sigma) and incubated in a humidified 9% CO_2 atmosphere at 37° C, according to the method of Green et al. [11].

After having grown to a subconfluent state (60%–80% confluence) in about 8 days, the keratinocytes were washed with EDTA (0,02% in PBS), dissociated again with trypsin/EDTA (0.05%/0.01% in PBS), counted, and seeded in 96-well microtiter plates (Nunc) with 3×10^4 cells/well onto a feeder layer of 1×10^4 3T3/well.

HaCaT Cultures

HaCaT cells (kindly provided by N. Fusenig) were seeded with a density of $6.5 \times 10^3/cm^2$ in tissue culture flasks (Nunc) and grown in DMEM (Gibco) supplemented with 10% fetal calf serum, glutamine (Gibco), penicillin (10

U/ml), and streptomycin (50 μg/ml) to a subconfluent stage. Then cells were dissociated with 0.1% trypsin/EDTA in PBS, counted, and seeded again in 96-well microtiter plates with a density of 2×10^4/well.

3T3 Cultures

3T3 fibroblasts were cultured according to the HaCaT protocol; the seeding density was 1×10^4/well.

Cytotoxicity Tests

Confluent first-passage multilayers (keratinocytes) or monolayers (HaCaT cells and 3T3 fibroblasts) were used for cytotoxicity testing; the cells were incubated for 24 h with various concentrations of the test agents and medium for controls, using eight wells per concentration of test agent.

Neutral Red Assay

The neutral red assay allows colorimetric quantitation of viable cells after their incubation with toxicants. It is based on the uptake of neutral red, a weakly cationic dye, and its accumulation in the lysosomes of viable uninjured cells [4, 5].

After the test agents had been removed from the cells, the cultures were washed with PBS (with Ca and Mg) and the medium was replaced with medium containing 50 μg/ml neutral red (Merck). After incubation for an additional 3 h at 37° C, the cells were washed with PBS and extracted with 1% acetic acid in 50% ethanol. Absorbance was determined spectrophotometrically at 540 nm using an ELISA reader (Titertek Multiskan Plus, Labsystem, Flow Laboratories).

Kenacid Blue Assay

The kenacid blue assay provides for the measurement of total cellular protein by a dye-binding method. Cells were washed with 1% acetic acid in 50% ethanol; then kenacid blue was added to the cells (0.4 mg/ml in 12% acetic acid 25% ethanol) for 10 min. After removal of the dye, cells were washed two times with 5% acetic acid in 10% ethanol and the remaining kenacid blue was extracted with desorbing solution (1 *M* potassium acetate in 70% ethanol). Absorbance was determined spectrophotometrically at 570 nm. The data represent the mean values of three independent repeats of tests, using eight wells for each concentration of test agent. Cytotoxicity data in all assays were standardized by expressing absorbance data in the presence of toxicants as a percentage of that in control cell culture. For each compound the midpoint toxicity NR 50, which is the concentration of toxicant that causes a 50% decrease in absorbance, was determined by extrapolation.

Results and Discussion

The cytotoxic effect of 12 chemicals with different skin- and eye-irritating potential was determined by neutral red uptake of keratinocytes in comparison with 3T3 mouse fibroblasts and HaCaT cells. Table 1 shows cytotoxicity

Table 1. In vitro cytotoxicity of 12 substances on keratinocytes, 3T3 fibroblasts and HaCaT cells (expressed as NR 50 concentration determined with the neutral red test). Comparison of the in vivo data of the Draize test with the in vitro results

Chemical substance	In vivo skin/eye irritation potential	In vitro keratinocyte culture NR 50 (m*M*)	In vitro 3T3 fibroblast culture NR 50 (m*M*)	In vitro HaCaT culture NR 50 (m*M*)	Sensitivity factor NR 50 keratinocyte/ NR 50 3T3 fibroblasts
Dimethylsulfoxide	Mild	2340.00	564.00	n.t.	4.1
Propylene glycol	Mild	1125.00	900.00	1317	1.25
Acetone	Mild-moderate (eye)	1123.00	670.00	n.t.	1.67
Phthalic acid anhydride	Severe, moderate (skin)	16.00	19.00	18	0.84
Acetaldehyde	Severe (eye), moderate (skin)	22.50	3.60	n.t.	6.25
Ascorbic acid	Severe	12.00	4.40	6.3	2.7
Allyl alcohol	Severe (eye)	10.30	1.30	n.t.	7.9
Benzyl alcohol	Severe (eye)	0.32	0.87	n.t.	0.36
Remolgan PM	Severe (eye), not irritant (skin), R41	0.22	0.10	0.1	2.2
Lactic acid	Severe	44.00	54.00	53	0.81
Sodium dodecyl sulfate	Severe, moderate	2.20	1.60	1.7	1.37
Tretinoin	Severe	0.02		n.t.	0.03

n.t., Not tested

data of the 12 test agents with a wide range of different toxic potencies, listed next to the in vivo data of the Draize rabbit-eye irritancy test. As shown in the table, 3T3 fibroblasts and HaCaT cells reveal for all tested chemicals almost identical NR 50 values as well as identical response curves for the dose-response relationships (Fig. 1).

Comparing the NR 50 values with the in vivo data from Draize tests, there is a good correlation between decreasing NR 50 values and increasing irritation potential in vivo, as we have already shown [12]. The multilayered keratinocyte cultures showed in most cases a significantly lower sensitivity to the test chemicals than 3T3 fibroblasts and HaCaT cells. Nevertheless, concerning their ranking, the in vitro data of keratinocytes correlate with in vivo data as do the in vitro data of 3T3 fibroblasts and HaCaT cells.

As a cause for the minor sensitivity, the stratified three-dimensional structure of the cell formation which complicates the penetration of the test chemical into the lower cell layers might be responsible, along with begin-

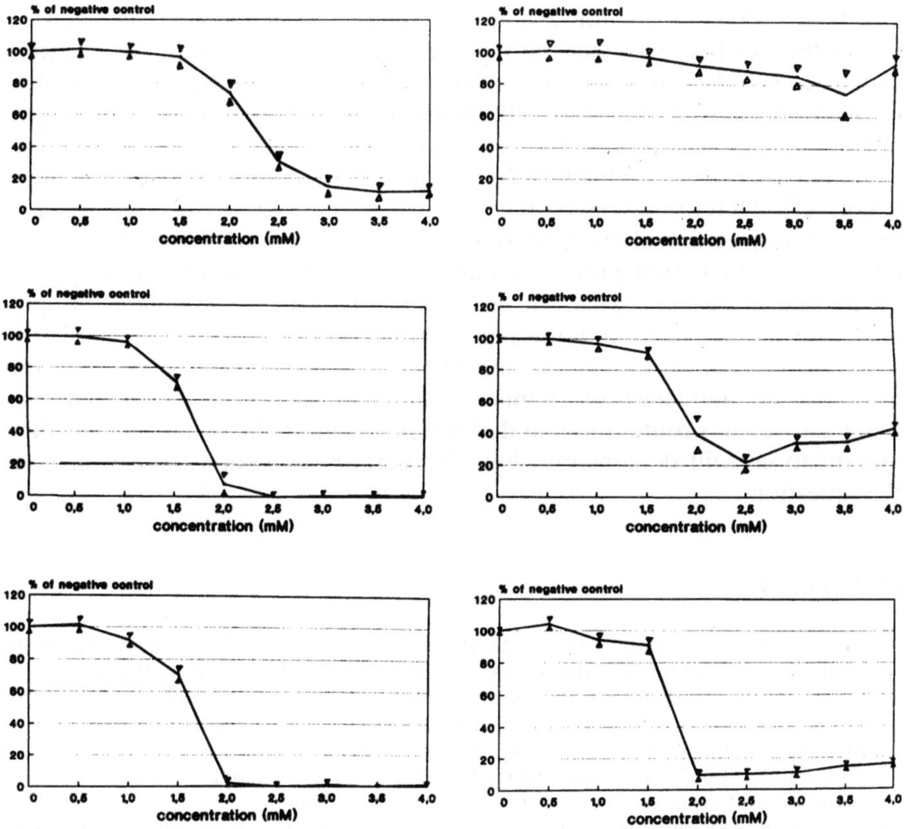

Fig. 1. Dose-response curves for sodium dodecyl effects on keratinocytes (above), HaCaT cells *(middle)*, and 3T3 fibroblasts *(below)*. *Left:* neutral red assay; *right:* kenacid blue assay

ning differentiation, and therefore a kind of barrier function of the upper cell layers.

To compare the variations of cytotoxicity in different cell types we calculated a sensitivity factor which expresses the ratio between NR 50 of keratinocytes and NR 50 of fibroblasts. The determination of this factor reveals remarkable differences between the individual chemicals. This tends to indicate that the different NR 50 values are due to cell-type-specific structures of the culture, but also that the nature of the chemical can result in varying penetration behavior and cytotoxicity toward different cells. However, if we compare the differences of sensitivity factor (maximum 7.9) with the differences in NR 50 values (maximum 10000) between a weakly and a severely irritating chemical, we can conclude that the different sensitivities of the investigated cell types have no influence on the ranking of the chemical. Accordingly, the choice of cell type is not crucial for an assessment of the primary effect of toxic substances and for gross classification of the substances analogous to the Draize test.

For the identification of tissue-specific toxic and inflammation-inducing effects (as they occur, for example, with skin irritation), a test system with other cell-type-specific end points and a corresponding exogenous substance-metabolizing system is imaginable. Keratinocytes are capable of secreting a wide range of inflammatory cytokines with immunologic, inflammatory, and proliferating properties, and the secretion of these cytokines can be modulated by various exogenous and endogenous agents [1]. Determination of the release of these mediators might be a more appropriate tool for assessing the skin irritation potential, particularly the inflammation processes that may be caused by chemical agents. Furthermore, keratinocytes are characterized by their expression of specific proliferation and differentiation markers such as certain cytokeratines, involucrin, filaggrin, or loricrin [9], and thus may be used for assessing the influence of chemicals or other biological compounds on the balance of proliferation and differentiation in human skin. Further investigations will demonstrate how far keratinocytes in vitro can answer this question.

References

1. Ansel J, Perry P, Brown J, Damm D, Phan T, Hart C, Luger T, Hefeneider S (1990) Cytokine modulation of keratinocyte cytokines. J Invest Dermatol 94:101S–107S
2. Babich H, Borenfreund E (1987) Structure-activity relationship (SAR) models established in vitro with the neutral red cytotoxicity assay. Toxicol In Vitro 1:3
3. Bartnik F, Künstler K (1987) Validierung von Ersatzmethoden für Tierversuche zur Prüfung auf lokale Verträgkichkeit. I. In vitro Untersuchungen mit Zellkulturen. Bundesministerium für Forschung und Technologie, Bonn
4. Borenfreund E, Borrero O (1984) In vitro cytotoxicity assays. Potential alternatives to the Draize ocular allergy test. Cell Biol Toxicol 1:55–56
5. Borenfreund E, Puerner JA (1985) Toxicity determined in vitro by morphological alterations and neutral red absorption. Toxicol Lett 24:119–124

6. Borenfreund E, Babich H, Martin-Alguacil N (1988) Comparisons of two in vitro cytotoxicity assays – the neutral red (NR) and tetrazolium MTT test. Toxicol In Vitro 2:1–6
7. Boukamp P, Petrusevska RT, Breitkreuz D, Hornung J, Markham A, Fusening NE (1988) Normal keratinization in a spontaneously immortalized aneuploid keratinocyte cell line. J Cell Biol 106:761–771
8. Clothier RH, Hulme L, Ahmed AB, Reeves HL, Smith M, Balls M (1988) In vitro cytotoxicity of 150 chemicals to 3T3-L1 cells, assessed by the FRAME Kenacid blue method. ATLA 16:84–95
9. Fuchs E (1990) Epidermal differentiation: the bare essentials. J Cell Biol 111:2807–2813
10. Gajjar L, Benford DJ (1990) Comparison of cultured keratinocytes and fibroblasts as models for irritancy testing in vitro. Toxicol In Vitro 4:280–283
11. Green H, Kehinde O, Thomas J (1979) Growth of cultured epidermal cells into multiple epithelia suitable for grafting. PNAS 76:5665
12. Hoh A, Maier K, Dreher RM (1987) Multilayered keratinocyte culture used for in vitro toxicology. J Mol Toxicol 1:537–546
13. Hopley JP, Benford DJ, Bridges JW (1985) Toxicity of polychlorinated biphenyl in a differentiating skin cell-culture system (Abstr). Food Chem Toxicol 23:325
14. Kalweit S, Besoke R, Gerner I, Spielmann H (1990) A national validation project of alternative methods to the Draize rabbit eye test. Toxicol In Vitro 4:702–706
15. Mosmann T (1983) Rapid colorimetric assay for cellular growth and survival: application to proliferation and cytotoxicity assays. J Immunol Methods 65:55–63
16. Reinhardt CA, Pelli DA, Zbinden G (1985) Interpretation of cell toxicity data for the estimation of potential irritation. Food Chem Toxicol 23:247
17. Rowan A, Goldberg A (1985) Perspectives on alternatives to current animal testing techniques in preclinical toxicology. Annu Rev Pharmacol Toxicol 25:225–247

SDS-Induced Activation of the Cutaneous Plasminogen Activator System: A Molecular Pathway Possibly Involved in SDS-Induced Skin Irritation

M. D. Kramer, M. Müller-Bardorff, P. J. Frosch,
and *G. Brunner*

Abstract

We characterized on the protein and enzyme level the components of the plasminogen activator (PA) system in cutaneous suction blister fluid (CBF) derived from normal human skin. By enzyme-linked immunosorbent assay, immunoblotting, and zymography, CBF was found to contain plasminogen, free plasminogen activator inhibitor-1 (PAI-1), no free tissue-type plasminogen activator (tPA), but two different high-molecular-weight tPA complexes (88 and 160 kDa). Immunoblotting with anti-PAI-1 antibodies yielded evidence that these complexes were composed of tPA and PAI-1. PA activity of one of these complexes (88 kDa) was partially restored by sodium dodecyl sulfate (SDS) during zymography. In radial fibrinolysis assays we found that plasmin activity could be induced in CBF from normal skin when SDS was added to the fluid in vitro. In similar experiments we demonstrated active PAs and plasmin in CBF induced on SDS-irritated skin areas. In conclusion, we suggest that plasmin is generated in SDS-irritated skin and that reactivation of PA activity from high-molecular-weight tPA/PAI-1 complexes might be a molecular pathway underlying this phenomenon.

Introduction

The plasminogen activator (PA) system is a major system of extracellular proteolysis. Its activation depends on the PA-mediated conversion of plasminogen to the trypsin-like, broad-specific serine proteinase plasmin [2, 11]. The two major types of human PAs are urokinase-type PA (uPA) and tissue-type PA (tPA). Keratinocytes have been shown to produce PAs of both types in vitro [14]. Zymographic analyses disclosed uPA and tPA activity in extracts of normal human epidermis [8]. Using a polyvalent antiserum for immunohistology, uPA was localized to the granular layer of the normal epidermis [12]. An independent study with monoclonal antibodies yielded conflicting results: Neither uPA nor tPA was detectable in normal skin [3, 4]. The PA substrate plasminogen, which is present in plasma and interstitial fluid, was immunohistologically localized to the basal layer of normal human epidermis [7, 9].

A. Bernd, J. Bereiter-Hahn,
F. Hevert, H. Holzmann (Eds.)
Cell and Tissue Culture Models
in Dermatological Research

The assumption that active plasmin possesses proinflammatory capacity was suggested by several findings:

a) the observation that the PA system is activated under diverse inflammatory conditions of human skin, such as tissue remodeling during wound healing [4], acantholysis during blister formation induced by anti-cell surface pemphigus antibodies [5, 13], and in psoriasis [3];
b) the finding that in skin organ cultures plasmin induces acantholysis [6]; and
c) the observation that active plasmin appears in tear fluid during chronic corneal ulceration and that local application of a plasmin inhibitor can lead to clinical improvement [16].

The potent skin irritant sodium dodecyl sulfate (SDS) can activate PA activity via reactivation of functional PA from PA/PA inhibitor (PAI) complexes [15]. In the present study we asked whether local activation of the PA system at the site of SDS application, possibly via reactivation of PA/PAI complexes, might be a molecular pathway involved in this type of skin irritation. For our experiments we employed the technique of cutaneous suction blister induction. The fluid contained in the blisters represents dermal interstitial fluid, which is in part derived as an ultrafiltrate from plasma [17] to which secretions of epidermal and dermal cells are added. The fluid can serve to explore the composition of extracellular interstitial fluid of human skin. In addition, comparison of cutaneous suction blister fluid (CBF) from nonirritated and irritated skin allows us to explore the locally occurring molecular alterations induced by the respective irritant.

We characterized the components of the PA system in CBF derived from normal human skin on the protein and enzyme activity level. By ELISA, immunoblotting, and zymography, CBF was found to contain plasminogen, free PAI-1, no free tPA, but two different high-molecular-weight tPA complexes (88 and 160 kDa). Immunoblotting with anti-PAI-1 antibodies yielded evidence that these complexes were composed of tPA and PAI-1. PA activity of one of the tPA complexes (the 88 kDa form) was partially restored by SDS during zymography. In radial fibrinolysis assays we found that plasmin activity could be induced in CBF from normal skin when SDS was added to the fluid in vitro. In similar experiments we demonstrated active PAs and plasmin in CBF induced on SDS-irritated skin areas. In conclusion, we suggest that plasmin generation appears in SDS-mediated skin irritation and that reactivation of PA activity from high-molecular-weight tPA/PAI-1 complexes might be a molecular pathway underlying this phenomenon.

Results and Discussion

Immunological Analysis of the Components of the PA System in CBF

The components of the PA system in CBF derived from normal and SDS-irritated human skin were analyzed by immunological methods. ELISA testing of 59 individual CBF samples revealed that the PA substrate plasminogen is present in a concentration range from 0.015 to 0.16 g/l (not shown; Müller-Bardorff, unpublished observations). Immunoblotting analysis of CBF using a plasminogen/plasmin-specific mAb (see Fig. 1, lane A) revealed a triple protein band at about M_r 92000, which corresponds to the known molecular weight of plasminogen. In radial fibrinolysis assays we found that plasminogen in CBF could be activated to plasmin by an excess of exogenously added uPA, which indicates the functional integrity of the plasminogen molecules (not shown). tPA was detected in CBF by immunoblotting using tPA-specific polyclonal antibodies (Fig. 1, lane B). Two immunoreactive proteins of a molecular weight of 88 and 160 kDa were detected. This suggests complex formation of intact tPA (68 kDa) and of a proteolytically derived low-molecular-weight form thereof (≈ 35 kDa) [2] with a PA inhibitor. In fact, immunoblotting experiments using a monoclonal antibody recognizing PAI-1 revealed the presence of this inhibitor in these high-molecular-weight complexes (not shown). In similar experiments, uPA was not detectable in CBF.

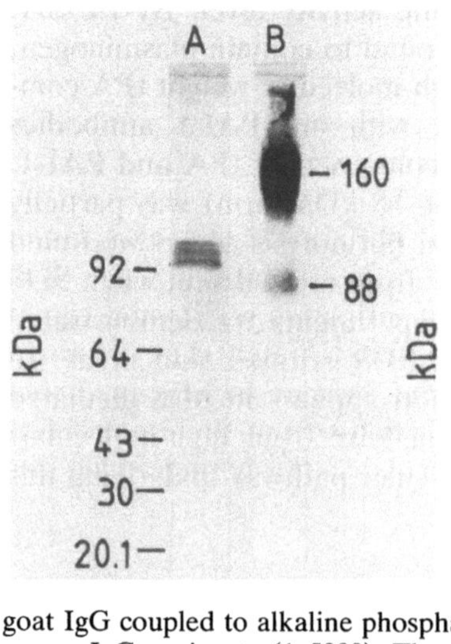

Fig. 1. Detection of plasminogen and tPA in CBF by immunoblotting. CBF (10 μl of a 1:15 dilution in SDS sample buffer per lane) was subjected to SDS-PAGE in 7.5%–15% polyacrylamide gradient gels under nonreducing conditions followed by electrophoretic transfer of the separated proteins onto nitrocellulose filters and immunostaining. *A*, Murine anti-plasminogen mAb PG-19 (own production; unpublished data), 10 μg/ml; *B*, goat anti-human tPA IgG-(American Diagnostica, Cat no. 387; 1:1000). Antibody binding was visualized by rabbit anti-goat IgG coupled to alkaline phosphatase (1:10000) or by the respective rabbit anti-mouse IgG conjugate (1:5000). The M_r of the molecular-weight markers and immunostained proteins is given in kDa. Anti-plasminogen mAb stained a triple band in the M_r region of 92 kDa. Anti-tPA stained two major compounds of 88 kDa and ≈ 160 kDa, respectively

Functional State of the PA System in CBF: Influence of SDS In Vitro and In Vivo

In CBF from normal human skin no plasmin activity was detectable by radial fibrinolysis (Fig. 3, I). Similarly, in zymography PA activity at the molecular position of free tPA (68 kDa) was not found in CBF (Fig. 2, lane B). The absence of free tPA and the presence of putative tPA/PAI complexes indicated an excess of PAI in CBF. In fact, reverse zymography of CBF revealed large amounts of an inhibitory protein, probably representing free PAI-1 (54 kDa; see clamp in Fig. 2, lane C). PA activity of one of the high-molecular-weight tPA/PAI-1 complexes (88 kDa) in CBF was partially restored in zymography in the presence of SDS (open arrow; Fig. 2, lane C).

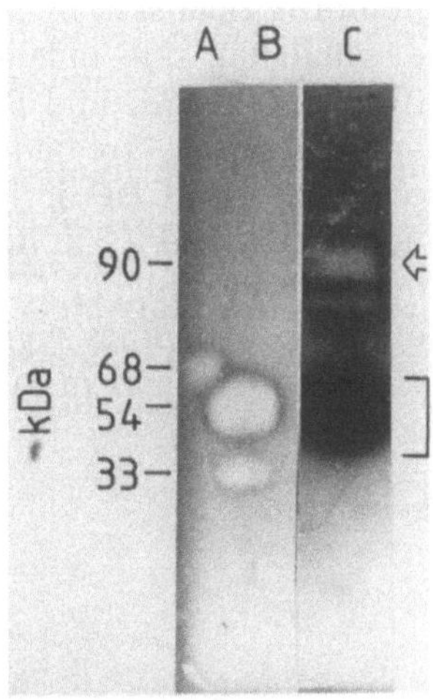

Fig. 2. Zymography of CBF. CBF samples were subjected to SDS-PAGE in 7.5%–15% polyacrylamide gradient gels under nonreducing conditions followed by zymography [1]. *A,* tPA (2 μg/lane); *B,* uPA (2 μg/lane); *C,* CBF (5 μl/lane). The M_r of the bands exhibiting PA activity is given in kDa: tPA (68 kDa) and uPA (high-M_r form: 54 kDa; low-M_r form: 33 kDa). *Open arrow,* PA activity in CBF; *clamp,* PA inhibitory activity in CBF. In this area reduced lysis is observed

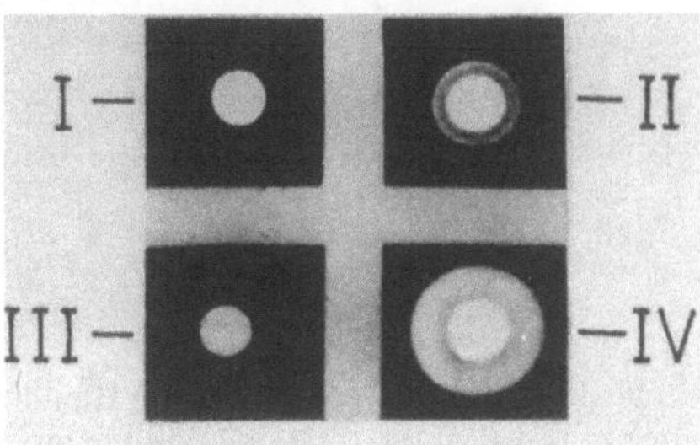

Fig. 3. Radial fibrinolysis assay of CBF from normal skin. CBF derived from normal human skin (for method see legend to Fig. 4) was mixed with SDS (1% w/v final concentration). A 1:5 dilution of SDS-treated *(II, IV)* and nontreated (I, III) CBF in assay buffer (30 m*M* Tris-HCl, pH 7.4, 60 m*M* NaCl, 0.05% NaN_3, 0.1% Triton X-100) was filled into the wells that had been punched into the fibrin agar indicator gel. The gel was prepared without *(I, II)* and with *(III, IV)* the addition of plasminogen (34 μg/ml finalconcentration). The radial lysis assay has been described previously [15]. Circular lysis zones were observed only when SDS-treated CBF *(II, IV)* was tested. Lysis zones in the indicator gels prepared without plasminogen *(II)* are evidence for plasmin activity generated from plasminogen present in CBF. The lysis zones were larger when plasminogen had been included into the fibrin gel *(IV),* indicating the presence of PA activity in CBF

This is in accordance with previous findings that SDS may partially reactivate the functional activity of PA/PAI complexes [15]. The 160-kDa tPA/PAI-1 complex did not exhibit detectable PA activity in zymography. These data provide evidence that SDS might reactivate PA from an 88-kDa tPA/PAI-1 complex in CBF.

To explore whether SDS can also activate the PA system in nonfractionated CBF, radial fibrinolysis assays were performed with CBF in the presence of SDS in vitro. As shown in Fig. 3, addition of SDS to CBF induced fibrinolysis, probably via reactivation of the 88-kDa tPA/PAI-1 complex followed by activation of plasminogen present in CBF (Fig. 3, II). Fibrinolysis was enhanced by incorporation of exogenous plasminogen into the fibrin gel (Fig. 3, IV), providing evidence for an SDS-inducible generation of PA and plasmin activity in CBF.

In further experiments we addressed the question of whether a similar activation of the PA system could occur in vivo during experimental irritation of normal skin by SDS. Abdominal skin of volunteers with no signs of disseminated skin inflammation was treated with graded amounts of SDS (0%, 1%, 2.5%, and 5% w/v in water) overnight. Thereafter, CBF was induced on the treated skin areas. As shown in Fig. 4, fibrinolysis was observed with CBF derived from skin areas previously irritated by application of 1% and 2.5% SDS but was not detectable in CBF from skin areas irritated with higher (5%) concentrations of the detergent or with water only. These findings suggest that SDS might act through comparable mechanisms in vivo and in vitro, leading to the generation of PA and plasmin activity in CBF.

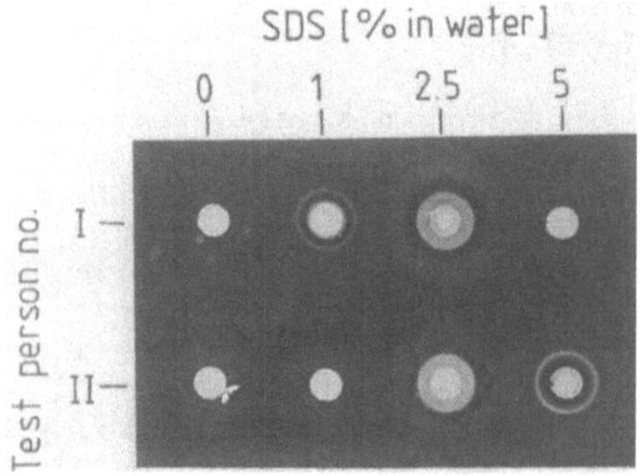

Fig. 4. Radial fibrinolysis assay of CBF from SDS-irritated skin. A 1:5 dilution in assay buffer of CBF derived from human skin 7 h after irritation with graded concentrations of SDS was tested in radial fibrinolysis in plasminogen-containing fibrin indicator gels. For experimental skin irritation 5 × 5 cm patches of nonwoven cotton (Cuticell; Beiersdorf AG; Hamburg, FRG) were saturated with graded concentrations of SDS (1%, 2.5%, and 5% w/v in water; negative control: water only). The patches were then applied under occlusive conditions for 24 h using Blenderm tape (3M company; Neuss, FRG). One hour after removal of the tape, suction blisters (6 mm diameter, 5 per patch area) were induced in the respective areas by the use of a Dermovac device according to the method of Kiistala [10]. Negative pressures of 100–350 mm Hg were applied for 2–3 h, carefully avoiding capillary bleeding. Samples contaminated with blood were excluded from further studies. The data obtained for two test persons are depicted. No fibrinolysis was observed in CBF derived from water-treated skin *(0)*. In CBF from SDS-treated skin the observed fibrinolysis is dependent on the SDS concentration: Optimal fibrinolysis was observed at intermediate concentrations of SDS, i.e., *2.5%*

In conclusion, the data provide evidence that the local activation of the PA system may be a molecular event that occurs during SDS-induced skin irritation. As yet, nothing is known about the origin of the different components of the PA system identified in CBF. It is of particular interest to explore whether PAs, PAIs, and/or respective PA/PAI complexes derived from keratinocytes, the primary cellular target after SDS application to the skin, are involved in this process. Moreover, nothing is known about the causal involvement of PA induction in the sequelae of the inflammatory reaction induced by SDS. The possible relevance of PA induction for the pathogenesis of SDS-induced skin irritation could be explored by the topical application of low-molecular-weight PA or plasmin inhibitors prior to SDS application.

Acknowledgements. This work was supported in part by the *Deutsche Forschungsgemeinschaft* (DFG Kr 931/2-2) and the *Bundesministerium für Forschung und Technologie* (01 KG 8808). The authors are indebted to Ms. U. Schirmer and Ms. S. Jobstmann for their expert technical assistance.

References

1. Brunner G, Schirrmacher V (1988) Fibrin autography of plasminogen activator by electrophoretic transfer into fibrin agar gels. Anal Biochem 168:411–416
2. Dano K, Andreasen PA, Grondahl-Hansen J, Kristensen P, Nielsen LS, Skriver L (1985) Plasminogen activators, tissue degradation and cancer. Adv Cancer Res 44:139–266
3. Grondahl-Hansen J, Ralfkiaer E, Nielsen LS, Kristensen P, Frentz G, Dano K (1987) Immunohistochemical localization of urokinase- and tissue-type plasminogen activators in psoriatic skin. J Invest Dermatol 88:28–32
4. Grondahl-Hansen J, Lund LR, Ralfkiaer E, Ottevanger V, Dano K (1988) Urokinase- and tissue-type plasminogen activators in keratinocytes during wound reepithelialization in vivo. J Invest Dermatol 90:790–795
5. Hashimoto K, Wun T-C, Baird J, Lazarus GS, Jensen P (1989) Characterization of keratinocyte plasminogen activator inhibitors and demonstration of the prevention of pemphigus IgG-induced acantholysis by a purified plasminogen activator inhibitor. J Invest Dermatol 92:310–315
6. Hunziker T, Vassalli J-D (1987) Plasmin induces acantholysis in skin organ cultures. Arch Dermatol Res 279:341–346
7. Isseroff RR, Rifkin DB (1983) Plasminogen is present in the basal layer of the epidermis. J Invest Dermatol 80:297–299
8. Jensen PJ, Baird J, Morioka S, Lessin S, Lazarus GS (1988) Epidermal plasminogen activator is abnormal in cutaneous lesions. J Invest Dermatol 90:777–782
9. Justus C, Müller S, Kramer MD (1987) A monoclonal antibody recognizing plasminogen and plasmin-altered reactivity in psoriatic lesions. Br J Dermatol 117:687–694
10. Kiistala U, Mustakallio KK (1964) In vivo separation of epidermis by production of suction blisters. Lancet :1444–1445
11. Mayer M (1990) Biochemical and biological aspects of the plasminogen activation system. Clin Biochem 23:197–211
12. Morioka S, Jensen PJ, Lazarus GS (1984) Localisation of human keratinocyte plasminogen activator with anti-urokinase antibody (Abstr). J Invest Dermatol 82:393

13. Morioka S, Lazarus GS, Jensen PJ (1987) Involvement of urokinase-type plasminogen activator in acantholysis induced by pemphigus IgG. J Invest Dermatol 89:474–477
14. Reinartz J, Boukamp P, Schickel E, Fusenig NE, Kramer MD (1991) Activation of the plasminogen activator system in a keratinocyte cell line (HaCaT) by alkyl sulfates. Biomed Biochim Acta 50:743–748
15. Saksela O (1981) Radial caseinolysis in agarose: a simple method for detection of plasminogen activator in the presence of inhibitory substances and serum. Anal Biochem 111:276–282
16. Salonen E-M, Tervo T, Törmä E, Takkanen A, Vaheri A (1987) Plasmin in tear fluid of patients with corneal ulcers: basis for new therapy. Acta Ophthalmol 65:3–12
17. Volden G, Kavli G, Haugen HF (1983) The pattern of lysosomal, cytosolic and plasma membrane enzymes in epidermis, dermis and suction blister fluid after ultraviolet radiation. Br J Dermatol 109:68–71

Subject Index

Springer-Verlag and the Environment

We at Springer-Verlag firmly believe that an international science publisher has a special obligation to the environment, and our corporate policies consistently reflect this conviction.

We also expect our business partners – paper mills, printers, packaging manufacturers, etc. – to commit themselves to using environmentally friendly materials and production processes.

The paper in this book is made from low- or no-chlorine pulp and is acid free, in conformance with international standards for paper permanency.

MIX
Papier aus verantwortungsvollen Quellen
Paper from responsible sources
FSC® C105338

If you have any concerns about our products,
you can contact us on
ProductSafety@springernature.com

In case Publisher is established outside the EU,
the EU authorized representative is:
Springer Nature Customer Service Center GmbH
Europaplatz 3, 69115 Heidelberg, Germany

Printed by Libri Plureos GmbH
in Hamburg, Germany